Electronics Cal   Handbook

D1452613

About the author

Daniel McBrearty is a freelance electronics engineer specializing in audio systems. He was an RAF electronics technician for several years before studying for an HND at Luton College of Higher Education. Thereafter he worked for various companies including Soundcraft Electronics and Drake Electronics. He is currently involved in the redevelopment of the Royal Opera House, Covent Garden. His other interests include playing the saxophone.

Electronics Calculations
Data Handbook

Daniel McBrearty

Newnes

OXFORD   BOSTON   JOHANNESBURG   MELBOURNE   NEW DELHI   SINGAPORE

Newnes
An imprint of Butterworth-Heinemann
Linacre House, Jordan Hill, Oxford OX2 8DP
225 Wildwood Avenue, Woburn, MA 01801-2041
A division of Reed Educational and Professional Publishing Ltd

 A member of the Reed Elsevier plc group

First published 1998
Transferred to digital printing 2004
© Daniel McBrearty 1998

British Library Cataloguing in Publication Data
A catalogue record for this book is available from the British Library.

ISBN 07506 3744 7

Library of Congress Cataloguing in Publication Data
A catalogue record for this book is available from the Library of Congress.

PLANT A
TREE
British Trust for
Conservation Volunteers

FOR EVERY TITLE THAT WE PUBLISH, BUTTERWORTH-HEINEMANN
WILL PAY FOR BTCV TO PLANT AND CARE FOR A TREE.

Typeset by Jayvee Computer Services and Exports Pvt. Ltd., India

Contents

Preface

This book began life as a selection of tables for electronics engineers. Dealing with such common requirements as setting op-amp gains, forming non-standard resistor values from standard pairs and setting CR time constants, it was my hope that such a book would find common favour.

In writing explanatory notes to accompany these tables, I found myself constantly wondering how much knowledge to assume on the part of the reader. After all, the book should be as useful to a beginner as to a professional with 20 years of experience.

The time came to make a decision, and Part One was born. In it I have attempted to go from first principles to the point where the reader can hope to understand the workings of the circuits discussed in Parts Two and Three with little prior knowledge. My intention has been to be non-mathematical, logical and concise. I have drawn a line at explaining the workings of electronic devices as there is not time or space, and this book only deals with them in the final chapter. An overview of the op-amp is given there. I hope that Part One provides a useful foundation for the beginner and a point of reference for the more experienced.

How well all this works as a whole will not be seen until the book is subjected to the final test by its readership. It is my hope that it will prove a valuable aid to you, whatever your level of experience. I will be happy to receive comments, criticism or suggestions for future improvement (and also to hear about any places where my spellchecker replaced 'squarewave' with 'scurvy' without my asking it to).

by mail:
Dan McBrearty
c/o Butterworth-Heinemann
Linacre House
Jordan Hill
Oxford OX2 8DP
UK

or by email:
dan@danmcb.demon.co.uk

Daniel McBrearty
December 1997

Acknowledgements

Thanks are due to Jim and Sybil McBrearty, for love and support in this; Eric Pressley and John Charlton, from whom I have learnt a great deal; all the staff at Butterworth-Heinemann for making the book a reality; William L. Bahn for helping me to convice myself that tuned circuits really should do what they do; and Hilary and Fi for putting up with my moods while I wrote!

Part One

Basic Concepts

Introduction to Part One

In this part of the book we take a look at the founding principles of electronics, from voltage, current and Ohm's Law to some basic ways to analyse circuits mathematically. If you know all of this then skip it, using it for reference as required. If not then read on.

I have tried to arrange ideas so that you are introduced to them sequentially, but this is a bit impractical in places. So if you are a real newcomer to electronics I'd like to suggest two readings; the first to get an overview of each topic, the second to cross-refer between them and to begin to see how they relate to each other.

1 Fundamentals

To the beginner

Before getting into the technical stuff, we might wish to ask ourselves what electricity is. My first recollection of thinking about it (though I did not know that I was) is of being a child and dismantling a prized radio to find out that no one was inside; just a mind-boggling collection of small coloured objects which evidently were not sweets, though some looked like them. My first direct experience of electricity was more sudden; an electric shock from a bar fire while trying to melt some plastic on it. I knew it was hot, but hadn't expected that! Later I came across the manual for a record player and amplifier in the house, which was old-fashioned enough to come with a circuit diagram. I was fascinated by it. It meant something to someone, but these hieroglyphics were like no language that I could figure out.

Later I took a technician's course, and a confusing set of concepts was presented to me as explanation for this unseen and magical force. Electricity, I had now realised, is used in a huge range of ways; recording sound, reproducing pictures, lighting our darkness and a lot else besides. All this was, they told me, due to unseen little balls which whiz around in some materials, creating equally unseen lines which can cause little balls in other places to whiz around as well. And this the mental territory of staid, rational looking people who would probably claim that they don't believe in magic.

I have to confess that, some years later, I have still not seen the little balls or the lines which they fling about, and I'm not very sure that I really understand them. But I have managed a reasonable career as a technician and an engineer, and I believe that I would have a fair chance of fixing my cassette machine, TV or house wiring should they misfunction. On the whole, I have not thought about the antics of little balls very much (though I suppose that it helps to know that they are there). Electricity looks, in my mind, more like water, wires like pipes, resistors like very thin pipes, capacitors like pairs of half-filled balloons which squash against each other, and inductors like – well something else. (I never said that the analogy was complete or accurate.)

I say all this as, I hope, a source of comfort to the new student of the subject who may be suffering the same pangs of insecurity about atoms, electrons and so on that I went through. Unless you feel strongly to do otherwise, let them live their own strange lives. You can be a very competent and effective user of the principles of electricity and electronics without needing to know much about them (though the same might not be said of completing your college course). Leave it to the physicists. What helps a great deal is having a good and clear grasp of the laws of electricity, and being able, as much as is possible, to prove them to yourself.

What is electricity?

We are never concerned with the behaviour of an individual atom in electronics; but with their movement in vast numbers, as they fling themselves, lemming-like, around the wires of your circuit. And yet we hardly think of them at all. We think a great deal about 'voltage', 'current' and about the golden rule of electronics, Ohm's Law. You will meet these soon.

Actually it is not atoms that move, but mostly bits of them, even tinier particles called electrons which separate themselves somehow from the larger atom which is their usual home. These electrons are now called 'free electrons'. Though un-imaginably small, they are pretty powerful. They carry something which is called a 'negative charge' which is another way of saying that if, having been sepa-rated from the parent atom, they should ever encounter another with a vacancy then they will be in there like a shot and pretty difficult to move. And the feeling is mutual; once it has lost an electron the atom feels the loss and will try to bring close to it any waifs that it senses nearby. Being much bigger, it moves much less easily, or not at all. All of which physicists understate quite magnificently when they describe atoms which are an electron short as being 'holes'. Where a material has very few free electrons but many holes it is said to be 'positively charged'. The movement of charge is called 'electric current'.

If we can't see electricity, how do we measure it?

The answer is that we don't, directly. We measure voltage and current, we read the values written on the sides of components, and we work the rest out from there. But that's jumping the gun. Who built the meters that we use for measure-ment? What did they use?

Electricity is unseen. It is only through its ability to affect things physically that we know anything about it, be that people a hundred and more years ago, watching thunderstorms and making the legs of dead frogs jump, or us with our loudspeakers and computer screens. There are two ways of converting electrical energy to mechanical energy that we are concerned with.

MMF and current
MMF is most familiar to us as magnetism. The physical force that you feel be-tween magnets is called MMF, or 'magnetomotive force'. A less familiar mani-

festation occurs when two wires with electric current passing through them are placed near to each other. A physical force exists between them. (It's usually too small to feel but the effect is exploited in electric motors.) An MMF is a result of a magnetic field, and the study of magnetic fields is called electromagnetics, which we won't get involved in here.

If we were to make the wires a given length and separate them by a given distance we could increase the current until we had some given measurable force. Then we would have a unit value of electrical current, the ampere or amp. The amp is abbreviated to A. The symbol which we use for electrical current in equations is I.

This in turn gives rise to a unit of measurement for electric charge, the coulomb. The coulomb is defined as the amount of charge passing a point in a wire that carries one amp of current in one second. This gives us our first equation:

$$I = \frac{\Delta Q}{\Delta t} \tag{1.1}$$

where I is current in amps, Q is charge in coulombs and t is time in seconds. (The symbol Δ means 'change in', so the term $\frac{\Delta Q}{\Delta t}$ means 'change in charge divided by change in time'–or the amount of charge which passed the point divided by the time period that charge was measured for.)

This tells us that a charge of one coulomb passing a point in one second equates to a current of one amp; two coulombs in a second, or half a coulomb in one second would be two amps; and so on. A coulomb is a lot of charge. Typically we deal in charges of thousandths of coulomb per second (mC/s), or mA. (The prefix 'm' denotes a thousandth. We will use these prefixes increasingly as we progress. They are described fully in Appendix 1.)

'Electron current' and 'conventional current'   There is something of a mix-up which has taken place over history with regard to the direction of current flow, and no one has ever bothered to sort it out. You would think from what we've said above that, as the electrons do most or all of the running around in circuits, the current flow would be from the negative end of a battery to the positive end. It is, but we have a convention of always arrowing diagrams with the direction of current flow reversed. When people were discovering the laws of electricity they had some idea about things moving in wires but no way of seeing in which direction, so they had to guess. They guessed wrongly. Later this was discovered, but by this time people had been labelling currents on circuit diagrams for ages, quite happily and without any problem. The numbers all work out the same; its like deciding to rewrite your bank statement with the signs reversed. A minus sign means you are in credit or have deposited cash, and a plus that you took money out or are in debt. So long as you know what the symbols mean there is no problem. So it was with current flow, and they left it at that. It was too much hassle to change the labels on everything. If you hear talk about 'electron current' (which goes from negative to positive) and 'conventional current'

(positive to negative) this is what it means. Most people, and this book, use conventional current.

EMF and voltage

Electric charges of the same polarity repel and of opposite polarity attract. The attraction is the second physical force, called the 'electromotive force' or EMF. This is how electrons and holes know about each others' presence and are caused to move towards each other. EMF's are caused by electric fields. The study of electric fields is electrostatics and, like electromagnetics, we won't get into it here.

EMF is often called 'potential difference' or 'voltage'. It is measured in volts, abbreviated to V. When we need a symbol for voltage in equations we will also use *V*.

Voltage is a comparison between two places that are charged. We say that one place on a circuit is '10 V positive with respect to' another or that a potential difference of 10 V exists. The important thing to remember is that the number is always a difference between two places, or it means nothing. Generally we take one point, the metal chassis of equipment, mains earth or perhaps a piece of wire connected to the power supply, and we call it '0 V' or 'earth'. This becomes our reference for other measurements, unless we state otherwise.

The volt is defined by its ability to make charge do work (by pushing it more or less quickly through something and generating another form of energy, like heat or light). The amount of potential required to make one coulomb of charge do one joule of work is one volt:

$$V = \frac{J}{Q} \tag{1.2}$$

where *V* is voltage in volts, *J* is work in joules and *Q* is charge in coulombs. A joule is a measurement of energy or of work done which is derived from physical quantities. We will not actually use this but we will develop it into something more useful in Chapter 2. For now it is enough to know that a volt is a definite measurable quantity.

How fast does electricity move?   When a current flows through a wire, the same electron doesn't go in one end and come out of the other; one bounces a little way along, until it is able to find a home, which it often does by dislodging another and sending it on its way down the wire. (In doing so it generates heat.) The average speed of electron movement in a wire is extremely low, fractions of millimetres per second, even for currents which are, by our standards, large.

However, what we generally mean by this question is something very different; when we apply a voltage to one end of a wire, how long is it before the voltage is felt at the other end of the wire? The answer is that the impulse travels at the speed of light – instantly for all our purposes.

An analogy is a string of marbles in a narrow tube. If you shove the end marble very hard, the impulse (the voltage) travels along the tube like a wave and reaches the other end quickly, say one metre in a second. In that time, any one marble may have only moved 1 cm. So the average marble speed (the current) is 1 cm/s while the impulse speed is 1 m/s, 100 times greater. The average marble speed changes at the other end of the tube just as quickly – they start falling out of the tube at a rate of 1 cm/s, a second later. With electrons the effect is the same, but unimaginably more exaggerated.

If that seems a bit confusing, please don't worry – it is to most people. The important thing is that changes in voltage or current travel instantly but electrons have different physical speeds. (The second fact is only relevant if you want to understand how power is dissipated in components. Even if you don't, you can get by quite well just knowing that it is. We look at this in Chapter 3.)

2 Circuits and components

Conductors and insulators

Why do some materials conduct electricity, while others don't? Some materials are populated by good and careful atoms which don't lose their electrons easily (and probably feel ashamed if they do). Hence the itinerant population is low. The few free electrons that exist tend to sit miserably in one place, having given up hope of ever finding a home. These are insulators, typically air, rubber or the fibreglass backing of a circuit board. In other materials atoms are positively feckless. There are many free electrons and should some charged material come near they move accordingly. How do they know? They feel the voltage. These are conductors. Most metals are conductors. We use copper a great deal, gold where corrosion could be a problem, and sometimes aluminium or other metals for housing equipment.

Resistance

Some materials fall into a kind of grey area between being conductors and insulators. We can use these to make resistors. It's as if the material was a kind of mud; if we make the mud thicker electrons move less easily. The resistance of a given piece of a material depends on 'how thick the mud is' (the 'resistivity' of the material) and its geometry. Making it thinner increases, and doubling or tripling length doubles or triples, its resistance.

Ohm's Law

The unit for resistance is the ohm, or, and the symbol in equations is R. Now we come to the most important equation in electronics:

$$V = IR \tag{2.1a}$$

with V in volts, I in amps, R in ohms. This is Ohm's Law. We will use it so much that it bears being rewritten in its two other forms:

$$I = \frac{V}{R} \tag{2.1b}$$

and:

$$R = \frac{V}{I} \tag{2.1c}$$

These are used so frequently that you need to know them by heart.

Ohm's Law tells us that placing a steady voltage across a material causes a proportional steady current to flow.

Even insulators and conductors have resistance. Insulators have very high resistance; tens or hundreds of $M\Omega$ or more, so that for the normal voltages that we use an unmeasurably small current flows. If the voltage is raised to sufficiently high levels, usually at least a kV, a condition called 'breakdown' occurs; electrons are blasted free from atoms within the structure of the material, and its essential nature is changed. You will know if this happens; there is a nasty burning smell, and things stop working. It is a condition which we try to avoid.

Conductors are the opposite; their resistance is very low, small fractions of an ohm typically. The commonest conductor that we use is copper. We cannot normally measure the voltage across a piece of copper wire, unless it is long. We can find out, from catalogues and so on, the resistance per metre of any given type of wire; this depends on the wire, but figures of around 0.1 Ω/m are typical. Occasionally we need to find this out and calculate the total resistance of a length of cable when our wires are very long, in a large system. The wire will also have a maximum current rating, which is what we can pass through it before the amount of heat generated in it becomes dangerous to its structure. The insulating material around it may melt, causing it to short to things that it shouldn't, or the conductor itself. We only need know that we should stay comfortably within the maximum rated current for a conductor. In essence, anything can be destroyed with too much voltage/current.

To summarize, we have three distinct classes of material; conductors, insulators and resistors. Conductors connect our circuit up and have very low resistance. We assume that the voltage is the same at both ends of them unless they are very long or current is very high. They are just vessels to carry current. Insulators are the separating material between different points in a circuit that need to be kept apart and have very high resistance. Resistors are electrical components having a known resistance. We can predict the current through them for any given voltage, or vice versa, using the all-powerful Ohm's Law, $V = IR$.

Circuits, diagrams and common expressions

A circuit is exactly what its name suggests; a loop around which electric current can flow. The loop is generally made up of conductors (bits of wire or otherwise), circuit elements or components (which we start to meet soon) and one or more sources of electrical energy – current or voltage.

Circuit diagrams, sometimes called schematic diagrams, are symbolic representations of the way that the components of a circuit are interconnected. They are the starting point for most designs, and invaluable aid in servicing or fault-finding equipment that already exists. Each element of the circuit, from a battery or a tiny resistor to a complex microprocessor chip, is represented by a symbol, with each connection to it represented by a line joining to it. Interconnections between elements are shown by lines drawn between components. In reality, these could be any form of conductor; a wire soldered directly to a leg of the component (not recommended for a microprocessor!), a copper track on a PCB, or a test lead with a crocodile clip at either end of it.

Sometimes a circuit diagram doesn't show all of the circuit, and you can't see all of the loop around which the current will flow. (For example, see Figure 2.2 later.) The inference here is that things start to happen when something else is connected to it, completing the circuit – in this case a DC voltage (as you will see).

'Block diagrams' are a special type of circuit diagram worthy of explanation. You generally find them with fairly complex electronic circuits which are either too large to see on one drawing, or more understandable by being broken into smaller parts. Each 'circuit block' is drawn as a box or some commonly understood symbol, with lines connecting the blocks. The lines represent 'signals' (in reality any means by which the individual circuits can communicate), generally one or more conductors carrying voltage or current. It is the job of the person drawing the block diagram to make it less vague than this definition!

Two more very common terms – 'short circuit' and 'open circuit'. A short circuit is what happens when you connect two points with a conductor. An open circuit between two points means that they are not connected at all (an insulator is between them).

Resistors

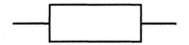

Figure 2.1 A resistor

We buy resistors as discrete electrical components. They come in values ranging from fractions of an ohm to tens of megohms, and are found everywhere in electronic circuits. We will see values between about 10 Ω and 1 M Ω most commonly in our circuits; values outside that do occur, but less commonly. They are identified by a series of coloured bands, each band signifying a number, or sometimes they just have the value written on. The colour code system is described in Appendix 2. The circuit symbol for a resistor is shown in Figure 2.1.

When we combine more than one resistor together in a circuit we can calculate an equivalent resistance for them, and treat them as a single component. There are two ways of doing this.

Resistors in series

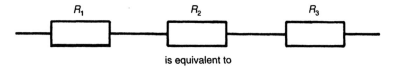

is equivalent to

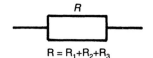

R = R₁+R₂+R₃

Figure 2.2 Resistors in series

Three resistors in series are shown in Figure 2.2. When we have this situation, we just add the values of the resistors to get their equivalent. We can do this for any number of resistors:

$$R = R_1 + R_2 + R_3 + \cdots + R_N \qquad (2.2)$$

Example 2.1: Three resistors of 1k, 4k7 and 680 Ω in series have an equivalent resistance of 6380 Ω

Resistors in parallel
Three resistors in parallel are shown in Figure 2.3. Here we must add the reciprocal of each to get the reciprocal of the equivalent resistance. Again this applies for any number of them:

$$\frac{1}{R} = \frac{1}{R_1} + \frac{1}{R_2} + \frac{1}{R_3} + \cdots + \frac{1}{R_N} \qquad (2.3a)$$

A special case of this worth remembering is when we have just two. Then we can calculate the equivalent in one step by dividing the product by the sum:

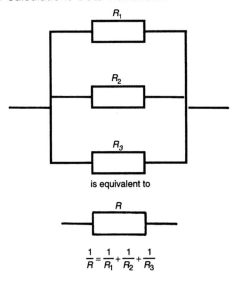

Figure 2.3 Resistors in parallel

Example 2.2: What is the equivalent parallel resistance of the three resistors from Example 2.1?

$$\frac{1}{R} = \frac{1}{1000} + \frac{1}{4700} + \frac{1}{680} = 2.68\text{m}$$

$$R = \frac{1}{2.68\text{m}} = 373\,\Omega$$

$$R = \frac{R_1 . R_2}{R_1 + R_2} \tag{2.3b}$$

We can always get some idea of equivalent resistance without calculating. Here are some rules of thumb:

1. For resistors in series the equivalent is always, obviously, larger than the largest resistor.
2. For resistors in parallel the equivalent is always smaller than the smallest.
3. Two same value resistors in parallel have a resistance which is half their value; three a third, and so on.

Are electrons psychic?

There are two ways in which electrons seem to know about what others are doing and be caused to move. One effect is due to EMF and the other to MMF. A full analysis requires excursions into the subjects of electrostatics and magnetics, but luckily we can get a good picture of these effects without all that. Both forces exist and have an application in electronics whether the charges are in the same conductor or two separate ones.

1. An EMF causes opposite charges to be attracted to each other, like ones to be repelled. If the two charges are in the same conductor this causes electrical current. If they are in two conductors separated by an insulator this gives rise to an effect known as capacitance. Capacitance is a property of a pair of isolated conductors which resists a change in the voltage between them.
2. Any electrical current creates a magnetic field, and any change in magnetic field creates a voltage in a conductor in the field. These two effects combine to create an effect known as inductance. There are two types of inductance. Self-inductance is a property of a conductor which opposes change in current flow. Mutual inductance is a property of isolated conductors by which a current flowing in one causes voltage in the other.

Capacitance and capacitors

When two conductors are separated by an insulator there is voltage between them if they have different amounts of charge. Changing that voltage requires charge to be moved, which requires energy. We can buy components specially made to have an appreciable amount of capacitance and use the effect in our circuits. Capacitors ('caps') consist of a pair of conductive plates separated by a special insulator called a 'dielectric', which magnifies the effect. When caps have no voltage across them we say that they are 'discharged' or, conversely, when they do that they are 'charged'.

The capacitance between the plates gets greater when:

1. they are moved closer together
2. they have greater surface area
3. the dielectric has a higher 'dielectric constant' – which, in English, means it is better at magnifying.

The unit of capacitance is the farad, or F, and the symbol used in equations is C. A farad is a lot of capacitance. A capacitor of one farad would have a potential of one volt across it when it had a charge of one volt on it:

$$Q = CV \quad \text{or} \quad C = \frac{Q}{V} \tag{2.4}$$

where Q is charge in coulombs, C is capacitance in farads and V is voltage in volts. This is more useful if we write it in terms of current, using Eqn(1.1):

$$I = \frac{\Delta Q}{\Delta t} = C \frac{\Delta V}{\Delta t} \tag{2.5}$$

This tells us that the rate of change of voltage across a capacitor is proportional to the current through it.

Capacitance exists everywhere, but is so small normally, say between two wires running side by side for a few feet, that we wouldn't bother much about it. (For cables containing more one conductor, separated by an insulator ('multicore cables') we can find out a value of 'typical capacitance between cores per metre', which we might take into account, if running signals through long cables, in the same way that we talked about for 'resistance/metre' earlier.)

Capacitors range in value from a few pF to 100 000 μF and more, with physical sizes getting larger with value. Figure 2.4 shows the circuit symbol for a capacitor.

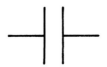

Figure 2.4 A capacitor

Capacitors come in a variety of types, with different characteristics. The most important of these are listed below:

1. Working voltage. If we put too much voltage across a capacitor, its dielectric, an insulator, will break down and current will flow from plate to plate, which it's not supposed to. The device will be destroyed. (They often, but not always, go short circuit.) The maximum working voltage tells us our safe limit. We try to design so that we never get too close to this.

2. Polarization. Large capacitors tend to be 'polarized' types – that is they use a special type of dielectric to squeeze more capacitance into a small volume, but the trade off is that you should only put voltage across them one way around. The most common type of polarized capacitor is the 'electrolytic'. They are marked with positive and negative terminals. If you disobey this by accident they go bang and bits of fluffy stuff come out of the top. Then you change them.

3. Tolerance. This varies greatly from type to type. Electrolytics have tolerances of something like −10%, +20%, because they are designed to be used in situations where you try to get as much capacitance into as small space as possible, and accuracy is not the prime consideration. At the other extreme, silvered mica types can be 1%, and give excellent stability. They are used in situations where the capacitor is used perhaps to set a time constant or a frequency (see Part Two) that must be accurately maintained.

4. Leakage current. Capacitors should be open circuit to a DC voltage. In practice you get a small current flowing, particularly for electrolytics.

There are other specifications which can come into play, but these are the most important. The value of a capacitor is usually written in numbers, in pF or in μF if no unit is specified. Unfortunately it is not always clear which, but it is usually possible to guess from the physical size, once you become familiar with them.

Capacitors in series and parallel

We can combine capacitors in series or parallel as we did with resistors. The law for caps in series is like that for resistors in parallel, and vice versa. Hence putting them in parallel creates a larger capacitor, and in series a smaller one. Here are the formulae:

In parallel:

$$C = C_1 + C_2 + C_3 + \ldots + C_N \qquad (2.6)$$

In series:

$$\frac{1}{C} = \frac{1}{C_1} + \frac{1}{C_2} + \frac{1}{C_3} + \ldots + \frac{1}{C_N} \qquad (2.7)$$

Self-inductance and inductors

Magnetic fields, when they move through a conductor, cause a voltage which can make electrons move. And electric current (or movement of electrons) causes a magnetic field. This adds up to an interaction of magnetic field, voltage and current, the net effect of which is this: if current flows in a wire there is a certain tendency for it to want to continue to do so, and not to change. (Sometimes we say this another way; when the current through an inductor changes, the inductor sets up a 'back EMF' which opposes the change in current.)

This effect is known as self-inductance (or just inductance). Like capacitance, it is not noticeable under normal circumstances. It can be made much stronger, however, if the wire is wound into a coil shape; the more turns the better. Inductance is also much increased by winding the coil on a core of some special material said to have a high 'permeability'; that is, it has the effect of concentrating the magnetic field. Such a device is called an inductor. Figure 2.5 is the circuit symbol.

The voltage across and current through an inductor are related as follows:

$$V = L \frac{\Delta I}{\Delta t} \qquad (2.8)$$

where L is the inductance in its standard unit, the henry (H).

This says that the voltage across an inductor is proportional to the rate of change of current through it. We can see that this is not unlike Eqn (2.5), for a capacitor. And, indeed, capacitors and inductors are in many ways the converse of each other. One resists change in voltage, the other change in current. As we look at these components again in the following chapters, we will see that in each circumstance they do opposite things to each other.

Figure 2.5 An inductor

Inductors are available with values from about 0.01 μH to 100 mH and more. Like capacitors, they get physically larger with value. As large value inductors consist of very many turns of wire on a heavy iron former, they also tend to get very heavy, one reason why they are not as popular as capacitors. They also tend to have a significant amount of resistance, which can be a disadvantage. The main specifications for an inductor are:

1. Internal resistance. This is caused by the large number of turns of wire, which can also be quite thin, to minimize the volume.
2. Maximum DC current. If too much DC current is passed through the wire of which the coil is made, it can overheat and melt, damaging or destroying the component. Even if the amount of DC is not so much as that, the core can 'saturate' – meaning that it doesn't work as well.

Despite their problems, they are quite common in some applications, and sometimes nothing else will do.

Inductors in series and parallel
Again, we can combine the inductance of series or parallel inductors and treat them as a single component. The laws are the same as those for resistors, and the inverse of those for capacitors.
In series:

$$L = L_1 + L_2 + L_3 + \ldots + L_N \tag{2.9}$$

In parallel:

$$\frac{1}{L} = \frac{1}{L_1} + \frac{1}{L_2} + \frac{1}{L_3} + \ldots + \frac{1}{L_N} \tag{2.10}$$

Mutual-inductance and transformers

Magnetic fields also cause electrons in unconnected conductors to influence each other. This effect is called mutual inductance. To make the effect strong, coils of wire are wound together on a core (which again has the effect of concentrating the magnetic field). This is a transformer. The symbol is shown in Figure 2.6.

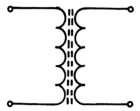

Figure 2.6 A transformer

The coil of a transformer into which you put current is called the 'primary'. The coil(s) from which you take current are the 'secondary(s)'. The 'turns ratio' of a transformer is the ratio of the number of turns on the primary to those on the secondary. When we apply AC voltage to the primary, we get AC from the secondary. And, conveniently:

$$\frac{T_p}{T_s} = \frac{V_p}{V_s} \tag{2.11}$$

where T_p and T_s are the number of turns on primary and secondary, respectively, and V_p and V_s the voltages.

Transformers are very useful in certain situations, the most common being the passing of a signal from one part of a circuit to another while keeping the two electrically isolated. Virtually all equipment connected to the AC mains supply is done so via a transformer. For safety reasons this is a special type, designed to have good electrical isolation. We *never* use anything but a proper mains transformer for this! Transformers are also very useful in many other circuits, and many other special types exist.

3 Direct current

AC and DC signals

We often hear electrical signals classified as 'AC' or 'DC'. These abbreviations stand for the terms 'alternating current' and 'direct current', but they are applied as much to voltages as to currents. In this chapter we will talk about DC. Given that we understand Ohm's Law, stated in Chapter 2, we have actually covered most of what ground there is. We will also introduce the term 'power' and consider its implications, which are important.

Direct current

A DC current is one which flows in one direction only. If we were to connect a resistor between the two terminals of a battery a DC current would result. For conventional current, it would flow from the positive terminal to the negative and have a magnitude, determined by Ohm's Law, of $I = V/R$. Similarly, we would refer to the voltage across that battery as a 'DC voltage'.

One application of DC is supplying power to an electronic circuit. The output from a power supply is often called the 'power rail' or just 'the rail'. The supply could be a separate circuit, operating from AC mains, or batteries. Virtually all circuits require one or more DC voltages to operate. A very common arrangement is to supply two rails with equal and opposite voltages, perhaps one of +15 V and one of –15 V. Then there will be three power rails; 0 V, used as a reference point for the whole circuit, and +/–15 V, used to supply power. Computer circuits often use a single rail of +5 V, perhaps with one or more other voltages used by the circuits which interface the computer to the outside world.

Circuit symbols for DC voltages and currents come in many flavours. Figure 3.1 shows some. You may just get lines, as in (a), with an indication of what the voltage or current is. (In a block diagram, like this example, the arrow may not follow the direction of current flow; it can just show that the supply is an 'input' to the circuit.) The symbol for a battery is shown at (b). The two general symbols for a voltage and current source are shown in (c). These would both indicate an unspecified device or circuit block; the intent is to indicate that a source exists at that point in the circuit, to help you see what is going on without actually telling you what physical devices to expect.

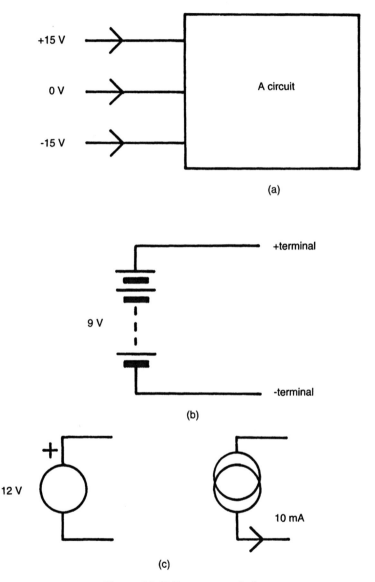

+15 V

0 V

-15 V

A circuit

(a)

+terminal

9 V

-terminal

(b)

12 V

+

10 mA

(c)

Figure 3.1 DC source symbols

Calculations on DC quantities

Addition, subtraction, multiplication and division can be performed on DC voltages and currents simply by performing the relevant operation on the numbers, paying due regard to their signs.

Resistors in DC circuits

There is hardly anything to say about this; they obey Ohm's Law, as already described. Also you should make sure that the rated power dissipation is not exceeded, as we discuss later.

Capacitors in DC circuits

If we put a DC voltage across a capacitor an instantaneous ('transient') current flows while it charges to the voltage. Thereafter no current flows as long as the voltage remains steady, with the exception of the leakage current that was discussed in Chapter 2. Capacitors are often used to isolate one part of a circuit from another to DC, while allowing AC signals to pass (as a 'DC block').

If we make a steady DC current flow into a capacitor (from a current source) the result is a ramp voltage across it. From Eqn (2.5), the ramp has a slope of:

$$\frac{\Delta V}{\Delta t} = \frac{I}{C} \qquad (3.1)$$

$\frac{\Delta V}{\Delta t}$ is the slope of the ramp in volts per second, I is the current in amps and C the capacitance in farads. This is a common method of making ramp generator circuits.

Example 3.1: A steady DC current of 1 mA is fed into a 680 pF capacitor; what is the slope of the voltage across its terminals? How long does it take to change by 10 V?

$$\text{slope} = \frac{I}{C} = \frac{1\text{m}}{680\,\text{p}} = 1.47\,\text{V}/\mu\text{s}$$
$$\text{Time taken to change 10 V is } \frac{10}{1.47} = 6.8\,\mu\text{s}$$

Inductors in DC circuits

An ideal inductor would be a short to DC. A real one appears to be a resistor with the value of the internal resistance of the inductor (hopefully very low). When DC current is passed through the inductor we get a transient voltage (the back EMF mentioned in Chapter 2). Thereafter the voltage will be small. Because of this, inductors are sometimes used to block AC but pass DC (as a 'choke').

Placing a steady DC voltage across an inductor causes the current waveform to be a ramp (in the same way that a DC current in a capacitor causes a ramp voltage). The slope of the current waveform is, from Eqn (2.8):

$$\frac{\Delta I}{\Delta t} = \frac{V}{L} \qquad\qquad (3.2)$$

(Did you notice how capacitors and inductors tend to do similar but opposite things to AC and DC voltages and currents?)

Power in DC circuits

Why do we worry about power dissipation?

As we saw in Chapter 2, connecting a voltage across a resistor causes a current to flow. The amount of the current depends on the value of the resistor; the lower the more current. In physical terms, this means that the average electron speed through the resistor is higher. Thus an electron making its way through strikes atoms in its path harder, and creates more heat. (We never have to worry about the electron speed, just the value of current; this is just a reminder of the physical process taking place.) The heat will be lost to the air, hopefully. How quickly the heat can dissipate depends mostly on physical structure of the resistor. If it cannot get away more quickly than it is generated, then the temperature will rise until it reaches a point where it destroys the component. Resistors go brown at this point, start to smell, and eventually go open circuit (sometimes catching fire in the process). Then you change them. Watching miniature resistors (1/8 W or 1/4 W types) catch fire isn't too bad, may even be mildly entertaining. (I have known heavy smoking engineers to use the trick to get a light in the middle of the night when the garage is shut, though I don't recommend the chemical content for your health.) Anything larger than that should definitely be avoided.

So that we can avoid this situation, resistors always have a 'maximum power rating'. Power is measured in watts(W). The power rating of a resistor might be anything from less than 0.1 W, for very small surface-mounted devices, to 100 W or more, for large, heavy devices intended to be bolted to 'heatsinks' (pieces of metal designed to radiate and make heat dissipate more quickly). On opening up pieces of electronic equipment we may sometimes see resistors which have obviously been overheating; the painted exterior may be brown and cracked. This may be the cause of the fault that we are investigating or an indication that some other fault is causing excessive current to flow. Sometimes it is an indication that the designer of the equipment wasn't as meticulous as he might have been. A resistor run continually at its maximum power may show signs of burning without misfunctioning, but the long-term reliability is likely to be reduced.

I have tried to follow a 'rule-of-thumb' which I learnt from an experienced engineer early in my career; to try not to run components continually at more than about one-third to one-half of their rated power. Running them at maximum for a short duration is generally OK; the device has time to cool off. If in doubt, prototype and use common sense. If it goes brown, you are being a bit ambitious.

Power calculation in DC circuits

How do we calculate the power dissipation in a resistor? Power is defined as the rate of energy dissipation, or:

$$P = \frac{J}{t} \tag{3.3a}$$

where P is the power in watts, J is the energy in joules and t is the time in seconds. From Eqn (1.2):

$$V = \frac{J}{Q} \quad \text{or} \quad J = VQ$$

Substituting into Eqn (3.3), we get:

$$P = V\frac{Q}{t} \tag{3.3b}$$

From Eqn (1.1):

$$I = \frac{Q}{t}$$

So:

$$P = VI \tag{3.4a}$$

Like Ohm's Law, this gets used a lot, and you need to know it. (The derivation is just for interest's sake.) The result is important enough to be restated, and learnt, in its two other forms:

$$I = \frac{P}{V} \tag{3.4b}$$

and:

$$V = \frac{P}{I}$$

Example 3.1: A resistor of 10 Ω has a voltage of 3 V across it. How much power does it consume?

Using Ohm's Law

$$I = \frac{V}{R} = 0.3 \text{ A}$$

and using Eqn (3.4a):

$$P = VI = 3 \times 0.3 = 0.9 \text{ W}$$

Often we use two equations which we can get from Ohm's Law and Eqn (3.4a) by a little algebraic substitution:

$$P = \frac{V^2}{R} \qquad\qquad (3.4d)$$

and

$$P = I^2 R \qquad\qquad (3.4e)$$

These enable us to answer questions like that posed in Example 3.1 in one step. Using Eqn (3.4d):

$$P = \frac{3^2}{10} = 0.9\,\text{W}$$

The business of using components within their maximum power rating quickly becomes fairly intuitive. For a circuit working from low voltages, say +/− 15 V DC, the maximum available voltage is 30 V. Our workshop resistor kit may contain miniature resistors all rated at 1/8 W. This enables us to calculate a value of resistor above which we cannot, no matter how we try, exceed the rated power.

Example 3.2: For the circumstances which we just described, what is the 'minimum' value, following the 'half-rated-power' rule-of-thumb?

$$P = \frac{V^2}{R} \quad \text{so} \quad R = \frac{V^2}{P}$$

P is half of 1/8:

$$\frac{0.125}{2} = 0.0625\,\text{W}$$

V = 30 V
Our threshold value of resistance is:

$$\frac{30^2}{0.0625} = 14400\,\Omega$$

− or say 15 kΩ.

This does not mean, of course, that we won't use any components of less than this value; but that above this value we can dot them around without worrying about power rating. At values around 15 k Ω, we only need worry if they spend a lot of time strapped across the power rails, which very few will. Even for low values, we will tend only to get the calculator out when we know that the resistor has a large voltage across it.

4 Alternating current

An AC voltage or current is any which is changing. It will very often be one which changes at a known rate. Where a DC voltage or current is easily described by its magnitude and sign ('3 V DC' or '−16.2 V DC' says it all, as long as we know where 0 V is), AC signals are a bit more complicated. We are going to start by considering the simplest type of AC signal. Then we will look at more complex ones.

The most ubiquitous AC signal is the sine wave. We will understand so much electronics in terms of how circuits respond to sine waves that it is worth our while spending some time describing and thinking about them.

The sine wave

If you haven't known about sine waves before, the word 'sine' probably brings back dimly remembered memories of school maths lessons, trigonometry, and all that goes with it. The sine is a mathematical function which describes the ratio of two sides of a right angled triangle in relation to one of the other angles, and a sine wave is a graph of the ratio against the angle. That's putting it mathematically. To many technicians and engineers, who try to do as much practical work as possible with the minimum use of mathematics, a sine wave is a very useful 'wiggle' which appears on the screens of an oscilloscope (a device which cleverly shows a graph of voltage against time) when they are working on equipment. We describe a great number of the essential properties of circuits in terms of what comes out when the input is a sine wave. The reasons for this are discussed later.

Figure 4.1 shows a sine wave. It is 2 V peak to peak (pk–pk or p–p) with a frequency of 1 Hz. (We'll define these terms soon.) The voltage begins at a level of 0 V, rises smoothly up to a level of 1 V, back down through 0 V to −1 V, and up to 0 V again. Each transition, 0 V to 1 V, 1 V to 0 V, 0 V to −1 V and −1 V to 0 V, takes the same amount of time, 0.25 s. The whole process is referred to as 'one cycle'.

We don't often come across sine waves which exist for one cycle only and then stop, however; usually the voltage oscillates continually while the circuit generating it is on. This may be for many hours. We call this a 'continuous signal'.

There are two things that we need to describe a sine wave unambiguously; how big it is, and how long it takes to complete one cycle. These two quantities we normally call 'amplitude' and either 'period' or 'frequency'.

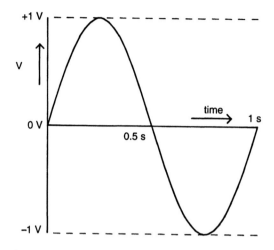

Figure 4.1 A sine wave

Amplitude

There are three ways of defining the amplitude of a sine wave:

1. Peak to peak amplitude is the difference between the positive and the negative peaks. Figure 4.1 is a sine wave of 2 V pk–pk. Pk–pk amplitude can be used to describe signals which are not sine waves as well.
2. Peak amplitude is the difference between the mid-point of the signal and the positive peak. Figure 4.1 is a sine wave of $(1–0) = 1$ V pk. We can say that:

$$\text{pk amplitude} = \frac{\text{pk} - \text{pk amplitude}}{2} \qquad (4.1)$$

3. Root mean square (RMS) amplitude is the level of a DC signal which has the same energy content as the AC signal. This odd way of specifying amplitude exists because it is needed for power calculations (on which more later in this chapter). For a sine wave:

$$\text{RMS amplitude} = \frac{\text{pk amplitude}}{\sqrt{2}}$$

or:

$$\text{RMS amplitude} = \text{pk amplitude} \times 0.707 \qquad (4.2a)$$

or again:

$$\text{pk amplitude} = \text{RMS amplitude} \times 1.414 \qquad (4.2b)$$

Which of these three we use depends on the situation, and we often need to convert between them. For instance, we might see a sine wave displayed on the screen of our oscilloscope, and wish to calculate the power which it causes to be dissipated in a certain component. Here we might read the pk–pk amplitude from the display, divide by two to get pk amplitude and then multiply by 0.707 to get the RMS amplitude. Then we can make a power calculation. Conversely, we might measure the amplitude of a sine wave using an RMS multimeter. If we needed for some reason to have a rough idea of the pk–pk value of the voltage, we could divide by 0.7 (multiply by 1.4) and then multiply by 2 in our heads – multiply by 2.8.

Period and frequency

The period of a sine wave is the time which it takes to complete one cycle. Its symbol in equations is T, and its unit is that of time, the second (s). For Figure 4.1, T is one second.

It is more usual to refer to the 'frequency' of a sine wave. The symbol for frequency in equations is f, and its unit is the hertz, abbreviated to Hz, meaning cycles per second. The frequency is the number of cycles completed in one second. Converting from frequency to period is very easy:

$$f = \frac{1}{T} \quad \text{or} \quad T = \frac{1}{f} \tag{4.3}$$

with f in Hz and T in s.

So our signal of Figure 4.1 has a frequency of 1 Hz.

Amplitude and frequency are independent of each other; a sine wave of 3 kHz will oscillate 3000 times per second whether it has a level of $1\,\mu V$ or 10 V. One of 10 V RMS will still be at 10 V RMS, and have the same heating effect through a given resistor whether it is at 10 Hz or 1 MHz.

(An interesting aside to this second case; if the frequency were 0.01 Hz, a period of 100 s, would this still be true? No; remember, from Chapter 3, the mechanism of heating and cooling of resistive components. In this case, the frequency is so low that we would do better to consider it as a slowly varying DC voltage. For a quarter of the period, 25 s, the resistor will have a voltage across it of 20 V or greater! This time is obviously long enough to destroy the resistor if it is not adequately rated. In our power calculations, we should budget for the pk amplitude of about 14 V. This is a very extreme example, and such slow moving signals are not often encountered. The statement in the above paragraph holds true for frequencies of a few Hz and above.)

Phase

We can describe a single sine wave completely by specifying its amplitude and frequency. When we are talking about two such signals the important question of phase arises.

Figure 4.2 shows the sine wave of Figure 4.1 again, labelled V_1, and a second nearly identical sine wave which starts its cycle at a later time t, labelled V_2. We say that there is a phase difference between V_1 and V_2.

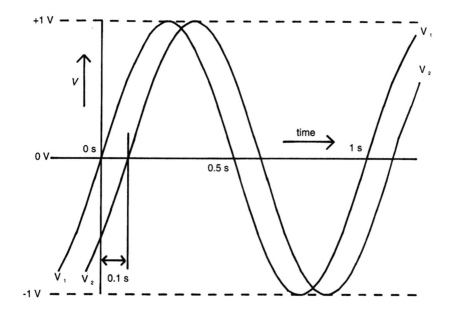

Figure 4.2 Illustration of phase

Phase is an angular quantity, and its units are degrees or radians. We use degrees in this book. There are two important points to bear in mind when talking about phase:

1. We only generally make phase comparisons between waveforms of the same frequency.
2. Phase, like voltage, is a relative quantity. It is meaningless to say that, for instance, 'the voltage has a phase angle of 30°' unless we specify that that 30° refers to some other waveform. A reference must be specified.

To calculate the phase difference between two waveforms, we use the following expression:

$$\phi = \frac{360t}{T} \qquad (4.4a)$$

We can, of course, calculate t from ϕ and T:

$$t = \frac{T\phi}{360} \qquad (4.4b)$$

where ϕ is the phase in degrees, t is the time difference in seconds and T is the period in seconds. For example, if t in Figure 3.2 is 100 ms, then $\phi = 36°$. V_2 starts 100 ms after V_1, so we can either say that V_2 lags V_1 by 36° or that V_1 leads V_2 by 36°. The two statements mean the same thing. We might, if we were being awkward, say that V_2 leads V_1 by $(360-36)°$ or 324°. This statement means that V_2 starts its cycle 324°, or 900 ms, before V_1 – just as true if they are continuous, but not the easiest way to say it. Normally we specify phase using the smallest possible angle, which is always less than 180°. Sometimes calculations give us numbers larger than this, and we convert them to a more sensible form.

Waveforms which are 90° out of phase are said to be in quadrature, while ones which are 180° out of phase are said to be in antiphase. Putting a waveform in antiphase is 'inverting' it. Waveforms which have no phase difference are simply 'in phase'.

Calculations on AC quantities

We can do calculations on AC quantities of the same frequency to analyse the behaviour of a circuit at that frequency. Every quantity that we use will have two components; magnitude (amplitude) and phase.

Mathematical operations on waveforms of any phase are slightly tricky and require the use of something called 'phasor mathematics'. Ideally, a mathematical technique called 'complex algebra' is used which makes use of a notional quantity called j, equal to $\sqrt{(-1)}$ to denote a phase shift of 90°. This technique allows not only voltages and currents but also combinations of reactances, resistances and even amplification to be treated under one umbrella. The behaviour of circuits can then be predicted at a fixed frequency.

The algebra involved is actually not much more difficult than the more conventional techniques which the next chapter uses, but a complete explanation of it is somewhat beyond the scope of this book. Should you already be familiar with it then Appendix 4 gives a quick restatement of its principle techniques.

Even if you are not (or don't want to bother with it) its not really a problem. There are some simple rules and special cases of phasor mathematics which are presented below. In Part Three we will consider what some basic circuits do under these conditions and hence get a lot of insight into their workings. You could take these ideas on board now, or refer back to this point when you get there. Anyway, here goes:

Phase Rule 1:   For waveforms in phase, magnitudes of quantities can be used in calculations directly. The phase of the result is the same as that of the 'inputs' to the calculation.

Phase Rule 2:   A positive value of phase means 'leading', a negative one 'lagging'.

Phase Rule 3:   A negative value of magnitude means an inversion of the quantity (i.e. it is in antiphase). So subtraction is addition with one wave-form inverted. We could say that $-V$ is V at an angle of $180°$.

Phase Rule 4: Adding two equal waveforms in antiphase gives a result of zero. This is called 'phase cancellation'.

Phase Rule 5: Multiplication of any pair of waveforms. The magnitudes are multiplied and the phase angles added together.

Phase Rule 6: Division of any pair of waveforms. The magnitudes are divided and the phase of the denominator subtracted from the phase of the numerator.

Resistors and sine waves

The terms pk, pk–pk and RMS can be used to refer to both voltages and currents. Provided that we are consistent we can do Ohm's Law calculations on resistors with any of them. So, for example:

1. A pk voltage of 10 V across a 1k resistor causes a pk current of $\frac{10}{1000} = 10\,\text{mA}$ to flow.
2. A pk–pk current of 3 mA through a 4k7 resistor causes a pk–pk voltage of $3 \times 4\text{k}7 = 14.1\,\text{V}$ across it.
3. An RMS voltage of 150 V across a 1M resistor causes an RMS current of $\frac{150}{1\text{M}} = 150\,\mu\text{A}$ to flow.

Current and voltage through a resistor are always in phase. In fact, the current waveform is always an exact replica of the voltage waveform, its magnitude at every point being determined by Ohm's Law.

Capacitors and sine waves

A capacitor with a sinusoidal voltage of frequency f across it will have a sinusoidal current flowing through it. The ratio of the voltage to the current is known as the 'reactance' of the capacitor at frequency f. The situation is analogous to that with a resistor, and the unit of reactance is again ohms. And Ohm's Law again applies:

$$V_C = I_C X_C \tag{4.5}$$

where X_C is capacitive reactance, V_C is the voltage across the capacitor and I_C is the current through it. As with resistors, amplitudes of voltage and current can be specified using pk, pk–pk or RMS conventions, so long as the same is used for both.

We can calculate the reactance of a capacitor at any particular frequency using the expression:

$$X_C = \frac{1}{2\pi f C} \qquad (4.6)$$

where C is the capacitance in farads and f is the frequency. We can see from this that the magnitude of the reactance of a capacitor decreases proportionally with frequency.

But hold on! Capacitors are more than 'frequency-dependent resistors'. They do something important to AC signals. The current through a capacitor always leads the voltage across it by 90°. This is the difference between a capacitor and a resistance with the same value as its reactance at that frequency.

We can often get a very good idea of what a circuit does by thinking of caps as 'frequency-dependent resistors', but we can't do calculations to solve CR networks (using techniques from Chapter 5) without taking account of phase. (A CR network is any consisting of capacitors and resistors – we'll meet CR, LR and LCR networks in Part Two.)

For calculation purposes, we consider that capacitive reactance has a phase angle of −90°, as passing a current with a phase angle of 0° through it results in a voltage of 90° lagging. (See Phase Rules 5 and 6 earlier in this chapter.) (It is this idea of assigning phase to reactance that makes otherwise incomprehensible complex numbers work.)

Now that we've met the terms resistance and reactance, this seems like the moment to mention the word 'impedance', which is used a lot. Strictly speaking, resistors have resistance, capacitors (and inductors) have reactance. Circuits have some combination of the two. If you want to be non-committal about which you mean, say 'impedance' – it can be either. And everyone should understand.

Inductors and sine waves

Like capacitors, the sinusoidal voltage across and current through an inductor are proportional at any given frequency. The ratio is again known as the reactance of the inductor, and here's Ohm's Law again:

$$V_L = I_L X_L \qquad (4.7)$$

with X_L as inductive reactance, V_L voltage across the inductor and I_L current through it. Again we may use pk, pk–pk or RMS amplitudes but must be consistent. For an inductor voltage leads current by 90°, so we consider that the reactance has a phase angle of 90°. The inductive reactance can be calculated using the expression:

$$X_L = 2\pi f L \qquad (4.8)$$

where L is the inductance in henries and f is the frequency in hertz. From this, inductive reactance increases proportionally with frequency.

(Again we see capacitors and inductors doing nice complementary things to each other, while resistors form a nice 'neutral background' against which to view it all.)

Why are sine waves so important?

In other words, why do we do so much of our analysis of AC circuits in terms of what they do when you put a sine wave through them?

All AC waveforms can be considered as being the sum of sine waves of various frequencies. The mathematical technique behind this is known as Fourier analysis. We are not going to get into the mathematics here, but we will consider the implications.

What the maths says is that we can treat any continuous periodic waveform (such as all the ones in Figure 4.3) as the sum of a number of sine waves of various frequencies. The frequencies will all be whole number multiples of the frequency of the repeating periodic waveform. So a 1 kHz triangle wave can be considered to be the sum of sine waves of 1 kHz, 2 kHz, 3 kHz . . . and so on. The amplitudes and phases of the sine waves will vary, generally getting less with frequency, but we could, if we knew them, and had the means to do it, reconstitute the waveform, like baking a cake to a recipe. For instance, the 'recipe' for a square wave of 1kHz is:

1 kHz sine wave, at 1 V RMS, plus ...

3 kHz sine wave, at $\frac{1}{3}$ V RMS, plus ...

5 kHz sine wave, at $\frac{1}{5}$ V RMS, plus ...

7 kHz sine wave, at $\frac{1}{7}$ V RMS, plus ...

. . .

n kHz sine wave, at $\frac{1}{n}$ V$_{RMS}$.

'n' can be any whole number as high as you care to keep going. The higher in frequency you continue, the better the shape of the resulting wave – that is the closer to the ideal, with the voltage changing from one voltage to the other almost instantaneously. Real square waves are not ideal, having a finite 'risetime' – the time taken to get from 10 to 90% of the value of the step.

You will notice in our square wave 'recipe' (known more properly as a Fourier series) that we only have the odd number multiples of 1 kHz. The 1 kHz waveform is known as the 'fundamental' of the series (i.e. the sine wave at the same frequency as the periodic waveform being analysed). The sine waves at 2 kHz, 3 kHz, and so on, are known as the 'harmonics' of the fundamental. 2 kHz would be the 2nd harmonic, 3 kHz the third, and so on. So we can say that 'a square wave consists only of the fundamental plus odd harmonics'. It contains no even harmonics. The exact amplitude of the result can be foretold, but we are not often worried about it.

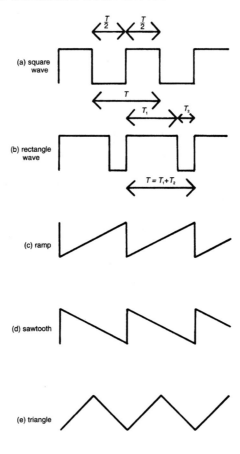

Figure 4.3 More AC waveforms

You can get tables showing the equations to write various waveforms in terms of their harmonics, but we rarely need them. Quite a lot can be done intuitively. Where a waveform has a sharp transition between two levels, that's where you get the high harmonics. The 'frequency response' of a circuit is the way that its output, relative to the input, varies with frequency. If a circuit, such as a low-pass filter (of which more in Part Two) tends to attenuate high frequencies then it will affect the edges of a square wave – they will be slower. If we know the frequency at which the filter begins to 'cut', then we can get an idea of how serious this is – for example if our square wave is at 1 kHz, and the filter starts to 'roll off' from 2 kHz, the effect will be quite noticeable. If the filter rolls off from 20 kHz, much less so. Conversely, the low frequencies of a waveform contribute most to the portions where it changes slowly, or not at all; the top and bottom of a square wave will tend to 'sag'. The time responses of 1st order circuits, which we look at in Chapter 9 are in fact extreme examples of this.

More AC waveforms

Figure 4.3 shows some types of AC signal which are commonly encountered. There is not too much to know about them, except to recognize their shapes and their names. With rectangle waves we sometimes hear talk of mark-to-space ratio. This is the ratio of duration of the more positive part to the more negative – $T_1 : T_2$ on the diagram. These waveforms are best specified, amplitude-wise, in terms of pk–pk level, again the difference between the positive and the negative voltage. Time-wise, we obviously refer to them by either their frequency or their period, with $f = 1/T$ once again.

AC waveforms with a DC level

It often happens that we come across signals which are a combination of AC and DC. Then we say that the waveform has a 'DC level'. The DC level of an AC waveform is just its average level. Hence if it has an equal area below 0 V as above it has a DC level of zero.

Sine waves, square waves, ramps, sawtooths and triangles have a DC level of zero if they move between equal positive and negative voltages. Rectangle waves with uneven mark-to-space ratios (if they didn't have they'd be square waves) will go more positive than negative (ratio less than 1:1) or more negative than positive (more than 1:1) to have no DC level. In fact changing their mark-to-space ratio changes their DC level too, assuming that they stay between the same two voltages.

Figure 4.4 shows some examples. The triangle wave goes between 1 and 9 V and so it has a DC level of 5 V. This is the case regardless of the duration of its ramps. (Sketch one of any shape on squared paper and then draw a line horizontally through its mid point. You should see that its area above the line is the same as that below. This is true for sines, squares, sawtooths and ramps as well.) So the DC level of any sine, square, ramp, sawtooth or triangle is:

$$V_{DC} = \frac{V_+ + V_-}{2} \tag{4.9}$$

For the triangle in Figure 4.4, $V_{DC} = (9 + 1)/2 = 5\,\text{V}$.
The rectangle wave goes between +5 V and −5 V. But it has a mark-to-space ratio of 3:1, and so has a positive DC level, as the area above 0 V is greater than that below. The DC level is given by:

$$V_{DC} = \frac{T_M}{T} V_M + \frac{T_S}{T} V_S \tag{4.10}$$

T_M is the duration of the mark, V_M its voltage, T_S and V_S the same for the space. T is the period, $T_M + T_S$. So for our example:

$$V_{DC} = 0.75 \times 5 - 0.25 \times 5 = 2.5\,\text{V}$$

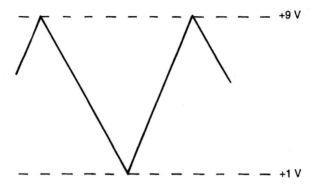

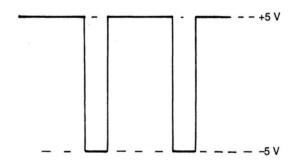

Figure 4.4 AC waveforms with DC level

Power calculations with AC waveforms

For power calculations involving AC waveforms, we must use the RMS value of
the signal. RMS stands for 'root-mean-square'. It is what we get when we take
the average of the square of the amplitude of a waveform over its cycle, and then
take the square root of that. If that sounds like a rigmarole, it is, but we never actually have to do it.

Sine waves
We already know how to find the RMS value of a sine wave from our discussions
on amplitude earlier in this chapter. When calculating the power that a sinu-

soidal signal develops in a load we must also take account of the phase angle between voltage across and current through that load.

The power dissipated by a circuit element which is passing a sinusoidal current (with no DC) is defined as:

$$P = VI \cos\phi \tag{4.11}$$

where P is the power in watts, V is the RMS voltage across the component in volts, I is the RMS current in amps and ϕ is the phase angle between voltage and current.

The $\cos\phi$ term varies between 1 when $\phi = 0°$ and 0 when $\phi = 90°$ (or $-90°$). Hence for resistors, for which $\phi = 0°$ always, we simply use the same power equations that were given for DC circuits in Chapter 3, of course using RMS values. Equally we can say that ideal capacitors and inductors, for which $\phi = 90°$ (or $-90°$), don't dissipate power. When we use circuits made up of combinations of these components, and ϕ is between 0 and 90°, the power will be somewhere between zero and 'volts-times-amps'. But all of that power will be dissipated in the resistive components of the circuit.

(An aside: the term 'peak power' is sometimes heard, but has no relationship to 'pk amplitude'. It is used in such things as sound amplification systems to describe the the maximum amount of electrical power delivered to a loudspeaker at any given moment. There is no such thing as pk – pk power. Thought I'd mention it.)

RMS values of other waveforms
Its not often that we have to calculate these, but if we do it's nice to know how. If you're a beginner you may well prefer to skip this. I just couldn't resist putting them in.

For any triangle, ramp or sawtooth wave with no DC level, and a positive excursion, V_{PK}:

$$RMS_{triangle} = \frac{V_{PK}}{\sqrt{3}} = 0.577\ V_{PK} \tag{4.12}$$

For any rectangle wave (of which a square wave is just a special case), if T_M is the duration of the mark, V_M its voltage, T_S the duration of the space, V_S its voltage and T the period ($T_M + T_S$):

$$RMS_{rectangle} = \sqrt{\left(V_M^2 \frac{T_M}{T} + V_S^2 \frac{T_S}{T} \right)} \tag{4.13}$$

with V_M the voltage of the mark and V_S that of the space. This takes into account the DC level.

For a signal with AC and DC components:

$$RMS_{AC+DC} = \sqrt{\left(V_{DC}^2 + V_{RMSAC}^2\right)} \qquad (4.14)$$

And finally, for a 'piecewise' signal with known RMS values over different parts of its period

$$RMS_{total} = \sqrt{\left(V1_{RMS}^2 \frac{t1}{T} + V2_{RMS}^2 \frac{t2}{T} + \cdots + Vn_{RMS}^2 \frac{tn}{T}\right)} \qquad (4.15)$$

We can see that Eqn (4.13) is just a special case of this.

By way of illustration:

Example 4.1: The triangle wave of Figure 4.4 has an AC component that goes between +4 V and −4 V. Its RMS value is therefore $0.577 \times 4 = 2.31$ V. It has a DC component of 5 V, so the true RMS value is:

$$RMS_{triangle} = \sqrt{(5^2 + 2.31^2)} = 5.5\,V$$

Example 4.2: A signal consists of a high frequency sine wave of 5 V pk for 400 ms with no DC, followed by a DC voltage of 2 V for 300 ms. What is the RMS value?

RMS value of the sine wave is $5 \times 0.707 = 3.54\,V$

$$RMS_{total} = \sqrt{\left(\frac{4}{7} \times 3.54^2 + \frac{3}{7} \times 2^2\right)} = 2.98\,V$$

---

5 A circuit analysis toolkit

Introduction

This chapter is intended as a quick guide to the basic techniques of network analysis. The mathematical requirements are minimal; you need to be able to transpose equations and solve simultaneous equations by substitution. We also get to consider some characteristics of real voltage and current sources, which ties in nicely with some of the analysis tools presented.

We will take DC circuits for all of our examples, as this is the easiest way of demonstrating the techniques. Everything said applies equally to AC circuits as well, but you generally need to use complex numbers (see Chapter 4 and Appendix 4) to take account of phase if there are capacitors and inductors present, or sources at different phases.

Basic topologies

There are two basic shapes that electric and electronic circuits come in; series and parallel. Each has associated with it one of two electrical laws which tell us how to go about calculating the voltages and currents in that circuit.

The series circuit and Kirchhoff's voltage law

A series circuit is represented in Figure 5.1. The components are an input voltage, V, 10 V DC, and three resistors: R_1, 1k; R_2, 2k2; and R_3, 4k7. We have also labelled on the drawing, for our convenience, the voltages across each resistor, V_1, V_2 and V_3. The voltages are labelled with arrows. With DC voltages we follow a convention of putting the head of the arrow to the more positive end of the component.

The key to analysing a series circuit is to remember that the current is identical through each component in the loop. (We can also say that connecting the components in a different order will not affect that current.) Hence the current in the loop is labelled just once, this time with an arrow on the loop, with the legend I next to it to identify it. The head of the arrow shows the direction of current flow.

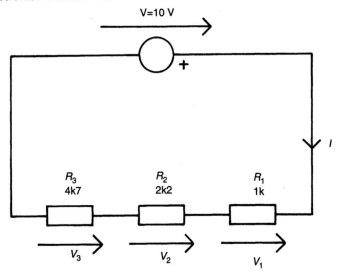

Figure 5.1 Series circuit

Example 5.1: Suppose that we wish to calculate V_1, V_2 and V_3. How do we do it? We need to find the value of current, I. Then we will be able to calculate each voltage using Ohm's Law.

We can find the total resistance as described in Chapter 2. In our case:

$$R_S = 1k + 2k2 + 4k7 = 7900\,\Omega$$

Now we can say that:

$$I = \frac{10}{7900} = 1.27\,\text{mA}$$

Now we can use Ohm's Law to find V_1, V_2 and V_3:

$$V_1 = 1.27\,\text{mA} \times 1k = 1.27\,\text{V}$$
$$V_2 = 1.27\,\text{mA} \times 2k2 = 2.78\,\text{V}$$
$$V_3 = 1.27\,\text{mA} \times 4k7 = 5.95\,\text{V}$$

Kirchhoff's voltage law states that, in any loop of voltages on a circuit, their sum will be zero if we take the clockwise ones as being positive, the anticlockwise ones as negative (or vice versa). Put another way, the clockwise ones equal the anticlockwise ones. This enables us to check our calculations: 1.27 + 2.78 + 5.95 = 10V – we did it right. In any mathematical analysis of a circuit, its always good to have a way to check our result, and using voltage loops is one way.

The parallel circuit and Kirchhoff's current law

Example 5.2: Figure 5.2 shows a basic parallel circuit made with the same components. With parallel circuits the rule is that the voltage across all the branches is identical. Hence there is only one voltage to be labelled, the source voltage. But there are four currents: I, the current leaving the voltage source, and I_1, I_2 and I_3, the currents through the resistors. We would like to calculate them.

We can see that the voltage across each resistor is the same; the 10 V source voltage. Therefore we can apply Ohm's Law directly to each branch of the circuit to find I_1, I_2 and I_3:

$$I_1 = \frac{10}{1k} = 10 \text{ mA}$$

$$I_2 = \frac{10}{2k2} = 4.55 \text{ mA}$$

$$I_3 = \frac{10}{4k7} = 2.13 \text{ mA}$$

Kirchhoff's current law states that the total of currents entering and leaving any 'node' (a node is any place where two or more conductors join) is zero. This enables us to calculate the current I:

$$I = I_1 + I_2 + I_3 = (10 + 4.55 + 2.13)\text{mA} = 16.68 \text{ mA}$$

We know that the resistance between any two points is the ratio of the voltage across them and the current passing between them. So in this case the resistance is:

$$R_P = \frac{10}{16.68 \text{ mA}} = 600 \, \Omega$$

We could have calculated the parallel resistance directly, by using the formula from Chapter 2:

$$\frac{1}{R_P} = \frac{1}{1k} + \frac{1}{2k2} + \frac{1}{4k7} = 1.67 \, \text{m} \Omega$$

So:

$$R_P = \frac{1}{1.67 \text{ mA}} = 600 \, \Omega$$

Again we have cross-checked our result, and it's fine.

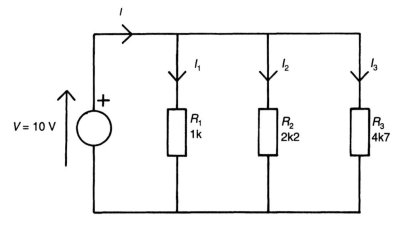

Figure 5.2 Parallel circuit

Arrow directions for voltages and currents

It is worth taking a moment to think about the rules that we apply when adding voltages and currents in using Kirchhoff's two theorems, as it can be surprisingly easy to confuse ourselves.

Here's the rules. If we follow these, things work:

1. The arrows for voltage across and current through a voltage or current source must point in the same direction.
2. The arrows for voltage across and current through a resistance must point in opposite directions.
3. For a voltage loop, the sum of arrows going clockwise equals the sum of arrows going anti-clockwise.
4. For a node, the sum of currents entering a node equals the sum of currents leaving the node, as denoted by the arrows' directions.

In the above examples, it was obvious which end of the resistors would be more positive and in which direction currents would be flowing. Hence drawing the arrows was easy.

In other instances it will not be obvious, when labelling currents and voltages at the start of the analysis, which end of the component is the more positive. Such a situation is shown in Figure 5.3. We wish to know the voltage V_2, but this voltage is affected by both voltage and current sources which have a tendency to create voltages of opposite polarity across R_2. The way to approach this sort of problem is not to worry about the polarity of the voltage; if we label it in the wrong direction, the voltage will come out as negative in solving the equations. We simply get on with labelling the voltages and currents (following rules 1 and 2), then we can write equations for voltage loops (using rule 3, the voltage law) or nodes (with rule 4, the current law).

Example 5.3: Solve Figure 5.3.

For node A, we can write:

$I_1 = I_2 + 10\,\text{mA}$
$I_2 = I_1 - 10\,\text{mA}$ (call this Eqn (A))

For the loop on the left we can write:

$V_{\text{IN}} = V_1 + V_2$
$V_2 = V_{\text{IN}} - V_1$

(but $V_1 = I_1.R_1 = 10\,\text{k}.I_1$ and $V_2 = I_2.R_2 = 2\,\text{k2}.I_2$)

so:

$2\,\text{k2}.I_2 = 5 - 10\,\text{k}.I_1$
$I_2 = \left(\frac{5-10\,\text{k}.I_1}{2\text{k2}}\right)$
$I_2 = 2.27\,\text{mA} - 4.55.I_1$

We can substitute this back into Eqn (A), giving:

$2.27\,\text{mA} - 4.55.I_1 = I_1 - 10\,\text{mA}$
$I_1(1 + 4.55) = 2.27\,\text{mA} + 10\,\text{mA}$
$I_1 = \frac{12.27\,\text{mA}}{5.55} = 2.21\,\text{mA}$

So, again from Equation A:

$I_2 = 2.21\,\text{mA} - 10\,\text{mA} = -7.79\,\text{mA}$

and therefore

$V_2 = I_2.R_2 = -17.1\,\text{V}$

and

$V_1 = 2.21\,\text{mA} \times 10\text{k} = 22.1\,\text{V}$

Our solution is: $V_1 = 22.1\,\text{V}$, $I_1 = 2.21\,\text{mA}$, $V_2 = -17.1\,\text{V}$, $I_2 = -7.79\,\text{mA}$. We can now redraw the circuit, reversing the direction of the arrows where our solution gave a minus sign (Figure 5.4). A few seconds of calculator-punching demonstrates that nothing went wrong with the number crunching, and the solution obeys the voltage and current laws. There are quicker ways to get to this result, but as a 'swiss army knife' technique, this is a good one to have. Soon we get some better ones still.

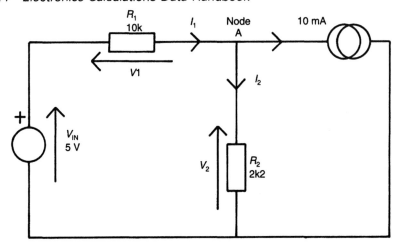

Figure 5.3 Example 5.3

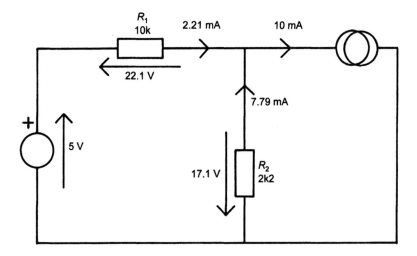

Figure 5.4 Solution to Example 5.3

More about voltage and current sources

The voltage and current sources that we considered above were 'ideal' – that is, we consider that they will supply their correct voltage or current, regardless of what load is placed on them. (As such, they have no real existence.) For the ideal voltage source, we could say that the change in voltage across it is zero for any change in output current. This means that a notional quantity which we call its 'internal resistance' is also zero, by Ohm's Law. For the current source we could say that the change in output current is zero for any change in output voltage. Hence its internal resistance is infinite. Real voltage and current sources are not ideal, and we have a way to represent this in our circuits (see Figure 5.5).

We give our ideal voltage source a series resistor, R_S. The closer this resistor is to a short circuit the closer the source is to ideal. If we connect a real voltage source to too low a load resistance the voltage across the load will drop (and the source could be damaged).

The current source is given a parallel resistor, R_P; the closer this is to an open circuit, the better the source. If we connect a real current source to too high a load resistance the current through the load will drop.

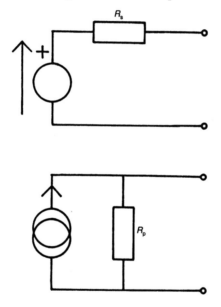

Figure 5.5 Representation of non-ideal sources

These circuit representations of real voltage and current sources are invaluable for circuit analysis, as we will now see. It is possible for any network of voltage sources, current sources and resistances to be replaced by either a single ideal voltage source with a series resistor, or a single ideal current source with a parallel resistor.

Thévenin's and Norton's equivalent circuits
The techniques that we used in the last example of circuit analysis are fine for instances where we can get a solution for the circuit by writing two or three

simultaneous equations and solving them. If the network gets much more complicated than that, however, the amount of maths involved can get pretty tedious, and it can get time consuming to arrive at the correct result without making a numerical error. Also, we have only solved the network for one value of load resistance; if we want to see what happens with a different value, we must do the calculations again. This is where replacing whole lumps of circuit with an equivalent non-ideal current or voltage source is great.

Here's how we go about finding the right values for our equivalent circuit:
Replace with a voltage source (Thévenin's equivalent circuit):

1. Cut the network at two places to isolate the section to be replaced.
2. Calculate the voltage at the load terminals with no load connected (called the 'open circuit voltage' or V_{TH}).
3. Calculate the equivalent resistance looking into the load terminals when any voltage sources are replaced with a short circuit and any current sources with an open circuit (called the 'output resistance' or R_{TH}).

Example 5.4: What is the voltage across R_L in Figure 5.6?

First we separate out the 12 V and the 3 mA source, and combine then into a single voltage source(– Figure 5.7). The points where we cut are labelled A and B. Replacing the 12 V source with a short and the 3 mA one with an open, we get $R_{TH} = 1 k + 1k = 2k$ (Figure 5.8). When we calculate the open circuit voltage, we get 3 mA $\times$ 1k = 3 V across R_2, and nothing across R_3, which has no current flowing in it. Hence V_{TH} is $12 - 3 = 9$ V.

The equivalent circuit is replaced into the original in Figure 5.9. Things are getting easier! Now we can cut and replace at the load terminals themselves, L_1 and L_2. This time we'll replace with a current source.

Looking into L_1 and L_2 with both sources replaced by a short, we get 2k in parallel with 2k, so R_N is 1k. (Figure 5.10).

To calculate short circuit current, we replace R_L with a short, and then calculate the current through the two resistors; 2.5 mA and 4.5 mA as shown on the drawing, taking care to label directions correctly. I_N is then clearly the difference between the two (by Kirchhoff's current law), 2 mA. So Figure 5.11 shows the equivalent circuit with R_1 now connected. If R_1 is 1 kΩ, parallel resistance with R_N is 500 Ω, and voltage across R_1 is 1 V, terminal L_1 being most positive.

Naturally we want to check, and the easiest way to do this is to redraw the original circuit, labelling in voltages and currents, using Ohm's Law and voltage and current laws, and working backwards from R_1. If there is a problem it will become apparent. We find we are OK (see Figure 5.12).

We now could find the current through any other load using our existing solution – no need to do all that number-crunching again.

The network can now be replaced with an ideal voltage source equivalent to V_{TH}, in series with a resistor equivalent to R_{TH}.

Replace with a current source (Norton's equivalent circuit):

1. Cut the network at two places to isolate the section to be replaced.
2. Calculate the current at the load terminals when they are short circuited (called the 'short circuit current' or I_N).
3. Calculate the equivalent resistance looking into the load terminals when any voltage sources are replaced with a short circuit and any current sources with an open circuit (called the 'output resistance' or R_N).

The network can now be replaced with an ideal current equivalent to I_N, in parallel with a resistor equivalent to R_N.

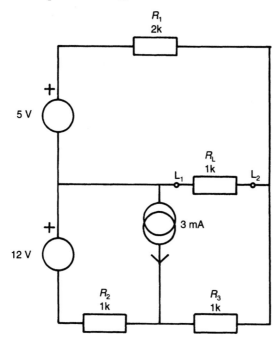

Figure 5.6   Example 5.4

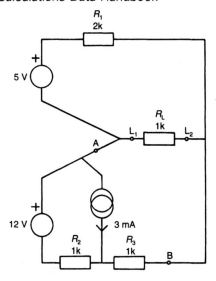

Figure 5.7 Figure 5.6 redrawn before simplifying

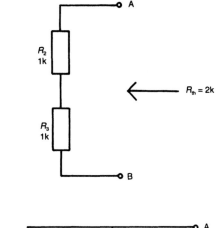

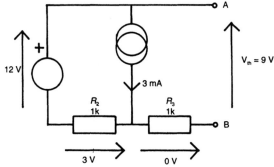

Figure 5.8 Finding R_{th} and V_{th}

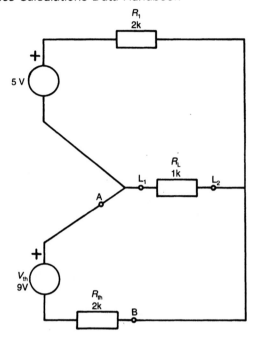

Figure 5.9 V_{th} and R_{th} replaced

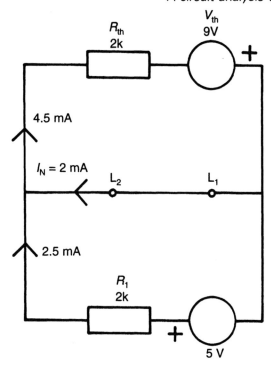

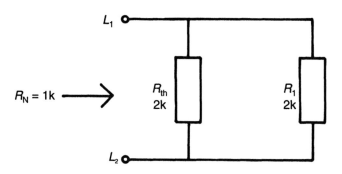

Figure 5.10 Finding I_N and R_N

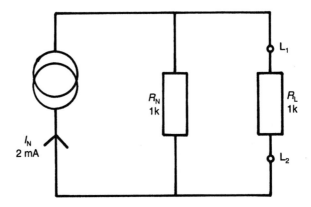

Figure 5.11 Equivalent circuit to Figure 5.6

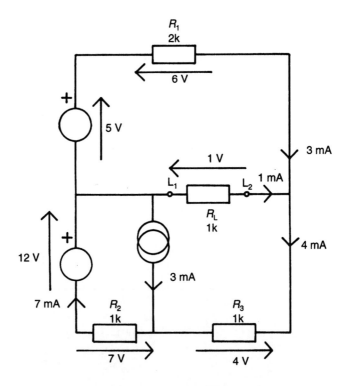

Figure 5.12 Solution of Figure 5.6

Part Two

Resistive Circuits

6 Equivalent resistances

Introduction

Resistors can be bought in certain standard values. The ranges of values are commonly called the E6, the E12 and the E24 series, the numbers denoting the quantity of values per decade of resistance. For building or repairing low voltage electronic circuits most workshops will stock a full range of E12 or E24 miniature resistors (i.e. 1/8 W or 1/4 W power rating), usually 1 or 2% tolerance, plus a selection of resistors rated at higher powers, which are available in combinations of E6 and E12 values. An additional range, E96, is also available, but these are more expensive, 0.1% tolerance, and few workshops will stock them as standard due to the large number of values. Also they are hard to obtain below 100 Ω or over about 250k.

The standard values are almost always fine, but once in a while you need a value that lies between two of them. This can be achieved by combining two components in series or parallel, their values being chosen so that the overall equivalent is near enough.

It is evident that choosing the best pair of resistors for this purpose might involve a fair amount of trial and error with a calculator, and that several near solutions might exist. For this purpose, three 'Tables of equivalent resistance' are provided in the rest of this chapter. Table 6.1 is for E6, Table 6.2 for E12, and Table 6.3 for E24. You should use the appropriate one for the range of components which you intend to use.

Using the Tables for resistors

Each table shows all achievable values between 1k and 10k with two resistors. s denotes series connection and p denotes parallel. The tables are sorted in order of ascending equivalent resistance.

Example 6.1: Choose a pair of E24 resistors to give an equivalent resistance of
7k7.

From Table 6.1:

R_E	R_1	s/p	R_2	P_{REL}
7700	200R	s	7k5	1.03
7700	1k5	s	6k2	1.24
7700	3k	s	4k7	1.64

There are plenty of others that would be close enough too.

To get a value of less than 1k, or greater than 10k, you should scale all values by
factors of ten as necessary.

Example 6.2: Choose a pair of E24 resistors to form a resistance of 125k.

R_E	R_1	s/p	R_2	P_{REL}
1250	1k5	p	7k5	1.2

125k is 1250 multiplied by 100, so resistors of 150k (1k5 times 100) and 750k
(7k5 times 100), connected in parallel, form a resistance of 125k.

Maximum power rating of the pair

Each line of the table gives a value of P_{REL}. Where each resistor of the pair is of
the same power rating, this can be used to find the overall power rating of the
pair. It is always a number between 1 and 2. For example, if a pair of 1/4 W resistors are used and P_{REL} has a value of 1.2 then the power rating of the pair is
equal to:

$$0.25 \times 1.2 = 0.3\,\text{W}$$

In this way, the pair of resistors can be lumped and considered as a single
circuit element.

Tolerance of components

If two components of the same tolerance are used the tolerance of the equivalent
resistance is the same.

Using the Tables for capacitors and inductors

Both inductors and capacitors are normally bought in combinations of the standard E6 and E12 ranges. Like resistors, they can be combined in series or parallel to obtain non-standard values. As was discussed in Chapter 2, inductors in series and parallel obey the same rules as resistors, while capacitors in series behave like resistors in parallel, and vice-versa.

Hence, the tables can be used to find equivalent capacitance or inductance. For capacitors, you should read p for s and s for p.

Example 6.3: To form a capacitor of 3nl using a range of 1%, E12 capacitors:

R_E	R_1	s/p	R_2	P_{REL}
3090	390R	s	2k7	1.14
3095	3k9	p	15k	1.26
3116	3k3	p	56k	1.06

This gives three alternatives: 2n7 p 390p, 3n9 s 15n and 3n3 s 56n. All three form a capacitor of 3nl, 1%. However, as capacitors (and inductors) tend to increase in physical size with increasing value the first might be preferable.

Table 6.1 *E6 equivalent resistances*

R_E	R_1	s/p	R_2	P_{REL}	R_E	R_1	s/p	R_2	P_{REL}
1010	10R	s	1k	1.01	1662	2k2	p	6k8	1.32
1010	330R	s	680R	1.49	1680	680R	s	1k	1.68
1015	15R	s	1k	1.02	1720	220R	s	1k5	1.15
1022	22R	s	1k	1.02	1803	2k2	p	10k	1.22
1031	1k5	p	3k3	1.45	1830	330R	s	1k5	1.22
1033	33R	s	1k	1.03	1919	2k2	p	15k	1.15
1047	47R	s	1k	1.05	1939	3k3	p	4k7	1.7
1068	68R	s	1k	1.07	1970	470R	s	1k5	1.31
1100	100R	s	1k	1.1	2000	1k	s	1k	2
1100	2k2	p	2k2	2	2000	2k2	p	22k	1.1
1137	1k5	p	4k7	1.32	2063	2k2	p	33k	1.07
1150	150R	s	1k	1.15	2102	2k2	p	47k	1.05
1150	470R	s	680R	1.69	2131	2k2	p	68k	1.03
1220	220R	s	1k	1.22	2153	2k2	p	100k	1.02
1229	1k5	p	6k8	1.22	2168	2k2	p	150k	1.01
1304	1k5	p	10k	1.15	2178	2k2	p	220k	1.01
1320	2k2	p	3k3	1.67	2180	680R	s	1k5	1.45
1330	330R	s	1k	1.33	2222	22R	s	2k2	1.01
1360	680R	s	680R	2	2222	3k3	p	6k8	1.49
1364	1k5	p	15k	1.1	2233	33R	s	2k2	1.02
1404	1k5	p	22k	1.07	2247	47R	s	2k2	1.02
1435	1k5	p	33k	1.05	2268	68R	s	2k2	1.03
1454	1k5	p	47k	1.03	2300	100R	s	2k2	1.05
1468	1k5	p	68k	1.02	2350	150R	s	2k2	1.07
1470	470R	s	1k	1.47	2350	4k7	p	4k7	2
1478	1k5	p	100k	1.02	2420	220R	s	2k2	1.1
1485	1k5	p	150k	1.01	2481	3k3	p	10k	1.33
1499	2k2	p	4k7	1.47	2500	1k	s	1k5	1.67
1515	15R	s	1k5	1.01	2530	330R	s	2k2	1.15
1522	22R	s	1k5	1.01	2670	470R	s	2k2	1.21
1533	33R	s	1k5	1.02	2705	3k3	p	15k	1.22
1547	47R	s	1k5	1.03	2779	4k7	p	6k8	1.69
1568	68R	s	1k5	1.05	2870	3k3	p	22k	1.15
1600	100R	s	1k5	1.07	2880	680R	s	2k2	1.31
1650	150R	s	1k5	1.1	3000	1k5	s	1k5	2
1650	3k3	p	3k3	2	3000	3k3	p	33k	1.1

Table 6.1 E6 equivalent resistances 59

R_E	R_1	s/p	R_2	P_{REL}	R_E	R_1	s/p	R_2	P_{REL}
3083	3k3	p	47k	1.07	4850	150R	s	4k7	1.03
3147	3k3	p	68k	1.05	4920	220R	s	4k7	1.05
3195	3k3	p	100k	1.03	5000	10k	p	10k	2
3197	4k7	p	10k	1.47	5030	330R	s	4k7	1.07
3200	1k	s	2k2	1.45	5170	470R	s	4k7	1.1
3229	3k3	p	150k	1.02	5194	6k8	p	22k	1.31
3251	3k3	p	220k	1.02	5380	680R	s	4k7	1.14
3267	3k3	p	330k	1.01	5500	2k2	s	3k3	1.67
3333	33R	s	3k3	1.01	5638	6k8	p	33k	1.21
3347	47R	s	3k3	1.01	5700	1k	s	4k7	1.21
3368	68R	s	3k3	1.02	5941	6k8	p	47k	1.14
3400	100R	s	3k3	1.03	6000	10k	p	15k	1.67
3400	6k8	p	6k8	2	6182	6k8	p	68k	1.1
3450	150R	s	3k3	1.05	6200	1k5	s	4k7	1.32
3520	220R	s	3k3	1.07	6367	6k8	p	100k	1.07
3579	4k7	p	15k	1.31	6505	6k8	p	150k	1.05
3630	330R	s	3k3	1.1	6596	6k8	p	220k	1.03
3700	1k5	s	2k2	1.68	6600	3k3	s	3k3	2
3770	470R	s	3k3	1.14	6663	6k8	p	330k	1.02
3873	4k7	p	22k	1.21	6703	6k8	p	470k	1.01
3980	680R	s	3k3	1.21	6733	6k8	p	680k	1.01
4048	6k8	p	10k	1.68	6868	68R	s	6k8	1.01
4114	4k7	p	33k	1.14	6875	10k	p	22k	1.45
4273	4k7	p	47k	1.1	6900	100R	s	6k8	1.01
4300	1k	s	3k3	1.3	6900	2k2	s	4k7	1.47
4396	4k7	p	68k	1.07	6950	150R	s	6k8	1.02
4400	2k2	s	2k2	2	7020	220R	s	6k8	1.03
4489	4k7	p	100k	1.05	7130	330R	s	6k8	1.05
4557	4k7	p	150k	1.03	7270	470R	s	6k8	1.07
4602	4k7	p	220k	1.02	7480	680R	s	6k8	1.1
4634	4k7	p	330k	1.01	7500	15k	p	15k	2
4653	4k7	p	470k	1.01	7674	10k	p	33k	1.3
4679	6k8	p	15k	1.45	7800	1k	s	6k8	1.15
4747	47R	s	4k7	1.01	8000	3k3	s	4k7	1.7
4768	68R	s	4k7	1.01	8246	10k	p	47k	1.21
4800	100R	s	4k7	1.02	8300	1k5	s	6k8	1.22
4800	1k5	s	3k3	1.45	8718	10k	p	68k	1.15

R_E	R_1	s/p	R_2	P_{REL}	R_E	R_1	s/p	R_2	P_{REL}
8919	15k	p	22k	1.68	9565	10k	p	220k	1.05
9000	2k2	s	6k8	1.32	9706	10k	p	330k	1.03
9091	10k	p	100k	1.1	9792	10k	p	470k	1.02
9375	10k	p	150k	1.07	9855	10k	p	680k	1.01
9400	4k7	s	4k7	2	9901	10k	p	1M	1.01

Table 6.2 *E12 equivalent resistances*

R_E	R_1	s/p	R_2	P_{REL}	R_E	R_1	s/p	R_2	P_{REL}
1010	10R	s	1k	1.01	1158	1k2	p	33k	1.04
1010	330R	s	680R	1.49	1164	1k2	p	39k	1.03
1012	12R	s	1k	1.01	1165	1k8	p	3k3	1.55
1015	15R	s	1k	1.02	1170	1k2	p	47k	1.03
1018	18R	s	1k	1.02	1175	1k2	p	56k	1.02
1020	1k2	p	6k8	1.18	1179	1k2	p	68k	1.02
1022	22R	s	1k	1.02	1180	180R	s	1k	1.18
1027	27R	s	1k	1.03	1183	1k2	p	82k	1.01
1030	470R	s	560R	1.84	1183	1k5	p	5k6	1.27
1031	1k5	p	3k3	1.45	1186	1k2	p	100k	1.01
1033	33R	s	1k	1.03	1188	1k2	p	120k	1.01
1039	39R	s	1k	1.04	1210	390R	s	820R	1.48
1040	220R	s	820R	1.27	1212	12R	s	1k2	1.01
1047	47R	s	1k	1.05	1212	2k2	p	2k7	1.81
1047	1k2	p	8k2	1.15	1215	15R	s	1k2	1.01
1056	56R	s	1k	1.06	1218	18R	s	1k2	1.02
1068	68R	s	1k	1.07	1220	220R	s	1k	1.22
1070	390R	s	680R	1.57	1222	22R	s	1k2	1.02
1071	1k2	p	10k	1.12	1227	27R	s	1k2	1.02
1080	1k8	p	2k7	1.67	1229	1k5	p	6k8	1.22
1082	82R	s	1k	1.08	1232	1k8	p	3k9	1.46
1083	1k5	p	3k9	1.38	1233	33R	s	1k2	1.03
1090	270R	s	820R	1.33	1239	39R	s	1k2	1.03
1091	1k2	p	12k	1.1	1240	560R	s	680R	1.82
1100	100R	s	1k	1.1	1247	47R	s	1k2	1.04
1100	2k2	p	2k2	2	1256	56R	s	1k2	1.05
1111	1k2	p	15k	1.08	1268	68R	s	1k2	1.06
1120	120R	s	1k	1.12	1268	1k5	p	8k2	1.18
1120	560R	s	560R	2	1270	270R	s	1k	1.27
1125	1k2	p	18k	1.07	1282	82R	s	1k2	1.07
1137	1k5	p	4k7	1.32	1290	470R	s	820R	1.57
1138	1k2	p	22k	1.05	1300	100R	s	1k2	1.08
1149	1k2	p	27k	1.04	1302	1k8	p	4k7	1.38
1150	150R	s	1k	1.15	1304	1k5	p	10k	1.15
1150	330R	s	820R	1.4	1320	120R	s	1k2	1.1
1150	470R	s	680R	1.69	1320	2k2	p	3k3	1.67

R_E	R_1	s/p	R_2	P_{REL}	R_E	R_1	s/p	R_2	P_{REL}
1330	330R	s	1k	1.33	1533	33R	s	1k5	1.02
1333	1k5	p	12k	1.13	1539	39R	s	1k5	1.03
1350	150R	s	1k2	1.13	1547	47R	s	1k5	1.03
1350	2k7	p	2k7	2	1556	56R	s	1k5	1.04
1360	680R	s	680R	2	1560	560R	s	1k	1.56
1362	1k8	p	5k6	1.32	1565	1k8	p	12k	1.15
1364	1k5	p	15k	1.1	1568	68R	s	1k5	1.05
1380	180R	s	1k2	1.15	1579	2k2	p	5k6	1.39
1380	560R	s	820R	1.68	1582	82R	s	1k5	1.05
1385	1k5	p	18k	1.08	1590	390R	s	1k2	1.33
1390	390R	s	1k	1.39	1595	2k7	p	3k9	1.69
1404	1k5	p	22k	1.07	1600	100R	s	1k5	1.07
1407	2k2	p	3k9	1.56	1607	1k8	p	15k	1.12
1420	220R	s	1k2	1.18	1620	120R	s	1k5	1.08
1421	1k5	p	27k	1.06	1636	1k8	p	18k	1.1
1423	1k8	p	6k8	1.26	1640	820R	s	820R	2
1435	1k5	p	33k	1.05	1650	150R	s	1k5	1.1
1444	1k5	p	39k	1.04	1650	3k3	p	3k3	2
1454	1k5	p	47k	1.03	1662	2k2	p	6k8	1.32
1461	1k5	p	56k	1.03	1664	1k8	p	22k	1.08
1468	1k5	p	68k	1.02	1670	470R	s	1k2	1.39
1470	270R	s	1k2	1.23	1680	180R	s	1k5	1.12
1470	470R	s	1k	1.47	1680	680R	s	1k	1.68
1473	1k5	p	82k	1.02	1688	1k8	p	27k	1.07
1476	1k8	p	8k2	1.22	1707	1k8	p	33k	1.05
1478	1k5	p	100k	1.02	1715	2k7	p	4k7	1.57
1481	1k5	p	120k	1.01	1720	220R	s	1k5	1.15
1485	1k5	p	150k	1.01	1721	1k8	p	39k	1.05
1485	2k7	p	3k3	1.82	1734	1k8	p	47k	1.04
1499	2k2	p	4k7	1.47	1735	2k2	p	8k2	1.27
1500	680R	s	820R	1.83	1744	1k8	p	56k	1.03
1515	15R	s	1k5	1.01	1754	1k8	p	68k	1.03
1518	18R	s	1k5	1.01	1760	560R	s	1k2	1.47
1522	22R	s	1k5	1.01	1761	1k8	p	82k	1.02
1525	1k8	p	10k	1.18	1768	1k8	p	100k	1.02
1527	27R	s	1k5	1.02	1770	270R	s	1k5	1.18
1530	330R	s	1k2	1.28	1773	1k8	p	120k	1.02

Table 6.2 E12 equivalent resistances 63

R_E	R_1	s/p	R_2	P_{REL}	R_E	R_1	s/p	R_2	P_{REL}
1779	1k8	p	150k	1.01	2070	270R	s	1k8	1.15
1782	1k8	p	180k	1.01	2076	3k3	p	5k6	1.59
1788	3k3	p	3k9	1.85	2083	2k2	p	39k	1.06
1803	2k2	p	10k	1.22	2102	2k2	p	47k	1.05
1818	18R	s	1k8	1.01	2117	2k2	p	56k	1.04
1820	820R	s	1k	1.82	2126	2k7	p	10k	1.27
1822	22R	s	1k8	1.01	2130	330R	s	1k8	1.18
1822	2k7	p	5k6	1.48	2131	2k2	p	68k	1.03
1827	27R	s	1k8	1.02	2131	3k9	p	4k7	1.83
1830	330R	s	1k5	1.22	2143	2k2	p	82k	1.03
1833	33R	s	1k8	1.02	2153	2k2	p	100k	1.02
1839	39R	s	1k8	1.02	2160	2k2	p	120k	1.02
1847	47R	s	1k8	1.03	2168	2k2	p	150k	1.01
1856	56R	s	1k8	1.03	2173	2k2	p	180k	1.01
1859	2k2	p	12k	1.18	2178	2k2	p	220k	1.01
1868	68R	s	1k8	1.04	2180	680R	s	1k5	1.45
1880	680R	s	1k2	1.57	2190	390R	s	1k8	1.22
1882	82R	s	1k8	1.05	2200	1k	s	1k2	1.83
1890	390R	s	1k5	1.26	2204	2k7	p	12k	1.23
1900	100R	s	1k8	1.06	2222	22R	s	2k2	1.01
1919	2k2	p	15k	1.15	2222	3k3	p	6k8	1.49
1920	120R	s	1k8	1.07	2227	27R	s	2k2	1.01
1933	2k7	p	6k8	1.4	2233	33R	s	2k2	1.02
1939	3k3	p	4k7	1.7	2239	39R	s	2k2	1.02
1950	150R	s	1k8	1.08	2247	47R	s	2k2	1.02
1950	3k9	p	3k9	2	2256	56R	s	2k2	1.03
1960	2k2	p	18k	1.12	2268	68R	s	2k2	1.03
1970	470R	s	1k5	1.31	2270	470R	s	1k8	1.26
1980	180R	s	1k8	1.1	2282	82R	s	2k2	1.04
2000	1k	s	1k	2	2288	2k7	p	15k	1.18
2000	2k2	p	22k	1.1	2299	3k9	p	5k6	1.7
2020	220R	s	1k8	1.12	2300	100R	s	2k2	1.05
2020	820R	s	1k2	1.68	2320	120R	s	2k2	1.05
2031	2k7	p	8k2	1.33	2320	820R	s	1k5	1.55
2034	2k2	p	27k	1.08	2348	2k7	p	18k	1.15
2060	560R	s	1k5	1.37	2350	150R	s	2k2	1.07
2063	2k2	p	33k	1.07	2350	4k7	p	4k7	2

R_E	R_1	s/p	R_2	P_{REL}	R_E	R_1	s/p	R_2	P_{REL}
2353	3k3	p	8k2	1.4	2756	56R	s	2k7	1.02
2360	560R	s	1k8	1.31	2760	560R	s	2k2	1.25
2380	180R	s	2k2	1.08	2768	68R	s	2k7	1.03
2400	1k2	s	1k2	2	2779	4k7	p	6k8	1.69
2405	2k7	p	22k	1.12	2782	82R	s	2k7	1.03
2420	220R	s	2k2	1.1	2789	3k3	p	18k	1.18
2455	2k7	p	27k	1.1	2800	100R	s	2k7	1.04
2470	270R	s	2k2	1.12	2800	1k	s	1k8	1.56
2479	3k9	p	6k8	1.57	2800	5k6	p	5k6	2
2480	680R	s	1k8	1.38	2806	3k9	p	10k	1.39
2481	3k3	p	10k	1.33	2820	120R	s	2k7	1.04
2496	2k7	p	33k	1.08	2850	150R	s	2k7	1.06
2500	1k	s	1k5	1.67	2870	3k3	p	22k	1.15
2525	2k7	p	39k	1.07	2880	180R	s	2k7	1.07
2530	330R	s	2k2	1.15	2880	680R	s	2k2	1.31
2553	2k7	p	47k	1.06	2920	220R	s	2k7	1.08
2555	4k7	p	5k6	1.84	2941	3k3	p	27k	1.12
2576	2k7	p	56k	1.05	2943	3k9	p	12k	1.33
2588	3k3	p	12k	1.28	2970	270R	s	2k7	1.1
2590	390R	s	2k2	1.18	2988	4k7	p	8k2	1.57
2597	2k7	p	68k	1.04	3000	1k2	s	1k8	1.67
2614	2k7	p	82k	1.03	3000	1k5	s	1k5	2
2620	820R	s	1k8	1.46	3000	3k3	p	33k	1.1
2629	2k7	p	100k	1.03	3020	820R	s	2k2	1.37
2641	2k7	p	120k	1.02	3030	330R	s	2k7	1.12
2643	3k9	p	8k2	1.48	3043	3k3	p	39k	1.08
2652	2k7	p	150k	1.02	3071	5k6	p	6k8	1.82
2660	2k7	p	180k	1.02	3083	3k3	p	47k	1.07
2667	2k7	p	220k	1.01	3090	390R	s	2k7	1.14
2670	470R	s	2k2	1.21	3095	3k9	p	15k	1.26
2673	2k7	p	270k	1.01	3116	3k3	p	56k	1.06
2700	1k2	s	1k5	1.8	3147	3k3	p	68k	1.05
2705	3k3	p	15k	1.22	3170	470R	s	2k7	1.17
2727	27R	s	2k7	1.01	3172	3k3	p	82k	1.04
2733	33R	s	2k7	1.01	3195	3k3	p	100k	1.03
2739	39R	s	2k7	1.01	3197	4k7	p	10k	1.47
2747	47R	s	2k7	1.02	3200	1k	s	2k2	1.45

Table 6.2 E12 equivalent resistances 65

R_E	R_1	s/p	R_2	P_{REL}	R_E	R_1	s/p	R_2	P_{REL}
3205	3k9	p	18k	1.22	3688	3k9	p	68k	1.06
3212	3k3	p	120k	1.03	3690	390R	s	3k3	1.12
3229	3k3	p	150k	1.02	3700	1k	s	2k7	1.37
3241	3k3	p	180k	1.02	3700	1k5	s	2k2	1.68
3251	3k3	p	220k	1.02	3717	6k8	p	8k2	1.83
3260	560R	s	2k7	1.21	3723	3k9	p	82k	1.05
3260	3k3	p	270k	1.01	3727	4k7	p	18k	1.26
3267	3k3	p	330k	1.01	3754	3k9	p	100k	1.04
3300	1k5	s	1k8	1.83	3770	470R	s	3k3	1.14
3313	3k9	p	22k	1.18	3777	3k9	p	120k	1.03
3328	5k6	p	8k2	1.68	3801	3k9	p	150k	1.03
3333	33R	s	3k3	1.01	3817	3k9	p	180k	1.02
3339	39R	s	3k3	1.01	3818	5k6	p	12k	1.47
3347	47R	s	3k3	1.01	3832	3k9	p	220k	1.02
3356	56R	s	3k3	1.02	3844	3k9	p	270k	1.01
3368	68R	s	3k3	1.02	3854	3k9	p	330k	1.01
3377	4k7	p	12k	1.39	3860	560R	s	3k3	1.17
3380	680R	s	2k7	1.25	3861	3k9	p	390k	1.01
3382	82R	s	3k3	1.02	3873	4k7	p	22k	1.21
3400	100R	s	3k3	1.03	3900	1k2	s	2k7	1.44
3400	1k2	s	2k2	1.55	3939	39R	s	3k9	1.01
3400	6k8	p	6k8	2	3947	47R	s	3k9	1.01
3408	3k9	p	27k	1.14	3956	56R	s	3k9	1.01
3420	120R	s	3k3	1.04	3968	68R	s	3k9	1.02
3450	150R	s	3k3	1.05	3980	680R	s	3k3	1.21
3480	180R	s	3k3	1.05	3982	82R	s	3k9	1.02
3488	3k9	p	33k	1.12	4000	100R	s	3k9	1.03
3520	220R	s	3k3	1.07	4000	1k8	s	2k2	1.82
3520	820R	s	2k7	1.3	4003	4k7	p	27k	1.17
3545	3k9	p	39k	1.1	4020	120R	s	3k9	1.03
3570	270R	s	3k3	1.08	4048	6k8	p	10k	1.68
3579	4k7	p	15k	1.31	4050	150R	s	3k9	1.04
3590	5k6	p	10k	1.56	4078	5k6	p	15k	1.37
3600	1k8	s	1k8	2	4080	180R	s	3k9	1.05
3601	3k9	p	47k	1.08	4100	8k2	p	8k2	2
3630	330R	s	3k3	1.1	4114	4k7	p	33k	1.14
3646	3k9	p	56k	1.07	4120	220R	s	3k9	1.06

R_E	R_1	s/p	R_2	P_{REL}	R_E	R_1	s/p	R_2	P_{REL}
4120	820R	s	3k3	1.25	4788	5k6	p	33k	1.17
4170	270R	s	3k9	1.07	4800	100R	s	4k7	1.02
4195	4k7	p	39k	1.12	4800	1k5	s	3k3	1.45
4200	1k5	s	2k7	1.56	4820	120R	s	4k7	1.03
4230	330R	s	3k9	1.08	4850	150R	s	4k7	1.03
4271	5k6	p	18k	1.31	4871	8k2	p	12k	1.68
4273	4k7	p	47k	1.1	4880	180R	s	4k7	1.04
4290	390R	s	3k9	1.1	4897	5k6	p	39k	1.14
4300	1k	s	3k3	1.3	4900	1k	s	3k9	1.26
4336	4k7	p	56k	1.08	4900	2k2	s	2k7	1.81
4340	6k8	p	12k	1.57	4920	220R	s	4k7	1.05
4370	470R	s	3k9	1.12	4935	6k8	p	18k	1.38
4396	4k7	p	68k	1.07	4970	270R	s	4k7	1.06
4400	2k2	s	2k2	2	5000	10k	p	10k	2
4445	4k7	p	82k	1.06	5004	5k6	p	47k	1.12
4460	560R	s	3k9	1.14	5030	330R	s	4k7	1.07
4464	5k6	p	22k	1.25	5090	390R	s	4k7	1.08
4489	4k7	p	100k	1.05	5091	5k6	p	56k	1.1
4500	1k2	s	3k3	1.36	5100	1k2	s	3k9	1.31
4500	1k8	s	2k7	1.67	5100	1k8	s	3k3	1.55
4505	8k2	p	10k	1.82	5170	470R	s	4k7	1.1
4523	4k7	p	120k	1.04	5174	5k6	p	68k	1.08
4557	4k7	p	150k	1.03	5194	6k8	p	22k	1.31
4580	680R	s	3k9	1.17	5242	5k6	p	82k	1.07
4580	4k7	p	180k	1.03	5260	560R	s	4k7	1.12
4602	4k7	p	220k	1.02	5302	8k2	p	15k	1.55
4620	4k7	p	270k	1.02	5303	5k6	p	100k	1.06
4634	4k7	p	330k	1.01	5350	5k6	p	120k	1.05
4638	5k6	p	27k	1.21	5380	680R	s	4k7	1.14
4644	4k7	p	390k	1.01	5398	5k6	p	150k	1.04
4653	4k7	p	470k	1.01	5400	1k5	s	3k9	1.38
4679	6k8	p	15k	1.45	5400	2k7	s	2k7	2
4720	820R	s	3k9	1.21	5431	5k6	p	180k	1.03
4747	47R	s	4k7	1.01	5432	6k8	p	27k	1.25
4756	56R	s	4k7	1.01	5455	10k	p	12k	1.83
4768	68R	s	4k7	1.01	5461	5k6	p	220k	1.03
4782	82R	s	4k7	1.02	5486	5k6	p	270k	1.02

Table 6.2 E12 equivalent resistances 67

R_E	R_1	s/p	R_2	P_{REL}	R_E	R_1	s/p	R_2	P_{REL}
5500	2k2	s	3k3	1.67	6367	6k8	p	100k	1.07
5507	5k6	p	330k	1.02	6420	820R	s	5k6	1.15
5520	820R	s	4k7	1.17	6429	10k	p	18k	1.56
5521	5k6	p	390k	1.01	6435	6k8	p	120k	1.06
5534	5k6	p	470k	1.01	6500	1k8	s	4k7	1.38
5545	5k6	p	560k	1.01	6505	6k8	p	150k	1.05
5634	8k2	p	18k	1.46	6552	6k8	p	180k	1.04
5638	6k8	p	33k	1.21	6568	8k2	p	33k	1.25
5656	56R	s	5k6	1.01	6596	6k8	p	220k	1.03
5668	68R	s	5k6	1.01	6600	1k	s	5k6	1.18
5682	82R	s	5k6	1.01	6600	2k7	s	3k9	1.69
5700	100R	s	5k6	1.02	6600	3k3	s	3k3	2
5700	1k	s	4k7	1.21	6633	6k8	p	270k	1.03
5700	1k8	s	3k9	1.46	6663	6k8	p	330k	1.02
5720	120R	s	5k6	1.02	6667	12k	p	15k	1.8
5750	150R	s	5k6	1.03	6683	6k8	p	390k	1.02
5780	180R	s	5k6	1.03	6703	6k8	p	470k	1.01
5790	6k8	p	39k	1.17	6718	6k8	p	560k	1.01
5820	220R	s	5k6	1.04	6733	6k8	p	680k	1.01
5870	270R	s	5k6	1.05	6775	8k2	p	39k	1.21
5900	1k2	s	4k7	1.26	6800	1k2	s	5k6	1.21
5930	330R	s	5k6	1.06	6868	68R	s	6k8	1.01
5941	6k8	p	47k	1.14	6875	10k	p	22k	1.45
5974	8k2	p	22k	1.37	6882	82R	s	6k8	1.01
5990	390R	s	5k6	1.07	6900	100R	s	6k8	1.01
6000	2k7	s	3k3	1.82	6900	2k2	s	4k7	1.47
6000	10k	p	15k	1.67	6920	120R	s	6k8	1.02
6000	12k	p	12k	2	6950	150R	s	6k8	1.02
6064	6k8	p	56k	1.12	6980	180R	s	6k8	1.03
6070	470R	s	5k6	1.08	6982	8k2	p	47k	1.17
6100	2k2	s	3k9	1.56	7020	220R	s	6k8	1.03
6160	560R	s	5k6	1.1	7070	270R	s	6k8	1.04
6182	6k8	p	68k	1.1	7100	1k5	s	5k6	1.27
6200	1k5	s	4k7	1.32	7130	330R	s	6k8	1.05
6279	6k8	p	82k	1.08	7153	8k2	p	56k	1.15
6280	680R	s	5k6	1.12	7190	390R	s	6k8	1.06
6290	8k2	p	27k	1.3	7200	3k3	s	3k9	1.85

R_E	R_1	s/p	R_2	P_{REL}	R_E	R_1	s/p	R_2	P_{REL}
7200	12k	p	18k	1.67	8308	12k	p	27k	1.44
7270	470R	s	6k8	1.07	8320	120R	s	8k2	1.01
7297	10k	p	27k	1.37	8350	150R	s	8k2	1.02
7318	8k2	p	68k	1.12	8380	180R	s	8k2	1.02
7360	560R	s	6k8	1.08	8420	220R	s	8k2	1.03
7400	1k8	s	5k6	1.32	8470	270R	s	8k2	1.03
7400	2k7	s	4k7	1.57	8485	10k	p	56k	1.18
7455	8k2	p	82k	1.1	8530	330R	s	8k2	1.04
7480	680R	s	6k8	1.1	8590	390R	s	8k2	1.05
7500	15k	p	15k	2	8600	1k8	s	6k8	1.26
7579	8k2	p	100k	1.08	8600	3k9	s	4k7	1.83
7620	820R	s	6k8	1.12	8670	470R	s	8k2	1.06
7674	10k	p	33k	1.3	8718	10k	p	68k	1.15
7676	8k2	p	120k	1.07	8760	560R	s	8k2	1.07
7765	12k	p	22k	1.55	8800	12k	p	33k	1.36
7775	8k2	p	150k	1.05	8880	680R	s	8k2	1.08
7800	1k	s	6k8	1.15	8900	3k3	s	5k6	1.59
7800	2k2	s	5k6	1.39	8913	10k	p	82k	1.12
7800	3k9	s	3k9	2	8919	15k	p	22k	1.68
7843	8k2	p	180k	1.05	9000	2k2	s	6k8	1.32
7905	8k2	p	220k	1.04	9000	18k	p	18k	2
7958	8k2	p	270k	1.03	9020	820R	s	8k2	1.1
7959	10k	p	39k	1.26	9091	10k	p	100k	1.1
8000	1k2	s	6k8	1.18	9176	12k	p	39k	1.31
8000	3k3	s	4k7	1.7	9200	1k	s	8k2	1.12
8001	8k2	p	330k	1.02	9231	10k	p	120k	1.08
8031	8k2	p	390k	1.02	9375	10k	p	150k	1.07
8059	8k2	p	470k	1.02	9400	1k2	s	8k2	1.15
8082	8k2	p	560k	1.01	9400	4k7	s	4k7	2
8102	8k2	p	680k	1.01	9474	10k	p	180k	1.06
8119	8k2	p	820k	1.01	9500	2k7	s	6k8	1.4
8182	15k	p	18k	1.83	9500	3k9	s	5k6	1.7
8246	10k	p	47k	1.21	9559	12k	p	47k	1.26
8282	82R	s	8k2	1.01	9565	10k	p	220k	1.05
8300	100R	s	8k2	1.01	9643	10k	p	270k	1.04
8300	1k5	s	6k8	1.22	9643	15k	p	27k	1.56
8300	2k7	s	5k6	1.48	9700	1k5	s	8k2	1.18

Table 6.2 E12 equivalent resistances 69

R_E	R_1	s/p	R_2	P_{REL}	R_E	R_1	s/p	R_2	P_{REL}
9706	10k	p	330k	1.03	9880	10k	p	820k	1.01
9750	10k	p	390k	1.03	9882	12k	p	56k	1.21
9792	10k	p	470k	1.02	9900	18k	p	22k	1.82
9825	10k	p	560k	1.02	9901	10k	p	1M	1.01
9855	10k	p	680k	1.01					

Table 6.3 *E24 equivalent resistances*

R_E	R_1	s/p	R_2	P_{REL}	R_E	R_1	s/p	R_2	P_{REL}
1001	91R	s	910R	1.1	1037	1k1	p	18k	1.06
1005	1k2	p	6k2	1.19	1039	39R	s	1k	1.04
1005	1k6	p	2k7	1.59	1040	130R	s	910R	1.14
1008	1k1	p	12k	1.09	1040	220R	s	820R	1.27
1010	10R	s	1k	1.01	1040	360R	s	680R	1.53
1010	100R	s	910R	1.11	1043	43R	s	1k	1.04
1010	330R	s	680R	1.49	1043	1k1	p	20k	1.06
1010	390R	s	620R	1.63	1043	1k6	p	3k	1.53
1011	11R	s	1k	1.01	1047	47R	s	1k	1.05
1012	12R	s	1k	1.01	1047	1k2	p	8k2	1.15
1013	13R	s	1k	1.01	1048	1k1	p	22k	1.05
1014	1k1	p	13k	1.08	1048	2k	p	2k2	1.91
1015	15R	s	1k	1.02	1050	300R	s	750R	1.4
1016	16R	s	1k	1.02	1050	430R	s	620R	1.69
1018	18R	s	1k	1.02	1051	51R	s	1k	1.05
1018	1k3	p	4k7	1.28	1052	1k1	p	24k	1.05
1020	20R	s	1k	1.02	1055	1k3	p	5k6	1.23
1020	110R	s	910R	1.12	1056	56R	s	1k	1.06
1020	200R	s	820R	1.24	1057	1k1	p	27k	1.04
1020	270R	s	750R	1.36	1059	1k5	p	3k6	1.42
1020	510R	s	510R	2	1060	150R	s	910R	1.16
1020	1k2	p	6k8	1.18	1060	240R	s	820R	1.29
1022	22R	s	1k	1.02	1060	1k2	p	9k1	1.13
1024	24R	s	1k	1.02	1061	1k1	p	30k	1.04
1025	1k1	p	15k	1.07	1062	62R	s	1k	1.06
1027	27R	s	1k	1.03	1065	1k1	p	33k	1.03
1029	1k1	p	16k	1.07	1067	1k1	p	36k	1.03
1029	1k8	p	2k4	1.75	1068	68R	s	1k	1.07
1030	30R	s	1k	1.03	1070	160R	s	910R	1.18
1030	120R	s	910R	1.13	1070	390R	s	680R	1.57
1030	470R	s	560R	1.84	1070	510R	s	560R	1.91
1031	1k5	p	3k3	1.45	1070	1k1	p	39k	1.03
1033	33R	s	1k	1.03	1071	1k2	p	10k	1.12
1034	1k2	p	7k5	1.16	1073	1k1	p	43k	1.03
1036	36R	s	1k	1.04	1075	75R	s	1k	1.08
1036	1k3	p	5k1	1.25	1075	1k1	p	47k	1.02

Table 6.3 E24 equivalent resistances 71

R_E	R_1	s/p	R_2	P_{REL}	R_E	R_1	s/p	R_2	P_{REL}
1075	1k3	p	6k2	1.21	1115	15R	s	1k1	1.01
1077	1k1	p	51k	1.02	1116	16R	s	1k1	1.01
1078	1k6	p	3k3	1.48	1116	1k2	p	16k	1.08
1079	1k1	p	56k	1.02	1118	18R	s	1k1	1.02
1080	330R	s	750R	1.44	1120	20R	s	1k1	1.02
1080	1k8	p	2k7	1.67	1120	120R	s	1k	1.12
1081	1k1	p	62k	1.02	1120	300R	s	820R	1.37
1082	82R	s	1k	1.08	1120	560R	s	560R	2
1082	1k1	p	68k	1.02	1122	22R	s	1k1	1.02
1082	1k2	p	11k	1.11	1122	1k3	p	8k2	1.16
1083	1k5	p	3k9	1.38	1124	24R	s	1k1	1.02
1084	1k1	p	75k	1.01	1125	1k2	p	18k	1.07
1085	1k1	p	82k	1.01	1125	1k8	p	3k	1.6
1087	1k1	p	91k	1.01	1127	27R	s	1k1	1.02
1088	1k1	p	100k	1.01	1130	30R	s	1k1	1.03
1089	1k1	p	110k	1.01	1130	130R	s	1k	1.13
1090	180R	s	910R	1.2	1130	220R	s	910R	1.24
1090	270R	s	820R	1.33	1130	510R	s	620R	1.82
1090	470R	s	620R	1.76	1132	1k2	p	20k	1.06
1091	91R	s	1k	1.09	1133	33R	s	1k1	1.03
1091	1k2	p	12k	1.1	1135	1k6	p	3k9	1.41
1091	1k3	p	6k8	1.19	1136	36R	s	1k1	1.03
1091	2k	p	2k4	1.83	1137	1k5	p	4k7	1.32
1099	1k2	p	13k	1.09	1138	1k2	p	22k	1.05
1100	100R	s	1k	1.1	1138	1k3	p	9k1	1.14
1100	2k2	p	2k2	2	1139	39R	s	1k1	1.04
1108	1k3	p	7k5	1.17	1140	390R	s	750R	1.52
1108	1k6	p	3k6	1.44	1143	43R	s	1k1	1.04
1110	110R	s	1k	1.11	1143	1k2	p	24k	1.05
1110	200R	s	910R	1.22	1147	47R	s	1k1	1.04
1110	360R	s	750R	1.48	1148	2k2	p	2k4	1.92
1110	430R	s	680R	1.63	1149	1k2	p	27k	1.04
1111	11R	s	1k1	1.01	1149	2k	p	2k7	1.74
1111	1k2	p	15k	1.08	1150	150R	s	1k	1.15
1112	12R	s	1k1	1.01	1150	240R	s	910R	1.26
1112	1k5	p	4k3	1.35	1150	330R	s	820R	1.4
1113	13R	s	1k1	1.01	1150	470R	s	680R	1.69

R_E	R_1	s/p	R_2	P_{REL}	R_E	R_1	s/p	R_2	P_{REL}
1150	1k3	p	10k	1.13	1191	91R	s	1k1	1.08
1151	51R	s	1k1	1.05	1194	1k6	p	4k7	1.34
1154	1k2	p	30k	1.04	1196	1k3	p	15k	1.09
1156	56R	s	1k1	1.05	1200	100R	s	1k1	1.09
1158	1k2	p	33k	1.04	1200	200R	s	1k	1.2
1159	1k5	p	5k1	1.29	1200	1k8	p	3k6	1.5
1160	160R	s	1k	1.16	1200	2k	p	3k	1.67
1161	1k2	p	36k	1.03	1200	2k4	p	2k4	2
1162	62R	s	1k1	1.06	1202	1k3	p	16k	1.08
1163	1k3	p	11k	1.12	1208	1k5	p	6k2	1.24
1164	1k2	p	39k	1.03	1210	110R	s	1k1	1.1
1165	1k8	p	3k3	1.55	1210	300R	s	910R	1.33
1166	1k6	p	4k3	1.37	1210	390R	s	820R	1.48
1167	1k2	p	43k	1.03	1212	12R	s	1k2	1.01
1168	68R	s	1k1	1.06	1212	1k3	p	18k	1.07
1170	1k2	p	47k	1.03	1212	2k2	p	2k7	1.81
1172	1k2	p	51k	1.02	1213	13R	s	1k2	1.01
1173	1k3	p	12k	1.11	1215	15R	s	1k2	1.01
1175	75R	s	1k1	1.07	1216	16R	s	1k2	1.01
1175	1k2	p	56k	1.02	1218	18R	s	1k2	1.02
1177	1k2	p	62k	1.02	1218	1k6	p	5k1	1.31
1179	1k2	p	68k	1.02	1220	20R	s	1k2	1.02
1180	180R	s	1k	1.18	1220	120R	s	1k1	1.11
1180	270R	s	910R	1.3	1220	220R	s	1k	1.22
1180	360R	s	820R	1.44	1220	470R	s	750R	1.63
1180	430R	s	750R	1.57	1221	1k3	p	20k	1.07
1180	560R	s	620R	1.9	1222	22R	s	1k2	1.02
1181	1k2	p	75k	1.02	1224	24R	s	1k2	1.02
1182	82R	s	1k1	1.07	1227	27R	s	1k2	1.02
1182	1k3	p	13k	1.1	1227	1k3	p	22k	1.06
1183	1k2	p	82k	1.01	1229	1k5	p	6k8	1.22
1183	1k5	p	5k6	1.27	1230	30R	s	1k2	1.03
1184	1k2	p	91k	1.01	1230	130R	s	1k1	1.12
1186	1k2	p	100k	1.01	1232	1k8	p	3k9	1.46
1187	1k2	p	110k	1.01	1233	33R	s	1k2	1.03
1188	1k2	p	120k	1.01	1233	1k3	p	24k	1.05
1190	510R	s	680R	1.75	1236	36R	s	1k2	1.03

Table 6.3 E24 equivalent resistances 73

R_E	R_1	s/p	R_2	P_{REL}	R_E	R_1	s/p	R_2	P_{REL}
1239	39R	s	1k2	1.03	1278	1k3	p	75k	1.02
1240	240R	s	1k	1.24	1280	180R	s	1k1	1.16
1240	330R	s	910R	1.36	1280	1k3	p	82k	1.02
1240	560R	s	680R	1.82	1282	82R	s	1k2	1.07
1240	620R	s	620R	2	1282	1k3	p	91k	1.01
1240	1k3	p	27k	1.05	1283	1k3	p	100k	1.01
1243	43R	s	1k2	1.04	1285	1k3	p	110k	1.01
1244	1k6	p	5k6	1.29	1286	1k3	p	120k	1.01
1245	2k	p	3k3	1.61	1286	2k	p	3k6	1.56
1246	1k3	p	30k	1.04	1287	1k3	p	130k	1.01
1247	47R	s	1k2	1.04	1288	1k5	p	9k1	1.16
1250	150R	s	1k1	1.14	1290	470R	s	820R	1.57
1250	430R	s	820R	1.52	1291	91R	s	1k2	1.08
1250	1k5	p	7k5	1.2	1295	1k6	p	6k8	1.24
1251	51R	s	1k2	1.04	1300	100R	s	1k2	1.08
1251	1k3	p	33k	1.04	1300	200R	s	1k1	1.18
1255	1k3	p	36k	1.04	1300	300R	s	1k	1.3
1256	56R	s	1k2	1.05	1300	390R	s	910R	1.43
1258	1k3	p	39k	1.03	1300	620R	s	680R	1.91
1260	160R	s	1k1	1.15	1302	1k8	p	4k7	1.38
1260	510R	s	750R	1.68	1304	1k5	p	10k	1.15
1262	62R	s	1k2	1.05	1310	110R	s	1k2	1.09
1262	1k3	p	43k	1.03	1310	560R	s	750R	1.75
1265	1k3	p	47k	1.03	1313	13R	s	1k3	1.01
1268	68R	s	1k2	1.06	1315	15R	s	1k3	1.01
1268	1k3	p	51k	1.03	1316	16R	s	1k3	1.01
1268	1k5	p	8k2	1.18	1318	18R	s	1k3	1.01
1269	1k8	p	4k3	1.42	1319	1k6	p	7k5	1.21
1269	2k2	p	3k	1.73	1320	20R	s	1k3	1.02
1270	270R	s	1k	1.27	1320	120R	s	1k2	1.1
1270	360R	s	910R	1.4	1320	220R	s	1k1	1.2
1271	1k3	p	56k	1.02	1320	1k5	p	11k	1.14
1271	2k4	p	2k7	1.89	1320	2k2	p	3k3	1.67
1272	1k6	p	6k2	1.26	1322	22R	s	1k3	1.02
1273	1k3	p	62k	1.02	1322	2k	p	3k9	1.51
1275	75R	s	1k2	1.06	1324	24R	s	1k3	1.02
1276	1k3	p	68k	1.02	1327	27R	s	1k3	1.02

R_E	R_1	s/p	R_2	P_{REL}	R_E	R_1	s/p	R_2	P_{REL}
1330	30R	s	1k3	1.02	1380	560R	s	820R	1.68
1330	130R	s	1k2	1.11	1382	82R	s	1k3	1.06
1330	330R	s	1k	1.33	1385	1k5	p	18k	1.08
1330	510R	s	820R	1.62	1389	2k4	p	3k3	1.73
1330	1k8	p	5k1	1.35	1390	390R	s	1k	1.39
1333	33R	s	1k3	1.03	1391	91R	s	1k3	1.07
1333	1k5	p	12k	1.13	1395	1k5	p	20k	1.08
1333	2k4	p	3k	1.8	1395	1k8	p	6k2	1.29
1336	36R	s	1k3	1.03	1397	1k6	p	11k	1.15
1339	39R	s	1k3	1.03	1400	100R	s	1k3	1.08
1339	1k6	p	8k2	1.2	1400	200R	s	1k2	1.17
1340	240R	s	1k1	1.22	1400	300R	s	1k1	1.27
1340	430R	s	910R	1.47	1403	2k	p	4k7	1.43
1343	43R	s	1k3	1.03	1404	1k5	p	22k	1.07
1345	1k5	p	13k	1.12	1407	2k2	p	3k9	1.56
1347	47R	s	1k3	1.04	1410	110R	s	1k3	1.08
1350	150R	s	1k2	1.13	1412	1k5	p	24k	1.06
1350	2k7	p	2k7	2	1412	1k6	p	12k	1.13
1351	51R	s	1k3	1.04	1420	120R	s	1k3	1.09
1356	56R	s	1k3	1.04	1420	220R	s	1k2	1.18
1360	160R	s	1k2	1.13	1420	510R	s	910R	1.56
1360	360R	s	1k	1.36	1421	1k5	p	27k	1.06
1360	680R	s	680R	2	1421	2k7	p	3k	1.9
1361	1k6	p	9k1	1.18	1423	1k8	p	6k8	1.26
1362	62R	s	1k3	1.05	1425	1k6	p	13k	1.12
1362	1k8	p	5k6	1.32	1429	1k5	p	30k	1.05
1364	1k5	p	15k	1.1	1430	130R	s	1k3	1.1
1365	2k	p	4k3	1.47	1430	330R	s	1k1	1.3
1366	2k2	p	3k6	1.61	1430	430R	s	1k	1.43
1368	68R	s	1k3	1.05	1430	680R	s	750R	1.91
1370	270R	s	1k1	1.25	1435	1k5	p	33k	1.05
1370	620R	s	750R	1.83	1437	2k	p	5k1	1.39
1371	1k5	p	16k	1.09	1440	240R	s	1k2	1.2
1375	75R	s	1k3	1.06	1440	620R	s	820R	1.76
1379	1k6	p	10k	1.16	1440	1k5	p	36k	1.04
1380	180R	s	1k2	1.15	1440	2k4	p	3k6	1.67
1380	470R	s	910R	1.52	1444	1k5	p	39k	1.04

Table 6.3 E24 equivalent resistances 75

R_E	R_1	s/p	R_2	P_{REL}	R_E	R_1	s/p	R_2	P_{REL}
1446	1k6	p	15k	1.11	1500	750R	s	750R	2
1449	1k5	p	43k	1.03	1500	1k6	p	24k	1.07
1450	150R	s	1k3	1.12	1500	3k	p	3k	2
1452	1k8	p	7k5	1.24	1503	1k8	p	9kl	1.2
1454	1k5	p	47k	1.03	1510	510R	s	1k	1.51
1455	1k6	p	16k	1.1	1510	1k6	p	27k	1.06
1455	2k2	p	4k3	1.51	1512	2k	p	6k2	1.32
1457	1k5	p	51k	1.03	1515	15R	s	1k5	1.01
1460	160R	s	1k3	1.12	1516	16R	s	1k5	1.01
1460	360R	s	1kl	1.33	1518	18R	s	1k5	1.01
1461	1k5	p	56k	1.03	1519	1k6	p	30k	1.05
1465	1k5	p	62k	1.02	1520	20R	s	1k5	1.01
1468	1k5	p	68k	1.02	1520	220R	s	1k3	1.17
1469	1k6	p	18k	1.09	1522	22R	s	1k5	1.01
1470	270R	s	1k2	1.23	1524	24R	s	1k5	1.02
1470	470R	s	1k	1.47	1525	1k8	p	10k	1.18
1470	560R	s	910R	1.62	1526	1k6	p	33k	1.05
1471	1k5	p	75k	1.02	1527	27R	s	1k5	1.02
1473	1k5	p	82k	1.02	1530	30R	s	1k5	1.02
1474	2k	p	5k6	1.36	1530	330R	s	1k2	1.28
1476	1k5	p	91k	1.02	1530	430R	s	1kl	1.39
1476	1k8	p	8k2	1.22	1530	620R	s	910R	1.68
1478	1k5	p	100k	1.02	1532	1k6	p	36k	1.04
1480	180R	s	1k3	1.14	1533	33R	s	1k5	1.02
1480	1k5	p	110k	1.01	1536	36R	s	1k5	1.02
1481	1k5	p	120k	1.01	1537	1k6	p	39k	1.04
1481	1k6	p	20k	1.08	1537	2k2	p	5kl	1.43
1483	1k5	p	130k	1.01	1539	39R	s	1k5	1.03
1485	1k5	p	150k	1.01	1540	240R	s	1k3	1.18
1485	2k7	p	3k3	1.82	1540	2k4	p	4k3	1.56
1486	2k4	p	3k9	1.62	1543	43R	s	1k5	1.03
1490	390R	s	1kl	1.35	1543	1k6	p	43k	1.04
1492	1k6	p	22k	1.07	1543	2k7	p	3k6	1.75
1499	2k2	p	4k7	1.47	1545	2k	p	6k8	1.29
1500	200R	s	1k3	1.15	1547	47R	s	1k5	1.03
1500	300R	s	1k2	1.25	1547	1k6	p	47k	1.03
1500	680R	s	820R	1.83	1547	1k8	p	11k	1.16

R_E	R_1	s/p	R_2	P_{REL}	R_E	R_1	s/p	R_2	P_{REL}
1551	51R	s	1k5	1.03	1608	2k	p	8k2	1.24
1551	1k6	p	51k	1.03	1610	110R	s	1k5	1.07
1556	56R	s	1k5	1.04	1610	510R	s	1k1	1.46
1556	1k6	p	56k	1.03	1616	16R	s	1k6	1.01
1560	360R	s	1k2	1.3	1618	18R	s	1k6	1.01
1560	560R	s	1k	1.56	1618	1k8	p	16k	1.11
1560	1k6	p	62k	1.03	1620	20R	s	1k6	1.01
1562	62R	s	1k5	1.04	1620	120R	s	1k5	1.08
1563	1k6	p	68k	1.02	1620	620R	s	1k	1.62
1565	1k8	p	12k	1.15	1622	22R	s	1k6	1.01
1567	1k6	p	75k	1.02	1624	24R	s	1k6	1.02
1568	68R	s	1k5	1.05	1624	2k2	p	6k2	1.35
1569	1k6	p	82k	1.02	1627	27R	s	1k6	1.02
1570	270R	s	1k3	1.21	1630	30R	s	1k6	1.02
1570	470R	s	1k1	1.43	1630	130R	s	1k5	1.09
1570	750R	s	820R	1.91	1630	330R	s	1k3	1.25
1571	3k	p	3k3	1.91	1630	430R	s	1k2	1.36
1572	1k6	p	91k	1.02	1632	2k4	p	5k1	1.47
1575	75R	s	1k5	1.05	1633	33R	s	1k6	1.02
1575	1k6	p	100k	1.02	1636	36R	s	1k6	1.02
1577	1k6	p	110k	1.01	1636	1k8	p	18k	1.1
1579	1k6	p	120k	1.01	1636	3k	p	3k6	1.83
1579	2k	p	7k5	1.27	1639	39R	s	1k6	1.02
1579	2k2	p	5k6	1.39	1640	820R	s	820R	2
1581	1k6	p	130k	1.01	1640	2k	p	9k1	1.22
1581	1k8	p	13k	1.14	1643	43R	s	1k6	1.03
1582	82R	s	1k5	1.05	1647	47R	s	1k6	1.03
1583	1k6	p	150k	1.01	1650	150R	s	1k5	1.1
1584	1k6	p	160k	1.01	1650	3k3	p	3k3	2
1589	2k4	p	4k7	1.51	1651	51R	s	1k6	1.03
1590	390R	s	1k2	1.33	1651	1k8	p	20k	1.09
1590	680R	s	910R	1.75	1656	56R	s	1k6	1.03
1591	91R	s	1k5	1.06	1659	2k7	p	4k3	1.63
1595	2k7	p	3k9	1.69	1660	160R	s	1k5	1.11
1600	100R	s	1k5	1.07	1660	360R	s	1k3	1.28
1600	300R	s	1k3	1.23	1660	560R	s	1k1	1.51
1607	1k8	p	15k	1.12	1660	750R	s	910R	1.82

Table 6.3 E24 equivalent resistances 77

R_E	R_1	s/p	R_2	P_{REL}	R_E	R_1	s/p	R_2	P_{REL}
1662	62R	s	1k6	1.04	1733	2k	p	13k	1.15
1662	2k2	p	6k8	1.32	1734	1k8	p	47k	1.04
1664	1k8	p	22k	1.08	1735	2k2	p	8k2	1.27
1667	2k	p	10k	1.2	1739	1k8	p	51k	1.04
1668	68R	s	1k6	1.04	1740	240R	s	1k5	1.16
1670	470R	s	1k2	1.39	1744	1k8	p	56k	1.03
1674	1k8	p	24k	1.08	1749	1k8	p	62k	1.03
1675	75R	s	1k6	1.05	1750	150R	s	1k6	1.09
1680	180R	s	1k5	1.12	1750	750R	s	1k	1.75
1680	680R	s	1k	1.68	1754	1k8	p	68k	1.03
1680	2k4	p	5k6	1.43	1758	1k8	p	75k	1.02
1682	82R	s	1k6	1.05	1760	160R	s	1k6	1.1
1688	1k8	p	27k	1.07	1760	560R	s	1k2	1.47
1690	390R	s	1k3	1.3	1761	1k8	p	82k	1.02
1691	91R	s	1k6	1.06	1765	1k8	p	91k	1.02
1692	2k	p	11k	1.18	1765	2k	p	15k	1.13
1696	3k	p	3k9	1.77	1765	2k7	p	5k1	1.53
1698	1k8	p	30k	1.06	1767	3k	p	4k3	1.7
1700	100R	s	1k6	1.06	1768	1k8	p	100k	1.02
1700	200R	s	1k5	1.13	1770	270R	s	1k5	1.18
1701	2k2	p	7k5	1.29	1770	470R	s	1k3	1.36
1707	1k8	p	33k	1.05	1771	1k8	p	110k	1.02
1710	110R	s	1k6	1.07	1772	2k2	p	9k1	1.24
1710	510R	s	1k2	1.43	1773	1k8	p	120k	1.02
1714	1k8	p	36k	1.05	1774	2k4	p	6k8	1.35
1714	2k	p	12k	1.17	1775	1k8	p	130k	1.01
1715	2k7	p	4k7	1.57	1778	2k	p	16k	1.13
1720	120R	s	1k6	1.08	1779	1k8	p	150k	1.01
1720	220R	s	1k5	1.15	1780	180R	s	1k6	1.11
1720	620R	s	1k1	1.56	1780	680R	s	1k1	1.62
1721	1k8	p	39k	1.05	1780	1k8	p	160k	1.01
1722	3k3	p	3k6	1.92	1782	1k8	p	180k	1.01
1728	1k8	p	43k	1.04	1788	3k3	p	3k9	1.85
1730	130R	s	1k6	1.08	1800	200R	s	1k6	1.13
1730	430R	s	1k3	1.33	1800	300R	s	1k5	1.2
1730	820R	s	910R	1.9	1800	2k	p	18k	1.11
1730	2k4	p	6k2	1.39	1800	3k6	p	3k6	2

R_E	R_1	s/p	R_2	P_{REL}	R_E	R_1	s/p	R_2	P_{REL}
1803	2k2	p	10k	1.22	1870	270R	s	1k6	1.17
1810	510R	s	1k3	1.39	1872	3k6	p	3k9	1.92
1818	18R	s	1k8	1.01	1875	75R	s	1k8	1.04
1818	2k	p	20k	1.1	1875	2k	p	30k	1.07
1818	2k4	p	7k5	1.32	1880	680R	s	1k2	1.57
1820	20R	s	1k8	1.01	1881	2k7	p	6k2	1.44
1820	220R	s	1k6	1.14	1882	82R	s	1k8	1.05
1820	620R	s	1k2	1.52	1882	2k2	p	13k	1.17
1820	820R	s	1k	1.82	1886	2k	p	33k	1.06
1820	910R	s	910R	2	1889	3k	p	5k1	1.59
1822	22R	s	1k8	1.01	1890	390R	s	1k5	1.26
1822	2k7	p	5k6	1.48	1891	91R	s	1k8	1.05
1824	24R	s	1k8	1.01	1895	2k	p	36k	1.06
1827	27R	s	1k8	1.02	1899	2k4	p	9k1	1.26
1830	30R	s	1k8	1.02	1900	100R	s	1k8	1.06
1830	330R	s	1k5	1.22	1900	300R	s	1k6	1.19
1831	3k	p	4k7	1.64	1902	2k	p	39k	1.05
1833	33R	s	1k8	1.02	1910	110R	s	1k8	1.06
1833	2k	p	22k	1.09	1910	910R	s	1k	1.91
1833	2k2	p	11k	1.2	1911	2k	p	43k	1.05
1836	36R	s	1k8	1.02	1918	2k	p	47k	1.04
1839	39R	s	1k8	1.02	1919	2k2	p	15k	1.15
1840	240R	s	1k6	1.15	1920	120R	s	1k8	1.07
1843	43R	s	1k8	1.02	1920	620R	s	1k3	1.48
1846	2k	p	24k	1.08	1920	820R	s	1k1	1.75
1847	47R	s	1k8	1.03	1925	2k	p	51k	1.04
1850	750R	s	1k1	1.68	1930	130R	s	1k8	1.07
1851	51R	s	1k8	1.03	1930	330R	s	1k6	1.21
1856	56R	s	1k8	1.03	1930	430R	s	1k5	1.29
1857	2k4	p	8k2	1.29	1931	2k	p	56k	1.04
1859	2k2	p	12k	1.18	1933	2k7	p	6k8	1.4
1860	360R	s	1k5	1.24	1934	2k2	p	16k	1.14
1860	560R	s	1k3	1.43	1935	2k4	p	10k	1.24
1862	62R	s	1k8	1.03	1938	2k	p	62k	1.03
1862	2k	p	27k	1.07	1939	3k3	p	4k7	1.7
1867	3k3	p	4k3	1.77	1943	2k	p	68k	1.03
1868	68R	s	1k8	1.04	1948	2k	p	75k	1.03

Table 6.3 E24 equivalent resistances 79

R_E	R_1	s/p	R_2	P_{REL}	R_E	R_1	s/p	R_2	P_{REL}
1950	150R	s	1k8	1.08	2022	3k	p	6k2	1.48
1950	750R	s	1k2	1.63	2024	24R	s	2k	1.01
1950	3k9	p	3k9	2	2026	2k4	p	13k	1.18
1952	2k	p	82k	1.02	2027	27R	s	2k	1.01
1953	3k	p	5k6	1.54	2030	30R	s	2k	1.02
1957	2k	p	91k	1.02	2030	430R	s	1k6	1.27
1959	3k6	p	4k3	1.84	2031	2k7	p	8k2	1.33
1960	160R	s	1k8	1.09	2033	33R	s	2k	1.02
1960	360R	s	1k6	1.23	2034	2k2	p	27k	1.08
1960	2k2	p	18k	1.12	2036	36R	s	2k	1.02
1961	2k	p	100k	1.02	2039	39R	s	2k	1.02
1964	2k	p	110k	1.02	2039	3k6	p	4k7	1.77
1967	2k	p	120k	1.02	2040	240R	s	1k8	1.13
1970	470R	s	1k5	1.31	2043	43R	s	2k	1.02
1970	2k	p	130k	1.02	2045	3k9	p	4k3	1.91
1970	2k4	p	11k	1.22	2047	47R	s	2k	1.02
1974	2k	p	150k	1.01	2050	750R	s	1k3	1.58
1975	2k	p	160k	1.01	2050	2k2	p	30k	1.07
1978	2k	p	180k	1.01	2051	51R	s	2k	1.03
1980	180R	s	1k8	1.1	2056	56R	s	2k	1.03
1980	680R	s	1k3	1.52	2060	560R	s	1k5	1.37
1980	2k	p	200k	1.01	2062	62R	s	2k	1.03
1982	2k2	p	20k	1.11	2063	2k2	p	33k	1.07
1985	2k7	p	7k5	1.36	2068	68R	s	2k	1.03
1990	390R	s	1k6	1.24	2069	2k4	p	15k	1.16
2000	200R	s	1k8	1.11	2070	270R	s	1k8	1.15
2000	1k	s	1k	2	2070	470R	s	1k6	1.29
2000	2k2	p	22k	1.1	2073	2k2	p	36k	1.06
2000	2k4	p	12k	1.2	2075	75R	s	2k	1.04
2004	3k3	p	5k1	1.65	2076	3k3	p	5k6	1.59
2010	510R	s	1k5	1.34	2082	82R	s	2k	1.04
2010	910R	s	1k1	1.83	2082	2k7	p	9k1	1.3
2015	2k2	p	24k	1.09	2082	3k	p	6k8	1.44
2020	20R	s	2k	1.01	2083	2k2	p	39k	1.06
2020	220R	s	1k8	1.12	2087	2k4	p	16k	1.15
2020	820R	s	1k2	1.68	2091	91R	s	2k	1.05
2022	22R	s	2k	1.01	2093	2k2	p	43k	1.05

R_E	R_1	s/p	R_2	P_{REL}	R_E	R_1	s/p	R_2	P_{REL}
2100	100R	s	2k	1.05	2168	2k7	p	11k	1.25
2100	300R	s	1k8	1.17	2170	2k2	p	160k	1.01
2100	1k	s	1k1	1.91	2173	2k2	p	180k	1.01
2102	2k2	p	47k	1.05	2176	2k2	p	200k	1.01
2109	2k2	p	51k	1.04	2178	2k2	p	220k	1.01
2110	110R	s	2k	1.06	2180	180R	s	2k	1.09
2110	510R	s	1k6	1.32	2180	680R	s	1k5	1.45
2110	910R	s	1k2	1.76	2182	2k4	p	24k	1.1
2110	3k6	p	5k1	1.71	2190	390R	s	1k8	1.22
2117	2k2	p	56k	1.04	2191	3k6	p	5k6	1.64
2118	2k4	p	18k	1.13	2196	3k	p	8k2	1.37
2120	120R	s	2k	1.06	2200	200R	s	2k	1.1
2120	620R	s	1k5	1.41	2200	1k	s	1k2	1.83
2120	820R	s	1k3	1.63	2200	1k1	s	1k1	2
2125	2k2	p	62k	1.04	2204	2k4	p	27k	1.09
2126	2k7	p	10k	1.27	2204	2k7	p	12k	1.23
2130	130R	s	2k	1.07	2210	910R	s	1k3	1.7
2130	330R	s	1k8	1.18	2210	3k9	p	5k1	1.76
2131	2k2	p	68k	1.03	2220	220R	s	2k	1.11
2131	3k9	p	4k7	1.83	2220	620R	s	1k6	1.39
2137	2k2	p	75k	1.03	2222	22R	s	2k2	1.01
2143	2k2	p	82k	1.03	2222	2k4	p	30k	1.08
2143	2k4	p	20k	1.12	2222	3k3	p	6k8	1.49
2143	3k	p	7k5	1.4	2224	24R	s	2k2	1.01
2148	2k2	p	91k	1.02	2227	27R	s	2k2	1.01
2150	150R	s	2k	1.08	2230	30R	s	2k2	1.01
2150	4k3	p	4k3	2	2230	430R	s	1k8	1.24
2153	2k2	p	100k	1.02	2233	33R	s	2k2	1.02
2154	3k3	p	6k2	1.53	2236	36R	s	2k2	1.02
2157	2k2	p	110k	1.02	2236	2k7	p	13k	1.21
2160	160R	s	2k	1.08	2237	2k4	p	33k	1.07
2160	360R	s	1k8	1.2	2239	39R	s	2k2	1.02
2160	560R	s	1k6	1.35	2240	240R	s	2k	1.12
2160	2k2	p	120k	1.02	2243	43R	s	2k2	1.02
2163	2k2	p	130k	1.02	2246	4k3	p	4k7	1.91
2164	2k4	p	22k	1.11	2247	47R	s	2k2	1.02
2168	2k2	p	150k	1.01	2250	750R	s	1k5	1.5

Table 6.3 E24 equivalent resistances 81

R_E	R_1	s/p	R_2	P_{REL}	R_E	R_1	s/p	R_2	P_{REL}
2250	2k4	p	36k	1.07	2333	4k3	p	5k1	1.84
2251	51R	s	2k2	1.02	2338	2k4	p	91k	1.03
2256	56R	s	2k2	1.03	2344	2k4	p	100k	1.02
2256	3k	p	9k1	1.33	2348	2k7	p	18k	1.15
2261	2k4	p	39k	1.06	2349	2k4	p	110k	1.02
2262	62R	s	2k2	1.03	2350	150R	s	2k2	1.07
2268	68R	s	2k2	1.03	2350	750R	s	1k6	1.47
2270	270R	s	2k	1.14	2350	4k7	p	4k7	2
2270	470R	s	1k8	1.26	2353	2k4	p	120k	1.02
2273	2k4	p	43k	1.06	2353	3k3	p	8k2	1.4
2275	75R	s	2k2	1.03	2354	3k6	p	6k8	1.53
2278	3k6	p	6k2	1.58	2356	2k4	p	130k	1.02
2280	680R	s	1k6	1.43	2357	3k	p	11k	1.27
2282	82R	s	2k2	1.04	2360	160R	s	2k2	1.07
2283	2k4	p	47k	1.05	2360	360R	s	2k	1.18
2288	2k7	p	15k	1.18	2360	560R	s	1k8	1.31
2291	91R	s	2k2	1.04	2362	2k4	p	150k	1.02
2292	2k4	p	51k	1.05	2365	2k4	p	160k	1.02
2292	3k3	p	7k5	1.44	2368	2k4	p	180k	1.01
2299	3k9	p	5k6	1.7	2372	2k4	p	200k	1.01
2300	100R	s	2k2	1.05	2374	2k4	p	220k	1.01
2300	300R	s	2k	1.15	2376	2k4	p	240k	1.01
2300	1k	s	1k3	1.77	2379	2k7	p	20k	1.14
2300	1k1	s	1k2	1.92	2380	180R	s	2k2	1.08
2301	2k4	p	56k	1.04	2390	390R	s	2k	1.2
2308	3k	p	10k	1.3	2394	3k9	p	6k2	1.63
2310	110R	s	2k2	1.05	2400	200R	s	2k2	1.09
2310	510R	s	1k8	1.28	2400	1k1	s	1k3	1.85
2310	2k7	p	16k	1.17	2400	1k2	s	1k2	2
2311	2k4	p	62k	1.04	2400	3k	p	12k	1.25
2318	2k4	p	68k	1.04	2405	2k7	p	22k	1.12
2320	120R	s	2k2	1.05	2410	910R	s	1k5	1.61
2320	820R	s	1k5	1.55	2420	220R	s	2k2	1.1
2326	2k4	p	75k	1.03	2420	620R	s	1k8	1.34
2330	130R	s	2k2	1.06	2420	820R	s	1k6	1.51
2330	330R	s	2k	1.17	2422	3k3	p	9k1	1.36
2332	2k4	p	82k	1.03	2424	24R	s	2k4	1.01

R_E	R_1	s/p	R_2	P_{REL}	R_E	R_1	s/p	R_2	P_{REL}
2427	27R	s	2k4	1.01	2510	910R	s	1k6	1.57
2427	2k7	p	24k	1.11	2512	2k7	p	36k	1.08
2430	30R	s	2k4	1.01	2520	120R	s	2k4	1.05
2430	430R	s	2k	1.22	2525	2k7	p	39k	1.07
2432	3k6	p	7k5	1.48	2526	3k	p	16k	1.19
2432	4k3	p	5k6	1.77	2530	130R	s	2k4	1.05
2433	33R	s	2k4	1.01	2530	330R	s	2k2	1.15
2436	36R	s	2k4	1.02	2538	3k3	p	11k	1.3
2438	3k	p	13k	1.23	2539	4k3	p	6k2	1.69
2439	39R	s	2k4	1.02	2540	2k7	p	43k	1.06
2440	240R	s	2k2	1.11	2550	150R	s	2k4	1.06
2443	43R	s	2k4	1.02	2550	750R	s	1k8	1.42
2446	4k7	p	5k1	1.92	2550	5k1	p	5k1	2
2447	47R	s	2k4	1.02	2553	2k7	p	47k	1.06
2451	51R	s	2k4	1.02	2555	4k7	p	5k6	1.84
2455	2k7	p	27k	1.1	2560	160R	s	2k4	1.07
2456	56R	s	2k4	1.02	2560	360R	s	2k2	1.16
2462	62R	s	2k4	1.03	2560	560R	s	2k	1.28
2468	68R	s	2k4	1.03	2564	2k7	p	51k	1.05
2470	270R	s	2k2	1.12	2566	3k9	p	7k5	1.52
2470	470R	s	2k	1.24	2571	3k	p	18k	1.17
2475	75R	s	2k4	1.03	2576	2k7	p	56k	1.05
2477	2k7	p	30k	1.09	2580	180R	s	2k4	1.08
2479	3k9	p	6k8	1.57	2580	3k6	p	9k1	1.4
2480	680R	s	1k8	1.38	2587	2k7	p	62k	1.04
2481	3k3	p	10k	1.33	2588	3k3	p	12k	1.28
2482	82R	s	2k4	1.03	2590	390R	s	2k2	1.18
2491	91R	s	2k4	1.04	2597	2k7	p	68k	1.04
2496	2k7	p	33k	1.08	2600	200R	s	2k4	1.08
2500	100R	s	2k4	1.04	2600	1k	s	1k6	1.63
2500	300R	s	2k2	1.14	2600	1k1	s	1k5	1.73
2500	1k	s	1k5	1.67	2600	1k3	s	1k3	2
2500	1k2	s	1k3	1.92	2606	2k7	p	75k	1.04
2500	3k	p	15k	1.2	2609	3k	p	20k	1.15
2502	3k6	p	8k2	1.44	2614	2k7	p	82k	1.03
2510	110R	s	2k4	1.05	2620	220R	s	2k4	1.09
2510	510R	s	2k	1.26	2620	620R	s	2k	1.31

Table 6.3 E24 equivalent resistances 83

R_E	R_1	s/p	R_2	P_{REL}	R_E	R_1	s/p	R_2	P_{REL}
2620	820R	s	1k8	1.46	2730	330R	s	2k4	1.14
2622	2k7	p	91k	1.03	2730	3k9	p	9k1	1.43
2629	2k7	p	100k	1.03	2733	33R	s	2k7	1.01
2630	430R	s	2k2	1.2	2733	4k3	p	7k5	1.57
2632	3k3	p	13k	1.25	2736	36R	s	2k7	1.01
2634	4k3	p	6k8	1.63	2736	3k3	p	16k	1.21
2635	2k7	p	110k	1.02	2739	39R	s	2k7	1.01
2640	240R	s	2k4	1.1	2743	43R	s	2k7	1.02
2640	3k	p	22k	1.14	2747	47R	s	2k7	1.02
2641	2k7	p	120k	1.02	2750	750R	s	2k	1.38
2643	3k9	p	8k2	1.48	2750	3k	p	33k	1.09
2645	2k7	p	130k	1.02	2751	51R	s	2k7	1.02
2647	3k6	p	10k	1.36	2756	56R	s	2k7	1.02
2652	2k7	p	150k	1.02	2760	360R	s	2k4	1.15
2655	2k7	p	160k	1.02	2760	560R	s	2k2	1.25
2660	2k7	p	180k	1.02	2762	62R	s	2k7	1.02
2664	2k7	p	200k	1.01	2768	68R	s	2k7	1.03
2667	2k7	p	220k	1.01	2769	3k	p	36k	1.08
2667	3k	p	24k	1.13	2769	3k6	p	12k	1.3
2669	5k1	p	5k6	1.91	2775	75R	s	2k7	1.03
2670	270R	s	2k4	1.11	2779	4k7	p	6k8	1.69
2670	470R	s	2k2	1.21	2782	82R	s	2k7	1.03
2670	2k7	p	240k	1.01	2786	3k	p	39k	1.08
2673	2k7	p	270k	1.01	2789	3k3	p	18k	1.18
2673	4k7	p	6k2	1.76	2790	390R	s	2k4	1.16
2680	680R	s	2k	1.34	2791	91R	s	2k7	1.03
2700	300R	s	2k4	1.13	2798	5k1	p	6k2	1.82
2700	1k1	s	1k6	1.69	2800	100R	s	2k7	1.04
2700	1k2	s	1k5	1.8	2800	1k	s	1k8	1.56
2700	3k	p	27k	1.11	2800	1k2	s	1k6	1.75
2705	3k3	p	15k	1.22	2800	1k3	s	1k5	1.87
2710	510R	s	2k2	1.23	2800	5k6	p	5k6	2
2710	910R	s	1k8	1.51	2804	3k	p	43k	1.07
2712	3k6	p	11k	1.33	2806	3k9	p	10k	1.39
2727	27R	s	2k7	1.01	2810	110R	s	2k7	1.04
2727	3k	p	30k	1.1	2819	3k6	p	13k	1.28
2730	30R	s	2k7	1.01	2820	120R	s	2k7	1.04

R_E	R_1	s/p	R_2	P_{REL}	R_E	R_1	s/p	R_2	P_{REL}
2820	620R	s	2k2	1.28	2940	240R	s	2k7	1.09
2820	820R	s	2k	1.41	2941	3k	p	150k	1.02
2820	3k	p	47k	1.06	2941	3k3	p	27k	1.12
2821	4k3	p	8k2	1.52	2942	5k6	p	6k2	1.9
2830	130R	s	2k7	1.05	2943	3k9	p	12k	1.33
2830	430R	s	2k4	1.18	2945	3k	p	160k	1.02
2833	3k	p	51k	1.06	2950	750R	s	2k2	1.34
2833	3k3	p	20k	1.17	2951	3k	p	180k	1.02
2847	3k	p	56k	1.05	2956	3k	p	200k	1.02
2850	150R	s	2k7	1.06	2960	560R	s	2k4	1.23
2860	160R	s	2k7	1.06	2960	3k	p	220k	1.01
2862	3k	p	62k	1.05	2963	3k	p	240k	1.01
2870	470R	s	2k4	1.2	2967	3k	p	270k	1.01
2870	3k3	p	22k	1.15	2970	270R	s	2k7	1.1
2873	3k	p	68k	1.04	2970	3k	p	300k	1.01
2879	3k9	p	11k	1.35	2973	3k3	p	30k	1.11
2880	180R	s	2k7	1.07	2988	4k7	p	8k2	1.57
2880	680R	s	2k2	1.31	3000	300R	s	2k7	1.11
2885	3k	p	75k	1.04	3000	1k	s	2k	1.5
2889	4k7	p	7k5	1.63	3000	1k2	s	1k8	1.67
2894	3k	p	82k	1.04	3000	1k5	s	1k5	2
2900	200R	s	2k7	1.07	3000	3k3	p	33k	1.1
2900	1k1	s	1k8	1.61	3000	3k6	p	18k	1.2
2900	1k3	s	1k6	1.81	3000	3k9	p	13k	1.3
2901	3k3	p	24k	1.14	3007	4k3	p	10k	1.43
2903	3k6	p	15k	1.24	3020	620R	s	2k4	1.26
2904	3k	p	91k	1.03	3020	820R	s	2k2	1.37
2910	510R	s	2k4	1.21	3023	3k3	p	36k	1.09
2910	910R	s	2k	1.46	3030	30R	s	3k	1.01
2913	3k	p	100k	1.03	3030	330R	s	2k7	1.12
2914	5k1	p	6k8	1.75	3033	33R	s	3k	1.01
2920	220R	s	2k7	1.08	3036	36R	s	3k	1.01
2920	3k	p	110k	1.03	3036	5k1	p	7k5	1.68
2920	4k3	p	9k1	1.47	3039	39R	s	3k	1.01
2927	3k	p	120k	1.03	3043	43R	s	3k	1.01
2932	3k	p	130k	1.02	3043	3k3	p	39k	1.08
2939	3k6	p	16k	1.23	3047	47R	s	3k	1.02

Table 6.3 E24 equivalent resistances 85

R_E	R_1	s/p	R_2	P_{REL}	R_E	R_1	s/p	R_2	P_{REL}
3051	51R	s	3k	1.02	3160	160R	s	3k	1.05
3051	3k6	p	20k	1.18	3161	3k3	p	75k	1.04
3056	56R	s	3k	1.02	3166	4k3	p	12k	1.36
3060	360R	s	2k7	1.13	3170	470R	s	2k7	1.17
3062	62R	s	3k	1.02	3172	3k3	p	82k	1.04
3065	3k3	p	43k	1.08	3176	3k6	p	27k	1.13
3068	68R	s	3k	1.02	3180	180R	s	3k	1.06
3071	5k6	p	6k8	1.82	3185	3k3	p	91k	1.04
3075	75R	s	3k	1.03	3195	3k3	p	100k	1.03
3080	680R	s	2k4	1.28	3197	4k7	p	10k	1.47
3082	82R	s	3k	1.03	3200	200R	s	3k	1.07
3083	3k3	p	47k	1.07	3200	1k	s	2k2	1.45
3090	390R	s	2k7	1.14	3200	1k2	s	2k	1.6
3091	91R	s	3k	1.03	3200	1k6	s	1k6	2
3092	4k3	p	11k	1.39	3204	3k3	p	110k	1.03
3094	3k6	p	22k	1.16	3205	3k9	p	18k	1.22
3095	3k9	p	15k	1.26	3206	5k6	p	7k5	1.75
3099	3k3	p	51k	1.06	3210	510R	s	2k7	1.19
3099	4k7	p	9k1	1.52	3212	3k3	p	120k	1.03
3100	100R	s	3k	1.03	3214	3k6	p	30k	1.12
3100	1k1	s	2k	1.55	3218	3k3	p	130k	1.03
3100	1k3	s	1k8	1.72	3220	220R	s	3k	1.07
3100	1k5	s	1k6	1.94	3220	820R	s	2k4	1.34
3100	6k2	p	6k2	2	3229	3k3	p	150k	1.02
3110	110R	s	3k	1.04	3231	4k3	p	13k	1.33
3110	910R	s	2k2	1.41	3233	3k3	p	160k	1.02
3116	3k3	p	56k	1.06	3240	240R	s	3k	1.08
3120	120R	s	3k	1.04	3241	3k3	p	180k	1.02
3130	130R	s	3k	1.04	3243	6k2	p	6k8	1.91
3130	430R	s	2k7	1.16	3246	3k3	p	200k	1.02
3130	3k6	p	24k	1.15	3246	3k6	p	33k	1.11
3133	3k3	p	62k	1.05	3251	3k3	p	220k	1.02
3136	3k9	p	16k	1.24	3255	3k3	p	240k	1.01
3144	5k1	p	8k2	1.62	3260	560R	s	2k7	1.21
3147	3k3	p	68k	1.05	3260	3k3	p	270k	1.01
3150	150R	s	3k	1.05	3264	3k3	p	300k	1.01
3150	750R	s	2k4	1.31	3264	3k9	p	20k	1.2

R_E	R_1	s/p	R_2	P_{REL}	R_E	R_1	s/p	R_2	P_{REL}
3267	3k3	p	330k	1.01	3390	390R	s	3k	1.13
3268	5k1	p	9k1	1.56	3391	91R	s	3k3	1.03
3270	270R	s	3k	1.09	3394	6k2	p	7k5	1.83
3273	3k6	p	36k	1.1	3400	100R	s	3k3	1.03
3293	4k7	p	11k	1.43	3400	1k	s	2k4	1.42
3296	3k6	p	39k	1.09	3400	1k2	s	2k2	1.55
3300	300R	s	3k	1.1	3400	1k6	s	1k8	1.89
3300	1k1	s	2k2	1.5	3400	6k8	p	6k8	2
3300	1k3	s	2k	1.65	3402	3k6	p	62k	1.06
3300	1k5	s	1k8	1.83	3408	3k9	p	27k	1.14
3310	910R	s	2k4	1.38	3410	110R	s	3k3	1.03
3313	3k9	p	22k	1.18	3419	3k6	p	68k	1.05
3320	620R	s	2k7	1.23	3420	120R	s	3k3	1.04
3322	3k6	p	43k	1.08	3430	130R	s	3k3	1.04
3328	5k6	p	8k2	1.68	3430	430R	s	3k	1.14
3330	330R	s	3k	1.11	3435	3k6	p	75k	1.05
3333	33R	s	3k3	1.01	3449	3k6	p	82k	1.04
3336	36R	s	3k3	1.01	3450	150R	s	3k3	1.05
3339	39R	s	3k3	1.01	3450	750R	s	2k7	1.28
3342	4k3	p	15k	1.29	3451	3k9	p	30k	1.13
3343	43R	s	3k3	1.01	3452	4k7	p	13k	1.36
3344	3k6	p	47k	1.08	3460	160R	s	3k3	1.05
3347	47R	s	3k3	1.01	3463	3k6	p	91k	1.04
3351	51R	s	3k3	1.02	3467	5k6	p	9k1	1.62
3355	3k9	p	24k	1.16	3470	470R	s	3k	1.16
3356	56R	s	3k3	1.02	3471	4k3	p	18k	1.24
3360	360R	s	3k	1.12	3475	3k6	p	100k	1.04
3362	62R	s	3k3	1.02	3480	180R	s	3k3	1.05
3363	3k6	p	51k	1.07	3484	5k1	p	11k	1.46
3368	68R	s	3k3	1.02	3486	3k6	p	110k	1.03
3375	75R	s	3k3	1.02	3488	3k9	p	33k	1.12
3377	4k7	p	12k	1.39	3495	3k6	p	120k	1.03
3377	5k1	p	10k	1.51	3500	200R	s	3k3	1.06
3380	680R	s	2k7	1.25	3500	1k1	s	2k4	1.46
3382	82R	s	3k3	1.02	3500	1k3	s	2k2	1.59
3383	3k6	p	56k	1.06	3500	1k5	s	2k	1.75
3389	4k3	p	16k	1.27	3503	3k6	p	130k	1.03

Table 6.3 E24 equivalent resistances 87

R_E	R_1	s/p	R_2	P_{REL}	R_E	R_1	s/p	R_2	P_{REL}
3510	510R	s	3k	1.17	3639	39R	s	3k6	1.01
3516	3k6	p	150k	1.02	3643	43R	s	3k6	1.01
3519	3k9	p	36k	1.11	3646	3k9	p	56k	1.07
3520	220R	s	3k3	1.07	3647	47R	s	3k6	1.01
3520	820R	s	2k7	1.3	3647	4k3	p	24k	1.18
3521	3k6	p	160k	1.02	3651	51R	s	3k6	1.01
3529	3k6	p	180k	1.02	3656	56R	s	3k6	1.02
3531	6k2	p	8k2	1.76	3660	360R	s	3k3	1.11
3536	3k6	p	200k	1.02	3662	62R	s	3k6	1.02
3539	4k3	p	20k	1.22	3663	5k1	p	13k	1.39
3540	240R	s	3k3	1.07	3668	68R	s	3k6	1.02
3542	3k6	p	220k	1.02	3669	3k9	p	62k	1.06
3545	3k9	p	39k	1.1	3675	75R	s	3k6	1.02
3547	3k6	p	240k	1.02	3680	680R	s	3k	1.23
3553	3k6	p	270k	1.01	3682	82R	s	3k6	1.02
3557	3k6	p	300k	1.01	3688	3k9	p	68k	1.06
3560	560R	s	3k	1.19	3688	6k2	p	9k1	1.68
3561	3k6	p	330k	1.01	3690	390R	s	3k3	1.12
3564	3k6	p	360k	1.01	3691	91R	s	3k6	1.03
3566	6k8	p	7k5	1.91	3700	100R	s	3k6	1.03
3570	270R	s	3k3	1.08	3700	1k	s	2k7	1.37
3576	3k9	p	43k	1.09	3700	1k3	s	2k4	1.54
3579	4k7	p	15k	1.31	3700	1k5	s	2k2	1.68
3579	5k1	p	12k	1.43	3707	3k9	p	75k	1.05
3590	5k6	p	10k	1.56	3709	4k3	p	27k	1.16
3597	4k3	p	22k	1.2	3710	110R	s	3k6	1.03
3600	300R	s	3k3	1.09	3711	5k6	p	11k	1.51
3600	1k2	s	2k4	1.5	3717	6k8	p	8k2	1.83
3600	1k6	s	2k	1.8	3720	120R	s	3k6	1.03
3600	1k8	s	1k8	2	3723	3k9	p	82k	1.05
3601	3k9	p	47k	1.08	3727	4k7	p	18k	1.26
3610	910R	s	2k7	1.34	3730	130R	s	3k6	1.04
3620	620R	s	3k	1.21	3730	430R	s	3k3	1.13
3623	3k9	p	51k	1.08	3740	3k9	p	91k	1.04
3630	330R	s	3k3	1.1	3750	150R	s	3k6	1.04
3633	4k7	p	16k	1.29	3750	750R	s	3k	1.25
3636	36R	s	3k6	1.01	3750	7k5	p	7k5	2

R_E	R_1	s/p	R_2	P_{REL}	R_E	R_1	s/p	R_2	P_{REL}
3754	3k9	p	100k	1.04	3873	4k7	p	22k	1.21
3760	160R	s	3k6	1.04	3892	6k8	p	9k1	1.75
3761	4k3	p	30k	1.14	3900	300R	s	3k6	1.08
3766	3k9	p	110k	1.04	3900	1k2	s	2k7	1.44
3770	470R	s	3k3	1.14	3900	1k5	s	2k4	1.63
3777	3k9	p	120k	1.03	3909	4k3	p	43k	1.1
3780	180R	s	3k6	1.05	3910	910R	s	3k	1.3
3786	3k9	p	130k	1.03	3914	5k6	p	13k	1.43
3800	200R	s	3k6	1.06	3917	7k5	p	8k2	1.91
3800	1k1	s	2k7	1.41	3920	620R	s	3k3	1.19
3800	1k6	s	2k2	1.73	3930	330R	s	3k6	1.09
3800	1k8	s	2k	1.9	3930	4k7	p	24k	1.2
3801	3k9	p	150k	1.03	3939	39R	s	3k9	1.01
3804	4k3	p	33k	1.13	3940	4k3	p	47k	1.09
3806	4k7	p	20k	1.24	3943	43R	s	3k9	1.01
3806	5k1	p	15k	1.34	3947	47R	s	3k9	1.01
3807	3k9	p	160k	1.02	3951	51R	s	3k9	1.01
3810	510R	s	3k3	1.15	3956	56R	s	3k9	1.01
3817	3k9	p	180k	1.02	3960	360R	s	3k6	1.1
3818	5k6	p	12k	1.47	3962	62R	s	3k9	1.02
3820	220R	s	3k6	1.06	3965	6k2	p	11k	1.56
3820	820R	s	3k	1.27	3966	4k3	p	51k	1.08
3825	3k9	p	200k	1.02	3968	68R	s	3k9	1.02
3827	6k2	p	10k	1.62	3974	5k1	p	18k	1.28
3832	3k9	p	220k	1.02	3975	75R	s	3k9	1.02
3838	3k9	p	240k	1.02	3980	680R	s	3k3	1.21
3840	240R	s	3k6	1.07	3982	82R	s	3k9	1.02
3841	4k3	p	36k	1.12	3990	390R	s	3k6	1.11
3844	3k9	p	270k	1.01	3991	91R	s	3k9	1.02
3850	3k9	p	300k	1.01	3993	4k3	p	56k	1.08
3854	3k9	p	330k	1.01	4000	100R	s	3k9	1.03
3858	3k9	p	360k	1.01	4000	1k	s	3k	1.33
3860	560R	s	3k3	1.17	4000	1k3	s	2k7	1.48
3861	3k9	p	390k	1.01	4000	1k6	s	2k4	1.67
3867	5k1	p	16k	1.32	4000	1k8	s	2k2	1.82
3870	270R	s	3k6	1.08	4000	2k	s	2k	2
3873	4k3	p	39k	1.11	4003	4k7	p	27k	1.17

Table 6.3 E24 equivalent resistances 89

R_E	R_1	s/p	R_2	P_{REL}	R_E	R_1	s/p	R_2	P_{REL}
4010	110R	s	3k9	1.03	4180	4k3	p	150k	1.03
4020	120R	s	3k9	1.03	4187	4k3	p	160k	1.03
4021	4k3	p	62k	1.07	4195	4k7	p	39k	1.12
4030	130R	s	3k9	1.03	4198	6k2	p	13k	1.48
4030	430R	s	3k6	1.12	4200	300R	s	3k9	1.08
4044	4k3	p	68k	1.06	4200	1k2	s	3k	1.4
4048	6k8	p	10k	1.68	4200	1k5	s	2k7	1.56
4050	150R	s	3k9	1.04	4200	1k8	s	2k4	1.75
4050	750R	s	3k3	1.23	4200	2k	s	2k2	1.91
4060	160R	s	3k9	1.04	4200	4k3	p	180k	1.02
4063	4k7	p	30k	1.16	4202	6k8	p	11k	1.62
4064	5k1	p	20k	1.26	4206	5k1	p	24k	1.21
4067	4k3	p	75k	1.06	4209	4k3	p	200k	1.02
4070	470R	s	3k6	1.13	4210	910R	s	3k3	1.28
4078	5k6	p	15k	1.37	4218	4k3	p	220k	1.02
4080	180R	s	3k9	1.05	4220	620R	s	3k6	1.17
4086	4k3	p	82k	1.05	4224	4k3	p	240k	1.02
4088	6k2	p	12k	1.52	4230	330R	s	3k9	1.08
4100	200R	s	3k9	1.05	4233	4k3	p	270k	1.02
4100	1k1	s	3k	1.37	4237	4k7	p	43k	1.11
4100	8k2	p	8k2	2	4239	4k3	p	300k	1.01
4106	4k3	p	91k	1.05	4245	4k3	p	330k	1.01
4110	510R	s	3k6	1.14	4249	4k3	p	360k	1.01
4111	7k5	p	9k1	1.82	4253	4k3	p	390k	1.01
4114	4k7	p	33k	1.14	4257	4k3	p	430k	1.01
4120	220R	s	3k9	1.06	4260	360R	s	3k9	1.09
4120	820R	s	3k3	1.25	4271	5k6	p	18k	1.31
4123	4k3	p	100k	1.04	4273	4k7	p	47k	1.1
4138	4k3	p	110k	1.04	4280	680R	s	3k6	1.19
4140	240R	s	3k9	1.06	4286	7k5	p	10k	1.75
4140	5k1	p	22k	1.23	4290	390R	s	3k9	1.1
4148	5k6	p	16k	1.35	4290	5k1	p	27k	1.19
4151	4k3	p	120k	1.04	4300	1k	s	3k3	1.3
4157	4k7	p	36k	1.13	4300	1k3	s	3k	1.43
4160	560R	s	3k6	1.16	4300	1k6	s	2k7	1.59
4162	4k3	p	130k	1.03	4303	4k7	p	51k	1.09
4170	270R	s	3k9	1.07	4313	8k2	p	9k1	1.9

R_E	R_1	s/p	R_2	P_{REL}	R_E	R_1	s/p	R_2	P_{REL}
4330	430R	s	3k9	1.11	4467	5k1	p	36k	1.14
4336	4k7	p	56k	1.08	4468	6k2	p	16k	1.39
4340	6k8	p	12k	1.57	4469	4k7	p	91k	1.05
4343	43R	s	4k3	1.01	4480	180R	s	4k3	1.04
4347	47R	s	4k3	1.01	4489	4k7	p	100k	1.05
4350	750R	s	3k6	1.21	4500	200R	s	4k3	1.05
4351	51R	s	4k3	1.01	4500	1k2	s	3k3	1.36
4356	56R	s	4k3	1.01	4500	1k5	s	3k	1.5
4359	5k1	p	30k	1.17	4500	1k8	s	2k7	1.67
4362	62R	s	4k3	1.01	4505	8k2	p	10k	1.82
4368	68R	s	4k3	1.02	4507	4k7	p	110k	1.04
4369	4k7	p	62k	1.08	4510	910R	s	3k6	1.25
4370	470R	s	3k9	1.12	4510	5k1	p	39k	1.13
4375	75R	s	4k3	1.02	4520	220R	s	4k3	1.05
4375	5k6	p	20k	1.28	4520	620R	s	3k9	1.16
4382	82R	s	4k3	1.02	4523	4k7	p	120k	1.04
4387	6k2	p	15k	1.41	4536	4k7	p	130k	1.04
4391	91R	s	4k3	1.02	4540	240R	s	4k3	1.06
4396	4k7	p	68k	1.07	4541	5k6	p	24k	1.23
4400	100R	s	4k3	1.02	4550	9k1	p	9k1	2
4400	1k1	s	3k3	1.33	4557	4k7	p	150k	1.03
4400	2k	s	2k4	1.83	4559	5k1	p	43k	1.12
4400	2k2	s	2k2	2	4566	4k7	p	160k	1.03
4410	110R	s	4k3	1.03	4570	270R	s	4k3	1.06
4410	510R	s	3k9	1.13	4580	680R	s	3k9	1.17
4417	5k1	p	33k	1.15	4580	4k7	p	180k	1.03
4420	120R	s	4k3	1.03	4592	4k7	p	200k	1.02
4420	820R	s	3k6	1.23	4600	300R	s	4k3	1.07
4423	4k7	p	75k	1.06	4600	1k	s	3k6	1.28
4430	130R	s	4k3	1.03	4600	1k3	s	3k3	1.39
4445	4k7	p	82k	1.06	4600	1k6	s	3k	1.53
4450	150R	s	4k3	1.03	4600	2k2	s	2k4	1.92
4459	7k5	p	11k	1.68	4601	5k1	p	47k	1.11
4460	160R	s	4k3	1.04	4602	4k7	p	220k	1.02
4460	560R	s	3k9	1.14	4610	4k7	p	240k	1.02
4464	5k6	p	22k	1.25	4612	6k2	p	18k	1.34
4465	6k8	p	13k	1.52	4615	7k5	p	12k	1.63

Table 6.3 E24 equivalent resistances 91

R_E	R_1	s/p	R_2	P_{REL}	R_E	R_1	s/p	R_2	P_{REL}
4620	4k7	p	270k	1.02	4791	91R	s	4k7	1.02
4628	4k7	p	300k	1.02	4800	100R	s	4k7	1.02
4630	330R	s	4k3	1.08	4800	1k2	s	3k6	1.33
4634	4k7	p	330k	1.01	4800	1k5	s	3k3	1.45
4636	5k1	p	51k	1.1	4800	1k8	s	3k	1.6
4638	5k6	p	27k	1.21	4800	2k4	s	2k4	2
4639	4k7	p	360k	1.01	4801	5k1	p	82k	1.06
4644	4k7	p	390k	1.01	4810	110R	s	4k7	1.02
4649	4k7	p	430k	1.01	4810	510R	s	4k3	1.12
4650	750R	s	3k9	1.19	4810	910R	s	3k9	1.23
4653	4k7	p	470k	1.01	4820	120R	s	4k7	1.03
4660	360R	s	4k3	1.08	4829	5k1	p	91k	1.06
4674	5k1	p	56k	1.09	4830	130R	s	4k7	1.03
4679	6k8	p	15k	1.45	4837	6k2	p	22k	1.28
4690	390R	s	4k3	1.09	4846	5k6	p	36k	1.16
4698	8k2	p	11k	1.75	4850	150R	s	4k7	1.03
4700	1k1	s	3k6	1.31	4853	5k1	p	100k	1.05
4700	2k	s	2k7	1.74	4860	160R	s	4k7	1.03
4712	5k1	p	62k	1.08	4860	560R	s	4k3	1.13
4719	5k6	p	30k	1.19	4871	8k2	p	12k	1.68
4720	820R	s	3k9	1.21	4874	5k1	p	110k	1.05
4730	430R	s	4k3	1.1	4880	180R	s	4k7	1.04
4733	6k2	p	20k	1.31	4892	5k1	p	120k	1.04
4744	5k1	p	68k	1.08	4897	5k6	p	39k	1.14
4747	47R	s	4k7	1.01	4900	200R	s	4k7	1.04
4751	51R	s	4k7	1.01	4900	1k	s	3k9	1.26
4756	56R	s	4k7	1.01	4900	1k3	s	3k6	1.36
4756	7k5	p	13k	1.58	4900	1k6	s	3k3	1.48
4762	62R	s	4k7	1.01	4900	2k2	s	2k7	1.81
4764	9k1	p	10k	1.91	4907	5k1	p	130k	1.04
4768	68R	s	4k7	1.01	4920	220R	s	4k7	1.05
4770	470R	s	4k3	1.11	4920	620R	s	4k3	1.14
4772	6k8	p	16k	1.43	4927	6k2	p	24k	1.26
4775	75R	s	4k7	1.02	4932	5k1	p	150k	1.03
4775	5k1	p	75k	1.07	4935	6k8	p	18k	1.38
4782	82R	s	4k7	1.02	4940	240R	s	4k7	1.05
4788	5k6	p	33k	1.17	4942	5k1	p	160k	1.03

R_E	R_1	s/p	R_2	P_{REL}	R_E	R_1	s/p	R_2	P_{REL}
4955	5k6	p	43k	1.13	5130	430R	s	4k7	1.09
4959	5k1	p	180k	1.03	5136	5k6	p	62k	1.09
4970	270R	s	4k7	1.06	5138	6k2	p	30k	1.21
4973	5k1	p	200k	1.03	5151	51R	s	5k1	1.01
4980	680R	s	4k3	1.16	5156	56R	s	5k1	1.01
4980	9k1	p	11k	1.83	5162	62R	s	5k1	1.01
4984	5k1	p	220k	1.02	5168	68R	s	5k1	1.01
4994	5k1	p	240k	1.02	5170	470R	s	4k7	1.1
5000	300R	s	4k7	1.06	5174	5k6	p	68k	1.08
5000	1k1	s	3k9	1.28	5175	75R	s	5k1	1.01
5000	2k	s	3k	1.67	5175	9k1	p	12k	1.76
5000	7k5	p	15k	1.5	5182	82R	s	5k1	1.02
5000	10k	p	10k	2	5191	91R	s	5k1	1.02
5004	5k6	p	47k	1.12	5194	6k8	p	22k	1.31
5005	5k1	p	270k	1.02	5200	100R	s	5k1	1.02
5015	5k1	p	300k	1.02	5200	1k3	s	3k9	1.33
5022	5k1	p	330k	1.02	5200	1k6	s	3k6	1.44
5028	8k2	p	13k	1.63	5200	2k2	s	3k	1.73
5029	5k1	p	360k	1.01	5210	110R	s	5k1	1.02
5030	330R	s	4k7	1.07	5210	510R	s	4k7	1.11
5034	5k1	p	390k	1.01	5210	910R	s	4k3	1.21
5040	5k1	p	430k	1.01	5211	5k6	p	75k	1.07
5042	6k2	p	27k	1.23	5219	6k2	p	33k	1.19
5045	5k1	p	470k	1.01	5220	120R	s	5k1	1.02
5046	5k6	p	51k	1.11	5230	130R	s	5k1	1.03
5050	750R	s	4k3	1.17	5238	10k	p	11k	1.91
5050	5k1	p	510k	1.01	5242	5k6	p	82k	1.07
5060	360R	s	4k7	1.08	5250	150R	s	5k1	1.03
5075	6k8	p	20k	1.34	5260	160R	s	5k1	1.03
5090	390R	s	4k7	1.08	5260	560R	s	4k7	1.12
5091	5k6	p	56k	1.1	5275	5k6	p	91k	1.06
5100	1k2	s	3k9	1.31	5280	180R	s	5k1	1.04
5100	1k5	s	3k6	1.42	5289	6k2	p	36k	1.17
5100	1k8	s	3k3	1.55	5294	7k5	p	18k	1.42
5100	2k4	s	2k7	1.89	5299	6k8	p	24k	1.28
5106	7k5	p	16k	1.47	5300	200R	s	5k1	1.04
5120	820R	s	4k3	1.19	5300	1k	s	4k3	1.23

Table 6.3 E24 equivalent resistances 93

R_E	R_1	s/p	R_2	P_{REL}	R_E	R_1	s/p	R_2	P_{REL}
5300	2k	s	3k3	1.61	5500	1k2	s	4k3	1.28
5302	8k2	p	15k	1.55	5500	1k6	s	3k9	1.41
5303	5k6	p	100k	1.06	5500	2k2	s	3k3	1.67
5320	220R	s	5k1	1.04	5500	11k	p	11k	2
5320	620R	s	4k7	1.13	5507	5k6	p	330k	1.02
5329	5k6	p	110k	1.05	5514	5k6	p	360k	1.02
5340	240R	s	5k1	1.05	5520	820R	s	4k7	1.17
5350	5k6	p	120k	1.05	5521	5k6	p	390k	1.01
5350	6k2	p	39k	1.16	5528	5k6	p	430k	1.01
5353	9k1	p	13k	1.7	5528	6k2	p	51k	1.12
5369	5k6	p	130k	1.04	5530	430R	s	5k1	1.08
5370	270R	s	5k1	1.05	5534	5k6	p	470k	1.01
5380	680R	s	4k7	1.14	5539	5k6	p	510k	1.01
5398	5k6	p	150k	1.04	5543	6k8	p	30k	1.23
5400	300R	s	5k1	1.06	5545	5k6	p	560k	1.01
5400	1k1	s	4k3	1.26	5570	470R	s	5k1	1.09
5400	1k5	s	3k9	1.38	5582	6k2	p	56k	1.11
5400	1k8	s	3k6	1.5	5593	7k5	p	22k	1.34
5400	2k4	s	3k	1.8	5600	1k3	s	4k3	1.3
5400	2k7	s	2k7	2	5600	2k	s	3k6	1.56
5411	5k6	p	160k	1.03	5610	510R	s	5k1	1.1
5419	6k2	p	43k	1.14	5610	910R	s	4k7	1.19
5421	8k2	p	16k	1.51	5634	8k2	p	18k	1.46
5430	330R	s	5k1	1.06	5636	6k2	p	62k	1.1
5431	5k6	p	180k	1.03	5638	6k8	p	33k	1.21
5432	6k8	p	27k	1.25	5652	10k	p	13k	1.77
5447	5k6	p	200k	1.03	5656	56R	s	5k6	1.01
5450	750R	s	4k7	1.16	5660	560R	s	5k1	1.11
5455	7k5	p	20k	1.38	5662	62R	s	5k6	1.01
5455	10k	p	12k	1.83	5664	9k1	p	15k	1.61
5460	360R	s	5k1	1.07	5668	68R	s	5k6	1.01
5461	5k6	p	220k	1.03	5675	75R	s	5k6	1.01
5472	5k6	p	240k	1.02	5682	82R	s	5k6	1.01
5477	6k2	p	47k	1.13	5682	6k2	p	68k	1.09
5486	5k6	p	270k	1.02	5691	91R	s	5k6	1.02
5490	390R	s	5k1	1.08	5700	100R	s	5k6	1.02
5497	5k6	p	300k	1.02	5700	1k	s	4k7	1.21

R_E	R_1	s/p	R_2	P_{REL}	R_E	R_1	s/p	R_2	P_{REL}
5700	1k8	s	3k9	1.46	5918	6k2	p	130k	1.05
5700	2k4	s	3k3	1.73	5920	820R	s	5k1	1.16
5700	2k7	s	3k	1.9	5930	330R	s	5k6	1.06
5710	110R	s	5k6	1.02	5941	6k8	p	47k	1.14
5714	7k5	p	24k	1.31	5954	6k2	p	150k	1.04
5720	120R	s	5k6	1.02	5958	11k	p	13k	1.85
5720	620R	s	5k1	1.12	5960	360R	s	5k6	1.06
5720	6k8	p	36k	1.19	5969	6k2	p	160k	1.04
5727	6k2	p	75k	1.08	5974	8k2	p	22k	1.37
5730	130R	s	5k6	1.02	5990	390R	s	5k6	1.07
5739	11k	p	12k	1.92	5994	6k2	p	180k	1.03
5750	150R	s	5k6	1.03	6000	1k3	s	4k7	1.28
5760	160R	s	5k6	1.03	6000	2k4	s	3k6	1.67
5764	6k2	p	82k	1.08	6000	2k7	s	3k3	1.82
5780	180R	s	5k6	1.03	6000	3k	s	3k	2
5780	680R	s	5k1	1.13	6000	6k8	p	51k	1.13
5790	6k8	p	39k	1.17	6000	7k5	p	30k	1.25
5800	200R	s	5k6	1.04	6000	10k	p	15k	1.67
5800	1k1	s	4k7	1.23	6000	12k	p	12k	2
5800	1k5	s	4k3	1.35	6010	910R	s	5k1	1.18
5800	2k2	s	3k6	1.61	6014	6k2	p	200k	1.03
5801	9k1	p	16k	1.57	6030	430R	s	5k6	1.08
5805	6k2	p	91k	1.07	6030	6k2	p	220k	1.03
5816	8k2	p	20k	1.41	6044	6k2	p	240k	1.03
5820	220R	s	5k6	1.04	6044	9k1	p	18k	1.51
5838	6k2	p	100k	1.06	6061	6k2	p	270k	1.02
5840	240R	s	5k6	1.04	6064	6k8	p	56k	1.12
5850	750R	s	5k1	1.15	6070	470R	s	5k6	1.08
5869	6k2	p	110k	1.06	6074	6k2	p	300k	1.02
5870	270R	s	5k6	1.05	6086	6k2	p	330k	1.02
5870	7k5	p	27k	1.28	6095	6k2	p	360k	1.02
5871	6k8	p	43k	1.16	6100	1k	s	5k1	1.2
5895	6k2	p	120k	1.05	6100	1k8	s	4k3	1.42
5900	300R	s	5k6	1.05	6100	2k2	s	3k9	1.56
5900	1k2	s	4k7	1.26	6103	6k2	p	390k	1.02
5900	1k6	s	4k3	1.37	6110	510R	s	5k6	1.09
5900	2k	s	3k9	1.51	6111	7k5	p	33k	1.23

Table 6.3 E24 equivalent resistances 95

R_E	R_1	s/p	R_2	P_{REL}	R_E	R_1	s/p	R_2	P_{REL}
6112	6k2	p	430k	1.01	6346	11k	p	15k	1.73
6112	8k2	p	24k	1.34	6350	150R	s	6k2	1.02
6119	6k2	p	470k	1.01	6350	750R	s	5k6	1.13
6126	6k2	p	510k	1.01	6360	160R	s	6k2	1.03
6128	6k8	p	62k	1.11	6367	6k8	p	100k	1.07
6132	6k2	p	560k	1.01	6380	180R	s	6k2	1.03
6139	6k2	p	620k	1.01	6386	7k5	p	43k	1.17
6154	10k	p	16k	1.63	6400	200R	s	6k2	1.03
6160	560R	s	5k6	1.1	6400	1k3	s	5k1	1.25
6182	6k8	p	68k	1.1	6404	6k8	p	110k	1.06
6200	1k1	s	5k1	1.22	6420	220R	s	6k2	1.04
6200	1k5	s	4k7	1.32	6420	820R	s	5k6	1.15
6207	7k5	p	36k	1.21	6429	10k	p	18k	1.56
6220	620R	s	5k6	1.11	6435	6k8	p	120k	1.06
6235	6k8	p	75k	1.09	6437	9k1	p	22k	1.41
6240	12k	p	13k	1.92	6440	240R	s	6k2	1.04
6254	9k1	p	20k	1.46	6440	8k2	p	30k	1.27
6262	62R	s	6k2	1.01	6462	6k8	p	130k	1.05
6268	68R	s	6k2	1.01	6468	7k5	p	47k	1.16
6275	75R	s	6k2	1.01	6470	270R	s	6k2	1.04
6279	6k8	p	82k	1.08	6500	300R	s	6k2	1.05
6280	680R	s	5k6	1.12	6500	1k8	s	4k7	1.38
6282	82R	s	6k2	1.01	6500	2k2	s	4k3	1.51
6290	7k5	p	39k	1.19	6500	13k	p	13k	2
6290	8k2	p	27k	1.3	6505	6k8	p	150k	1.05
6291	91R	s	6k2	1.01	6510	910R	s	5k6	1.16
6300	100R	s	6k2	1.02	6519	11k	p	16k	1.69
6300	1k2	s	5k1	1.24	6523	6k8	p	160k	1.04
6300	1k6	s	4k7	1.34	6530	330R	s	6k2	1.05
6300	2k	s	4k3	1.47	6538	7k5	p	51k	1.15
6300	2k4	s	3k9	1.62	6552	6k8	p	180k	1.04
6300	2k7	s	3k6	1.75	6560	360R	s	6k2	1.06
6300	3k	s	3k3	1.91	6568	8k2	p	33k	1.25
6310	110R	s	6k2	1.02	6576	6k8	p	200k	1.03
6320	120R	s	6k2	1.02	6590	390R	s	6k2	1.06
6327	6k8	p	91k	1.07	6596	6k8	p	220k	1.03
6330	130R	s	6k2	1.02	6598	9k1	p	24k	1.38

R_E	R_1	s/p	R_2	P_{REL}	R_E	R_1	s/p	R_2	P_{REL}
6600	1k	s	5k6	1.18	6857	12k	p	16k	1.75
6600	1k5	s	5k1	1.29	6868	68R	s	6k8	1.01
6600	2k7	s	3k9	1.69	6872	7k5	p	82k	1.09
6600	3k	s	3k6	1.83	6875	75R	s	6k8	1.01
6600	3k3	s	3k3	2	6875	10k	p	22k	1.45
6613	6k8	p	240k	1.03	6880	680R	s	6k2	1.11
6614	7k5	p	56k	1.13	6882	82R	s	6k8	1.01
6630	430R	s	6k2	1.07	6887	8k2	p	43k	1.19
6633	6k8	p	270k	1.03	6891	91R	s	6k8	1.01
6649	6k8	p	300k	1.02	6900	100R	s	6k8	1.01
6663	6k8	p	330k	1.02	6900	1k3	s	5k6	1.23
6667	10k	p	20k	1.5	6900	1k8	s	5k1	1.35
6667	12k	p	15k	1.8	6900	2k2	s	4k7	1.47
6670	470R	s	6k2	1.08	6900	3k	s	3k9	1.77
6674	6k8	p	360k	1.02	6900	3k3	s	3k6	1.92
6679	8k2	p	36k	1.23	6910	110R	s	6k8	1.02
6683	6k8	p	390k	1.02	6920	120R	s	6k8	1.02
6691	7k5	p	62k	1.12	6929	7k5	p	91k	1.08
6694	6k8	p	430k	1.02	6930	130R	s	6k8	1.02
6700	1k1	s	5k6	1.2	6950	150R	s	6k8	1.02
6700	1k6	s	5k1	1.31	6950	750R	s	6k2	1.12
6700	2k	s	4k7	1.43	6960	160R	s	6k8	1.02
6700	2k4	s	4k3	1.56	6964	13k	p	15k	1.87
6703	6k8	p	470k	1.01	6977	7k5	p	100k	1.08
6710	510R	s	6k2	1.08	6980	180R	s	6k8	1.03
6711	6k8	p	510k	1.01	6982	8k2	p	47k	1.17
6718	6k8	p	560k	1.01	6982	9k1	p	30k	1.3
6726	6k8	p	620k	1.01	7000	200R	s	6k8	1.03
6733	6k8	p	680k	1.01	7000	2k7	s	4k3	1.63
6755	7k5	p	68k	1.11	7020	220R	s	6k8	1.03
6760	560R	s	6k2	1.09	7020	820R	s	6k2	1.13
6775	8k2	p	39k	1.21	7021	7k5	p	110k	1.07
6800	1k2	s	5k6	1.21	7040	240R	s	6k8	1.04
6806	9k1	p	27k	1.34	7059	7k5	p	120k	1.06
6818	7k5	p	75k	1.1	7059	10k	p	24k	1.42
6820	620R	s	6k2	1.1	7064	8k2	p	51k	1.16
6828	11k	p	18k	1.61	7070	270R	s	6k8	1.04

Table 6.3 E24 equivalent resistances 97

R_E	R_1	s/p	R_2	P_{REL}	R_E	R_1	s/p	R_2	P_{REL}
7091	7k5	p	130k	1.06	7333	11k	p	22k	1.5
7097	11k	p	20k	1.55	7347	7k5	p	360k	1.02
7100	300R	s	6k8	1.04	7358	7k5	p	390k	1.02
7100	1k5	s	5k6	1.27	7360	560R	s	6k8	1.08
7100	2k	s	5k1	1.39	7371	7k5	p	430k	1.02
7100	2k4	s	4k7	1.51	7378	9k1	p	39k	1.23
7110	910R	s	6k2	1.15	7382	7k5	p	470k	1.02
7130	330R	s	6k8	1.05	7391	7k5	p	510k	1.01
7133	9k1	p	33k	1.28	7392	8k2	p	75k	1.11
7143	7k5	p	150k	1.05	7400	1k2	s	6k2	1.19
7153	8k2	p	56k	1.15	7400	1k8	s	5k6	1.32
7160	360R	s	6k8	1.05	7400	2k7	s	4k7	1.57
7164	7k5	p	160k	1.05	7401	7k5	p	560k	1.01
7172	13k	p	16k	1.81	7410	7k5	p	620k	1.01
7190	390R	s	6k8	1.06	7418	7k5	p	680k	1.01
7200	1k	s	6k2	1.16	7420	620R	s	6k8	1.09
7200	1k6	s	5k6	1.29	7426	7k5	p	750k	1.01
7200	3k3	s	3k9	1.85	7455	8k2	p	82k	1.1
7200	3k6	s	3k6	2	7480	680R	s	6k8	1.1
7200	7k5	p	180k	1.04	7500	1k3	s	6k2	1.21
7200	12k	p	18k	1.67	7500	2k4	s	5k1	1.47
7229	7k5	p	200k	1.04	7500	3k6	s	3k9	1.92
7230	430R	s	6k8	1.06	7500	10k	p	30k	1.33
7242	8k2	p	62k	1.13	7500	12k	p	20k	1.6
7253	7k5	p	220k	1.03	7500	15k	p	15k	2
7264	9k1	p	36k	1.25	7511	9k1	p	43k	1.21
7270	470R	s	6k8	1.07	7522	8k2	p	91k	1.09
7273	7k5	p	240k	1.03	7543	11k	p	24k	1.46
7297	7k5	p	270k	1.03	7548	13k	p	18k	1.72
7297	10k	p	27k	1.37	7550	750R	s	6k8	1.11
7300	1k1	s	6k2	1.18	7575	75R	s	7k5	1.01
7300	2k2	s	5k1	1.43	7579	8k2	p	100k	1.08
7300	3k	s	4k3	1.7	7582	82R	s	7k5	1.01
7310	510R	s	6k8	1.08	7591	91R	s	7k5	1.01
7317	7k5	p	300k	1.03	7600	100R	s	7k5	1.01
7318	8k2	p	68k	1.12	7600	2k	s	5k6	1.36
7333	7k5	p	330k	1.02	7600	3k3	s	4k3	1.77

R_E	R_1	s/p	R_2	P_{REL}	R_E	R_1	s/p	R_2	P_{REL}
7610	110R	s	7k5	1.01	7879	13k	p	20k	1.65
7620	120R	s	7k5	1.02	7890	390R	s	7k5	1.05
7620	820R	s	6k8	1.12	7900	1k1	s	6k8	1.16
7624	9k1	p	47k	1.19	7900	3k6	s	4k3	1.84
7630	130R	s	7k5	1.02	7905	8k2	p	220k	1.04
7631	8k2	p	110k	1.07	7929	8k2	p	240k	1.03
7650	150R	s	7k5	1.02	7930	430R	s	7k5	1.06
7660	160R	s	7k5	1.02	7935	9k1	p	62k	1.15
7674	10k	p	33k	1.3	7958	8k2	p	270k	1.03
7676	8k2	p	120k	1.07	7959	10k	p	39k	1.26
7680	180R	s	7k5	1.02	7970	470R	s	7k5	1.06
7700	200R	s	7k5	1.03	7982	8k2	p	300k	1.03
7700	1k5	s	6k2	1.24	8000	1k2	s	6k8	1.18
7700	3k	s	4k7	1.64	8000	1k8	s	6k2	1.29
7710	910R	s	6k8	1.13	8000	2k4	s	5k6	1.43
7713	8k2	p	130k	1.06	8000	3k3	s	4k7	1.7
7720	220R	s	7k5	1.03	8000	12k	p	24k	1.5
7722	9k1	p	51k	1.18	8000	16k	p	16k	2
7740	240R	s	7k5	1.03	8001	8k2	p	330k	1.02
7742	15k	p	16k	1.94	8010	510R	s	7k5	1.07
7765	12k	p	22k	1.55	8017	8k2	p	360k	1.02
7770	270R	s	7k5	1.04	8026	9k1	p	68k	1.13
7775	8k2	p	150k	1.05	8031	8k2	p	390k	1.02
7800	300R	s	7k5	1.04	8047	8k2	p	430k	1.02
7800	1k	s	6k8	1.15	8049	11k	p	30k	1.37
7800	1k6	s	6k2	1.26	8059	8k2	p	470k	1.02
7800	2k2	s	5k6	1.39	8060	560R	s	7k5	1.07
7800	2k7	s	5k1	1.53	8070	8k2	p	510k	1.02
7800	3k9	s	3k9	2	8082	8k2	p	560k	1.01
7800	8k2	p	160k	1.05	8093	8k2	p	620k	1.01
7816	11k	p	27k	1.41	8100	1k3	s	6k8	1.19
7826	10k	p	36k	1.28	8100	3k	s	5k1	1.59
7828	9k1	p	56k	1.16	8102	8k2	p	680k	1.01
7830	330R	s	7k5	1.04	8111	8k2	p	750k	1.01
7843	8k2	p	180k	1.05	8113	10k	p	43k	1.23
7860	360R	s	7k5	1.05	8115	9k1	p	75k	1.12
7877	8k2	p	200k	1.04	8119	8k2	p	820k	1.01

Table 6.3 E24 equivalent resistances 99

R_E	R_1	s/p	R_2	P_{REL}	R_E	R_1	s/p	R_2	P_{REL}
8120	620R	s	7k5	1.08	8459	9k1	p	120k	1.08
8171	13k	p	22k	1.59	8470	270R	s	8k2	1.03
8180	680R	s	7k5	1.09	8471	16k	p	18k	1.89
8182	15k	p	18k	1.83	8485	10k	p	56k	1.18
8191	9k1	p	82k	1.11	8500	300R	s	8k2	1.04
8200	2k	s	6k2	1.32	8500	1k	s	7k5	1.13
8200	3k9	s	4k3	1.91	8505	9k1	p	130k	1.07
8246	10k	p	47k	1.21	8530	330R	s	8k2	1.04
8250	750R	s	7k5	1.1	8560	360R	s	8k2	1.04
8250	11k	p	33k	1.33	8571	12k	p	30k	1.4
8273	9k1	p	91k	1.1	8571	15k	p	20k	1.75
8282	82R	s	8k2	1.01	8580	9k1	p	150k	1.06
8291	91R	s	8k2	1.01	8580	11k	p	39k	1.28
8300	100R	s	8k2	1.01	8590	390R	s	8k2	1.05
8300	1k5	s	6k8	1.22	8600	1k1	s	7k5	1.15
8300	2k7	s	5k6	1.48	8600	1k8	s	6k8	1.26
8300	3k6	s	4k7	1.77	8600	2k4	s	6k2	1.39
8308	12k	p	27k	1.44	8600	3k	s	5k6	1.54
8310	110R	s	8k2	1.01	8600	3k9	s	4k7	1.83
8320	120R	s	8k2	1.01	8600	4k3	s	4k3	2
8320	820R	s	7k5	1.11	8610	9k1	p	160k	1.06
8330	130R	s	8k2	1.02	8611	10k	p	62k	1.16
8341	9k1	p	100k	1.09	8630	430R	s	8k2	1.05
8350	150R	s	8k2	1.02	8662	9k1	p	180k	1.05
8360	160R	s	8k2	1.02	8670	470R	s	8k2	1.06
8361	10k	p	51k	1.2	8700	1k2	s	7k5	1.16
8380	180R	s	8k2	1.02	8700	3k6	s	5k1	1.71
8400	200R	s	8k2	1.02	8704	9k1	p	200k	1.05
8400	1k6	s	6k8	1.24	8710	510R	s	8k2	1.06
8400	2k2	s	6k2	1.35	8718	10k	p	68k	1.15
8400	3k3	s	5k1	1.65	8739	9k1	p	220k	1.04
8405	9k1	p	110k	1.08	8759	11k	p	43k	1.26
8410	910R	s	7k5	1.12	8760	560R	s	8k2	1.07
8420	220R	s	8k2	1.03	8768	9k1	p	240k	1.04
8426	11k	p	36k	1.31	8775	13k	p	27k	1.48
8432	13k	p	24k	1.54	8800	1k3	s	7k5	1.17
8440	240R	s	8k2	1.03	8800	2k	s	6k8	1.29

R_E	R_1	s/p	R_2	P_{REL}	R_E	R_1	s/p	R_2	P_{REL}
8800	12k	p	33k	1.36	9110	910R	s	8k2	1.11
8803	9k1	p	270k	1.03	9167	10k	p	110k	1.09
8820	620R	s	8k2	1.08	9176	12k	p	39k	1.31
8824	10k	p	75k	1.13	9191	91R	s	9k1	1.01
8832	9k1	p	300k	1.03	9194	11k	p	56k	1.2
8856	9k1	p	330k	1.03	9200	100R	s	9k1	1.01
8876	9k1	p	360k	1.03	9200	1k	s	8k2	1.12
8880	680R	s	8k2	1.08	9200	2k4	s	6k8	1.35
8889	16k	p	20k	1.8	9200	3k	s	6k2	1.48
8893	9k1	p	390k	1.02	9200	3k6	s	5k6	1.64
8900	2k7	s	6k2	1.44	9210	110R	s	9k1	1.01
8900	3k3	s	5k6	1.59	9220	120R	s	9k1	1.01
8911	9k1	p	430k	1.02	9230	130R	s	9k1	1.01
8913	10k	p	82k	1.12	9231	10k	p	120k	1.08
8914	11k	p	47k	1.23	9231	15k	p	24k	1.63
8919	15k	p	22k	1.68	9250	150R	s	9k1	1.02
8927	9k1	p	470k	1.02	9260	160R	s	9k1	1.02
8940	9k1	p	510k	1.02	9263	16k	p	22k	1.73
8950	750R	s	8k2	1.09	9280	180R	s	9k1	1.02
8954	9k1	p	560k	1.02	9286	10k	p	130k	1.08
8968	9k1	p	620k	1.01	9300	200R	s	9k1	1.02
8980	9k1	p	680k	1.01	9300	1k1	s	8k2	1.13
8991	9k1	p	750k	1.01	9300	1k8	s	7k5	1.24
9000	1k5	s	7k5	1.2	9320	220R	s	9k1	1.02
9000	2k2	s	6k8	1.32	9326	13k	p	33k	1.39
9000	3k9	s	5k1	1.76	9340	240R	s	9k1	1.03
9000	4k3	s	4k7	1.91	9342	11k	p	62k	1.18
9000	9k1	p	820k	1.01	9370	270R	s	9k1	1.03
9000	12k	p	36k	1.33	9375	10k	p	150k	1.07
9000	18k	p	18k	2	9382	12k	p	43k	1.28
9010	9k1	p	910k	1.01	9400	300R	s	9k1	1.03
9010	10k	p	91k	1.11	9400	1k2	s	8k2	1.15
9020	820R	s	8k2	1.1	9400	4k3	s	5k1	1.84
9048	11k	p	51k	1.22	9400	4k7	s	4k7	2
9070	13k	p	30k	1.43	9412	10k	p	160k	1.06
9091	10k	p	100k	1.1	9430	330R	s	9k1	1.04
9100	1k6	s	7k5	1.21	9460	360R	s	9k1	1.04

Table 6.3 E24 equivalent resistances 101

R_E	R_1	s/p	R_2	P_{REL}	R_E	R_1	s/p	R_2	P_{REL}
9468	11k	p	68k	1.16	9720	620R	s	9k1	1.07
9474	10k	p	180k	1.06	9730	10k	p	360k	1.03
9474	18k	p	20k	1.9	9750	10k	p	390k	1.03
9490	390R	s	9k1	1.04	9750	13k	p	39k	1.33
9500	1k3	s	8k2	1.16	9773	10k	p	430k	1.02
9500	2k	s	7k5	1.27	9780	680R	s	9k1	1.07
9500	2k7	s	6k8	1.4	9792	10k	p	470k	1.02
9500	3k3	s	6k2	1.53	9800	1k6	s	8k2	1.2
9500	3k9	s	5k6	1.7	9800	3k	s	6k8	1.44
9524	10k	p	200k	1.05	9800	3k6	s	6k2	1.58
9530	430R	s	9k1	1.05	9800	4k7	s	5k1	1.92
9551	13k	p	36k	1.36	9808	10k	p	510k	1.02
9559	12k	p	47k	1.26	9814	11k	p	91k	1.12
9565	10k	p	220k	1.05	9825	10k	p	560k	1.02
9570	470R	s	9k1	1.05	9841	10k	p	620k	1.02
9593	11k	p	75k	1.15	9850	750R	s	9k1	1.08
9600	10k	p	240k	1.04	9855	10k	p	680k	1.01
9600	16k	p	24k	1.67	9868	10k	p	750k	1.01
9610	510R	s	9k1	1.06	9880	10k	p	820k	1.01
9643	10k	p	270k	1.04	9882	12k	p	56k	1.21
9643	15k	p	27k	1.56	9891	10k	p	910k	1.01
9660	560R	s	9k1	1.06	9900	2k4	s	7k5	1.32
9677	10k	p	300k	1.03	9900	4k3	s	5k6	1.77
9699	11k	p	82k	1.13	9900	18k	p	22k	1.82
9700	1k5	s	8k2	1.18	9901	10k	p	1M	1.01
9700	2k2	s	7k5	1.29	9910	11k	p	100k	1.11
9706	10k	p	330k	1.03	9920	820R	s	9k1	1.09
9714	12k	p	51k	1.24	9982	13k	p	43k	1.3

7 Maximum powers for resistors

Tables 7.1–7.6 show the voltage across or current through standard value resistors which are equivalent to a selection of common rated maximum powers.

As discussed in Chapter 3, it is wise to avoid running components continuously at full rated power. A good rule of thumb is not to exceed half of rated power; that is roughly 0.7 times the value of voltage or current presented here.

Table 7.1 *Maximum power for resistors up to 0.25 W*

R	0.063 W	0.1 W	0.125 W	0.155 W	0.25 W	
0R10	79.4m	100m	112m	124m	158m	V
	794m	1.00	1.12	1.24	1.58	A
0R12	86.9m	110m	122m	136m	173m	V
	725m	913m	1.02	1.14	1.44	A
0R15	97.2m	122m	137m	152m	194m	V
	648m	816m	913m	1.02	1.29	A
0R18	106m	134m	150m	167m	212m	V
	592m	745m	833m	928m	1.18	A
0R22	118m	148m	166m	185m	235m	V
	535m	674m	754m	839m	1.07	A
0R27	130m	164m	184m	205m	260m	V
	483m	609m	680m	758m	962m	A
0R33	144m	182m	203m	226m	287m	V
	437m	550m	615m	685m	870m	A
0R39	157m	197m	221m	246m	312m	V
	402m	506m	566m	630m	801m	A
0R47	172m	217m	242m	270m	343m	V
	366m	461m	516m	574m	729m	A
0R56	188m	237m	265m	295m	374m	V
	335m	423m	472m	526m	668m	A
0R68	207m	261m	292m	325m	412m	V
	304m	383m	429m	477m	606m	A
0R82	227m	286m	320m	357m	453m	V
	277m	349m	390m	435m	552m	A
1R	251m	316m	354m	394m	500m	V
	251m	316m	354m	394m	500m	A
1R1	263m	332m	371m	413m	524m	V
	239m	302m	337m	375m	477m	A
1R2	275m	346m	387m	431m	548m	V
	229m	289m	323m	359m	456m	A
1R3	286m	361m	403m	449m	570m	V
	220m	277m	310m	345m	439m	A
1R5	307m	387m	433m	482m	612m	V
	205m	258m	289m	321m	408m	A
1R6	317m	400m	447m	498m	632m	V
	198m	250m	280m	311m	395m	A

R	0.063 W	0.1 W	0.125 W	0.155 W	0.25 W	
1R8	337m	424m	474m	528m	671m	V
	187m	236m	264m	293m	373m	A
2R	355m	447m	500m	557m	707m	V
	177m	224m	250m	278m	354m	A
2R2	372m	469m	524m	584m	742m	V
	169m	213m	238m	265m	337m	A
2R4	389m	490m	548m	610m	775m	V
	162m	204m	228m	254m	323m	A
2R7	412m	520m	581m	647m	822m	V
	153m	192m	215m	240m	304m	A
3R	435m	548m	612m	682m	866m	V
	145m	183m	204m	227m	289m	A
3R3	456m	574m	642m	715m	908m	V
	138m	174m	195m	217m	275m	A
3R6	476m	600m	671m	747m	949m	V
	132m	167m	186m	207m	264m	A
3R9	496m	624m	698m	777m	987m	V
	127m	160m	179m	199m	253m	A
4R3	520m	656m	733m	816m	1.04	V
	121m	152m	170m	190m	241m	A
4R7	544m	686m	766m	854m	1.08	V
	116m	146m	163m	182m	231m	A
5R1	567m	714m	798m	889m	1.13	V
	111m	140m	157m	174m	221m	A
5R6	594m	748m	837m	932m	1.18	V
	106m	134m	149m	166m	211m	A
6R2	625m	787m	880m	980m	1.24	V
	101m	127m	142m	158m	201m	A
6R8	655m	825m	922m	1.03	1.30	V
	96.3m	121m	136m	151m	192m	A
7R5	687m	866m	968m	1.08	1.37	V
	91.7m	115m	129m	144m	183m	A
8R2	719m	906m	1.01	1.13	1.43	V
	87.7m	110m	123m	137m	175m	A
9R1	757m	954m	1.07	1.19	1.51	V
	83.2m	105m	117m	131m	166m	A
10R	794m	1.00	1.12	1.24	1.58	V
	79.4m	100m	112m	124m	158m	A

Table 7.1 Maximum power for resistors up to 0.25 W   105

R	0.063 W	0.1 W	0.125 W	0.155 W	0.25 W	
11R	832m	1.05	1.17	1.31	1.66	V
	75.7m	95.3m	107m	119m	151m	A
12R	869m	1.10	1.22	1.36	1.73	V
	72.5m	91.3m	102m	114m	144m	A
13R	905m	1.14	1.27	1.42	1.80	V
	69.6m	87.7m	98.1m	109m	139m	A
15R	972m	1.22	1.37	1.52	1.94	V
	64.8m	81.6m	91.3m	102m	129m	A
16R	1.00	1.26	1.41	1.57	2.00	V
	62.7m	79.1m	88.4m	98.4m	125m	A
18R	1.06	1.34	1.50	1.67	2.12	V
	59.2m	74.5m	83.3m	92.8m	118m	A
20R	1.12	1.41	1.58	1.76	2.24	V
	56.1m	70.7m	79.1m	88.0m	112m	A
22R	1.18	1.48	1.66	1.85	2.35	V
	53.5m	67.4m	75.4m	83.9m	107m	A
24R	1.23	1.55	1.73	1.93	2.45	V
	51.2m	64.5m	72.2m	80.4m	102m	A
27R	1.30	1.64	1.84	2.05	2.60	V
	48.3m	60.9m	68.0m	75.8m	96.2m	A
30R	1.37	1.73	1.94	2.16	2.74	V
	45.8m	57.7m	64.5m	71.9m	91.3m	A
33R	1.44	1.82	2.03	2.26	2.87	V
	43.7m	55.0m	61.5m	68.5m	87.0m	A
36R	1.51	1.90	2.12	2.36	3.00	V
	41.8m	52.7m	58.9m	65.6m	83.3m	A
39R	1.57	1.97	2.21	2.46	3.12	V
	40.2m	50.6m	56.6m	63.0m	80.1m	A
43R	1.65	2.07	2.32	2.58	3.28	V
	38.3m	48.2m	53.9m	60.0m	76.2m	A
47R	1.72	2.17	2.42	2.70	3.43	V
	36.6m	46.1m	51.6m	57.4m	72.9m	A
51R	1.79	2.26	2.52	2.81	3.57	V
	35.1m	44.3m	49.5m	55.1m	70.0m	A
56R	1.88	2.37	2.65	2.95	3.74	V
	33.5m	42.3m	47.2m	52.6m	66.8m	A
62R	1.98	2.49	2.78	3.10	3.94	V
	31.9m	40.2m	44.9m	50.0m	63.5m	A

R	0.063 W	0.1 W	0.125 W	0.155 W	0.25 W	
68R	2.07	2.61	2.92	3.25	4.12	V
	30.4m	38.3m	42.9m	47.7m	60.6m	A
75R	2.17	2.74	3.06	3.41	4.33	V
	29.0m	36.5m	40.8m	45.5m	57.7m	A
82R	2.27	2.86	3.20	3.57	4.53	V
	27.7m	34.9m	39.0m	43.5m	55.2m	A
91R	2.39	3.02	3.37	3.76	4.77	V
	26.3m	33.1m	37.1m	41.3m	52.4m	A
100R	2.51	3.16	3.54	3.94	5.00	V
	25.1m	31.6m	35.4m	39.4m	50.0m	A
110R	2.63	3.32	3.71	4.13	5.24	V
	23.9m	30.2m	33.7m	37.5m	47.7m	A
120R	2.75	3.46	3.87	4.31	5.48	V
	22.9m	28.9m	32.3m	35.9m	45.6m	A
130R	2.86	3.61	4.03	4.49	5.70	V
	22.0m	27.7m	31.0m	34.5m	43.9m	A
150R	3.07	3.87	4.33	4.82	6.12	V
	20.5m	25.8m	28.9m	32.1m	40.8m	A
160R	3.17	4.00	4.47	4.98	6.32	V
	19.8m	25.0m	28.0m	31.1m	39.5m	A
180R	3.37	4.24	4.74	5.28	6.71	V
	18.7m	23.6m	26.4m	29.3m	37.3m	A
200R	3.55	4.47	5.00	5.57	7.07	V
	17.7m	22.4m	25.0m	27.8m	35.4m	A
220R	3.72	4.69	5.24	5.84	7.42	V
	16.9m	21.3m	23.8m	26.5m	33.7m	A
240R	3.89	4.90	5.48	6.10	7.75	V
	16.2m	20.4m	22.8m	25.4m	32.3m	A
270R	4.12	5.20	5.81	6.47	8.22	V
	15.3m	19.2m	21.5m	24.0m	30.4m	A
300R	4.35	5.48	6.12	6.82	8.66	V
	14.5m	18.3m	20.4m	22.7m	28.9m	A
330R	4.56	5.74	6.42	7.15	9.08	V
	13.8m	17.4m	19.5m	21.7m	27.5m	A
360R	4.76	6.00	6.71	7.47	9.49	V
	13.2m	16.7m	18.6m	20.7m	26.4m	A
390R	4.96	6.24	6.98	7.77	9.87	V
	12.7m	16.0m	17.9m	19.9m	25.3m	A

Table 7.1 Maximum power for resistors up to 0.25 W 107

R	0.063 W	0.1 W	0.125 W	0.155 W	0.25 W	
430R	5.20	6.56	7.33	8.16	10.4	V
	12.1m	15.2m	17.0m	19.0m	24.1m	A
470R	5.44	6.86	7.66	8.54	10.8	V
	11.6m	14.6m	16.3m	18.2m	23.1m	A
510R	5.67	7.14	7.98	8.89	11.3	V
	11.1m	14.0m	15.7m	17.4m	22.1m	A
560R	5.94	7.48	8.37	9.32	11.8	V
	10.6m	13.4m	14.9m	16.6m	21.1m	A
620R	6.25	7.87	8.80	9.80	12.4	V
	10.1m	12.7m	14.2m	15.8m	20.1m	A
680R	6.55	8.25	9.22	10.3	13.0	V
	9.63m	12.1m	13.6m	15.1m	19.2m	A
750R	6.87	8.66	9.68	10.8	13.7	V
	9.17m	11.5m	12.9m	14.4m	18.3m	A
820R	7.19	9.06	10.1	11.3	14.3	V
	8.77m	11.0m	12.3m	13.7m	17.5m	A
910R	7.57	9.54	10.7	11.9	15.1	V
	8.32m	10.5m	11.7m	13.1m	16.6m	A
1k	7.94	10.0	11.2	12.4	15.8	V
	7.94m	10.0m	11.2m	12.4m	15.8m	A
1k1	8.32	10.5	11.7	13.1	16.6	V
	7.57m	9.53m	10.7m	11.9m	15.1m	A
1k2	8.69	11.0	12.2	13.6	17.3	V
	7.25m	9.13m	10.2m	11.4m	14.4m	A
1k3	9.05	11.4	12.7	14.2	18.0	V
	6.96m	8.77m	9.81m	10.9m	13.9m	A
1k5	9.72	12.2	13.7	15.2	19.4	V
	6.48m	8.16m	9.13m	10.2m	12.9m	A
1k6	10.0	12.6	14.1	15.7	20.0	V
	6.27m	7.91m	8.84m	9.84m	12.5m	A
1k8	10.6	13.4	15.0	16.7	21.2	V
	5.92m	7.45m	8.33m	9.28m	11.8m	A
2k	11.2	14.1	15.8	17.6	22.4	V
	5.61m	7.07m	7.91m	8.80m	11.2m	A
2k2	11.8	14.8	16.6	18.5	23.5	V
	5.35m	6.74m	7.54m	8.39m	10.7m	A
2k4	12.3	15.5	17.3	19.3	24.5	V
	5.12m	6.45m	7.22m	8.04m	10.2m	A

R	0.063 W	0.1 W	0.125 W	0.155 W	0.25 W	
2k7	13.0	16.4	18.4	20.5	26.0	V
	4.83m	6.09m	6.80m	7.58m	9.62m	A
3k	13.7	17.3	19.4	21.6	27.4	V
	4.58m	5.77m	6.45m	7.19m	9.13m	A
3k3	14.4	18.2	20.3	22.6	28.7	V
	4.37m	5.50m	6.15m	6.85m	8.70m	A
3k6	15.1	19.0	21.2	23.6	30.0	V
	4.18m	5.27m	5.89m	6.56m	8.33m	A
3k9	15.7	19.7	22.1	24.6	31.2	V
	4.02m	5.06m	5.66m	6.30m	8.01m	A
4k3	16.5	20.7	23.2	25.8	32.8	V
	3.83m	4.82m	5.39m	6.00m	7.62m	A
4k7	17.2	21.7	24.2	27.0	34.3	V
	3.66m	4.61m	5.16m	5.74m	7.29m	A
5k1	17.9	22.6	25.2	28.1	35.7	V
	3.51m	4.43m	4.95m	5.51m	7.00m	A
5k6	18.8	23.7	26.5	29.5	37.4	V
	3.35m	4.23m	4.72m	5.26m	6.68m	A
6k2	19.8	24.9	27.8	31.0	39.4	V
	3.19m	4.02m	4.49m	5.00m	6.35m	A
6k8	20.7	26.1	29.2	32.5	41.2	V
	3.04m	3.83m	4.29m	4.77m	6.06m	A
7k5	21.7	27.4	30.6	34.1	43.3	V
	2.90m	3.65m	4.08m	4.55m	5.77m	A
8k2	22.7	28.6	32.0	35.7	45.3	V
	2.77m	3.49m	3.90m	4.35m	5.52m	A
9k1	23.9	30.2	33.7	37.6	47.7	V
	2.63m	3.31m	3.71m	4.13m	5.24m	A
10k	25.1	31.6	35.4	39.4	50.0	V
	2.51m	3.16m	3.54m	3.94m	5.00m	A
11k	26.3	33.2	37.1	41.3	52.4	V
	2.39m	3.02m	3.37m	3.75m	4.77m	A
12k	27.5	34.6	38.7	43.1	54.8	V
	2.29m	2.89m	3.23m	3.59m	4.56m	A
13k	28.6	36.1	40.3	44.9	57.0	V
	2.20m	2.77m	3.10m	3.45m	4.39m	A
15k	30.7	38.7	43.3	48.2	61.2	V
	2.05m	2.58m	2.89m	3.21m	4.08m	A

Table 7.1 Maximum power for resistors up to 0.25 W 109

R	0.063 W	0.1 W	0.125 W	0.155 W	0.25 W	
16k	31.7	40.0	44.7	49.8	63.2	V
	1.98m	2.50m	2.80m	3.11m	3.95m	A
18k	33.7	42.4	47.4	52.8	67.1	V
	1.87m	2.36m	2.64m	2.93m	3.73m	A
20k	35.5	44.7	50.0	55.7	70.7	V
	1.77m	2.24m	2.50m	2.78m	3.54m	A
22k	37.2	46.9	52.4	58.4	74.2	V
	1.69m	2.13m	2.38m	2.65m	3.37m	A
24k	38.9	49.0	54.8	61.0	77.5	V
	1.62m	2.04m	2.28m	2.54m	3.23m	A
27k	41.2	52.0	58.1	64.7	82.2	V
	1.53m	1.92m	2.15m	2.40m	3.04m	A
30k	43.5	54.8	61.2	68.2	86.6	V
	1.45m	1.83m	2.04m	2.27m	2.89m	A
33k	45.6	57.4	64.2	71.5	90.8	V
	1.38m	1.74m	1.95m	2.17m	2.75m	A
36k	47.6	60.0	67.1	74.7	94.9	V
	1.32m	1.67m	1.86m	2.07m	2.64m	A
39k	49.6	62.4	69.8	77.7	98.7	V
	1.27m	1.60m	1.79m	1.99m	2.53m	A
43k	52.0	65.6	73.3	81.6	104	V
	1.21m	1.52m	1.70m	1.90m	2.41m	A
47k	54.4	68.6	76.6	85.4	108	V
	1.16m	1.46m	1.63m	1.82m	2.31m	A
51k	56.7	71.4	79.8	88.9	113	V
	1.11m	1.40m	1.57m	1.74m	2.21m	A
56k	59.4	74.8	83.7	93.2	118	V
	1.06m	1.34m	1.49m	1.66m	2.11m	A
62k	62.5	78.7	88.0	98.0	124	V
	1.01m	1.27m	1.42m	1.58m	2.01m	A
68k	65.5	82.5	92.2	103	130	V
	963u	1.21m	1.36m	1.51m	1.92m	A
75k	68.7	86.6	96.8	108	137	V
	917u	1.15m	1.29m	1.44m	1.83m	A
82k	71.9	90.6	101	113	143	V
	877u	1.10m	1.23m	1.37m	1.75m	A
91k	75.7	95.4	107	119	151	V
	832u	1.05m	1.17m	1.31m	1.66m	A

R	0.063 W	0.1 W	0.125 W	0.155 W	0.25 W	
100k	79.4	100	112	124	158	V
	794u	1.00m	1.12m	1.24m	1.58m	A
110k	83.2	105	117	131	166	V
	757u	953u	1.07m	1.19m	1.51m	A
120k	86.9	110	122	136	173	V
	725u	913u	1.02m	1.14m	1.44m	A
130k	90.5	114	127	142	180	V
	696u	877u	981u	1.09m	1.39m	A
150k	97.2	122	137	152	194	V
	648u	816u	913u	1.02m	1.29m	A
160k	100	126	141	157	200	V
	627u	791u	884u	984u	1.25m	A
180k	106	134	150	167	212	V
	592u	745u	833u	928u	1.18m	A
200k	112	141	158	176	224	V
	561u	707u	791u	880u	1.12m	A
220k	118	148	166	185	235	V
	535u	674u	754u	839u	1.07m	A
240k	123	155	173	193	245	V
	512u	645u	722u	804u	1.02m	A
270k	130	164	184	205	260	V
	483u	609u	680u	758u	962u	A
300k	137	173	194	216	274	V
	458u	577u	645u	719u	913u	A
330k	144	182	203	226	287	V
	437u	550u	615u	685u	870u	A
360k	151	190	212	236	300	V
	418u	527u	589u	656u	833u	A
390k	157	197	221	246	312	V
	402u	506u	566u	630u	801u	A
430k	165	207	232	258	328	V
	383u	482u	539u	600u	762u	A
470k	172	217	242	270	343	V
	366u	461u	516u	574u	729u	A
510k	179	226	252	281	357	V
	351u	443u	495u	551u	700u	A
560k	188	237	265	295	374	V
	335u	423u	472u	526u	668u	A

Table 7.1 Maximum power for resistors up to 0.25 W   111

R	0.063 W	0.1 W	0.125 W	0.155 W	0.25 W	
620k	198	249	278	310	394	V
	319u	402u	449u	500u	635u	A
680k	207	261	292	325	412	V
	304u	383u	429u	477u	606u	A
750k	217	274	306	341	433	V
	290u	365u	408u	455u	577u	A
820k	227	286	320	357	453	V
	277u	349u	390u	435u	552u	A
910k	239	302	337	376	477	V
	263u	331u	371u	413u	524u	A
1M	251	316	354	394	500	V
	251u	316u	354u	394u	500u	A
1M2	275	346	387	431	548	V
	229u	289u	323u	359u	456u	A
1M5	307	387	433	482	612	V
	205u	258u	289u	321u	408u	A
1M8	337	424	474	528	671	V
	187u	236u	264u	293u	373u	A
2M2	372	469	524	584	742	V
	169u	213u	238u	265u	337u	A
2M7	412	520	581	647	822	V
	153u	192u	215u	240u	304u	A
3M3	456	574	642	715	908	V
	138u	174u	195u	217u	275u	A
3M9	496	624	698	777	987	V
	127u	160u	179u	199u	253u	A
4M7	544	686	766	854	1.08k	V
	116u	146u	163u	182u	231u	A
5M6	594	748	837	932	1.18k	V
	106u	134u	149u	166u	211u	A
6M8	655	825	922	1.03k	1.30k	V
	96.3u	121u	136u	151u	192u	A
8M2	719	906	1.01k	1.13k	1.43k	V
	87.7u	110u	123u	137u	175u	A
10M	794	1.00k	1.12k	1.24k	1.58k	V
	79.4u	100u	112u	124u	158u	A

Table 7.2 *Maximum power for resistors between 0.33 and 1 W*

R	0.33 W	0.4 W	0.5 W	0.66 W	1 W	
0R10	182m	200m	224m	257m	316m	V
	1.82	2.00	2.24	2.57	3.16	A
0R12	199m	219m	245m	281m	346m	V
	1.66	1.83	2.04	2.35	2.89	A
0R15	222m	245m	274m	315m	387m	V
	1.48	1.63	1.83	2.10	2.58	A
0R18	244m	268m	300m	345m	424m	V
	1.35	1.49	1.67	1.91	2.36	A
0R22	269m	297m	332m	381m	469m	V
	1.22	1.35	1.51	1.73	2.13	A
0R27	298m	329m	367m	422m	520m	V
	1.11	1.22	1.36	1.56	1.92	A
0R33	330m	363m	406m	467m	574m	V
	1.00	1.10	1.23	1.41	1.74	A
0R39	359m	395m	442m	507m	624m	V
	920m	1.01	1.13	1.30	1.60	A
0R47	394m	434m	485m	557m	686m	V
	838m	923m	1.03	1.19	1.46	A
0R56	430m	473m	529m	608m	748m	V
	768m	845m	945m	1.09	1.34	A
0R68	474m	522m	583m	670m	825m	V
	697m	767m	857m	985m	1.21	A
0R82	520m	573m	640m	736m	906m	V
	634m	698m	781m	897m	1.10	A
1R	574m	632m	707m	812m	1.00	V
	574m	632m	707m	812m	1.00	A
1R1	602m	663m	742m	852m	1.05	V
	548m	603m	674m	775m	953m	A
1R2	629m	693m	775m	890m	1.10	V
	524m	577m	645m	742m	913m	A
1R3	655m	721m	806m	926m	1.14	V
	504m	555m	620m	713m	877m	A
1R5	704m	775m	866m	995m	1.22	V
	469m	516m	577m	663m	816m	A
1R6	727m	800m	894m	1.03	1.26	V
	454m	500m	559m	642m	791m	A

Table 7.2 Maximum power for resistors between 0.33 and 1 W   113

R	0.33 W	0.4 W	0.5 W	0.66 W	1 W	
1R8	771m	849m	949m	1.09	1.34	V
	428m	471m	527m	606m	745m	A
2R	812m	894m	1.00	1.15	1.41	V
	406m	447m	500m	574m	707m	A
2R2	852m	938m	1.05	1.20	1.48	V
	387m	426m	477m	548m	674m	A
2R4	890m	980m	1.10	1.26	1.55	V
	371m	408m	456m	524m	645m	A
2R7	944m	1.04	1.16	1.33	1.64	V
	350m	385m	430m	494m	609m	A
3R	995m	1.10	1.22	1.41	1.73	V
	332m	365m	408m	469m	577m	A
3R3	1.04	1.15	1.28	1.48	1.82	V
	316m	348m	389m	447m	550m	A
3R6	1.09	1.20	1.34	1.54	1.90	V
	303m	333m	373m	428m	527m	A
3R9	1.13	1.25	1.40	1.60	1.97	V
	291m	320m	358m	411m	506m	A
4R3	1.19	1.31	1.47	1.68	2.07	V
	277m	305m	341m	392m	482m	A
4R7	1.25	1.37	1.53	1.76	2.17	V
	265m	292m	326m	375m	461m	A
5R1	1.30	1.43	1.60	1.83	2.26	V
	254m	280m	313m	360m	443m	A
5R6	1.36	1.50	1.67	1.92	2.37	V
	243m	267m	299m	343m	423m	A
6R2	1.43	1.57	1.76	2.02	2.49	V
	231m	254m	284m	326m	402m	A
6R8	1.50	1.65	1.84	2.12	2.61	V
	220m	243m	271m	312m	383m	A
7R5	1.57	1.73	1.94	2.22	2.74	V
	210m	231m	258m	297m	365m	A
8R2	1.64	1.81	2.02	2.33	2.86	V
	201m	221m	247m	284m	349m	A
9R1	1.73	1.91	2.13	2.45	3.02	V
	190m	210m	234m	269m	331m	A
10R	1.82	2.00	2.24	2.57	3.16	V
	182m	200m	224m	257m	316m	A

R	0.33 W	0.4 W	0.5 W	0.66 W	1 W	
11R	1.91	2.10	2.35	2.69	3.32	V
	173m	191m	213m	245m	302m	A
12R	1.99	2.19	2.45	2.81	3.46	V
	166m	183m	204m	235m	289m	A
13R	2.07	2.28	2.55	2.93	3.61	V
	159m	175m	196m	225m	277m	A
15R	2.22	2.45	2.74	3.15	3.87	V
	148m	163m	183m	210m	258m	A
16R	2.30	2.53	2.83	3.25	4.00	V
	144m	158m	177m	203m	250m	A
18R	2.44	2.68	3.00	3.45	4.24	V
	135m	149m	167m	191m	236m	A
20R	2.57	2.83	3.16	3.63	4.47	V
	128m	141m	158m	182m	224m	A
22R	2.69	2.97	3.32	3.81	4.69	V
	122m	135m	151m	173m	213m	A
24R	2.81	3.10	3.46	3.98	4.90	V
	117m	129m	144m	166m	204m	A
27R	2.98	3.29	3.67	4.22	5.20	V
	111m	122m	136m	156m	192m	A
30R	3.15	3.46	3.87	4.45	5.48	V
	105m	115m	129m	148m	183m	A
33R	3.30	3.63	4.06	4.67	5.74	V
	100m	110m	123m	141m	174m	A
36R	3.45	3.79	4.24	4.87	6.00	V
	95.7m	105m	118m	135m	167m	A
39R	3.59	3.95	4.42	5.07	6.24	V
	92.0m	101m	113m	130m	160m	A
43R	3.77	4.15	4.64	5.33	6.56	V
	87.6m	96.4m	108m	124m	152m	A
47R	3.94	4.34	4.85	5.57	6.86	V
	83.8m	92.3m	103m	119m	146m	A
51R	4.10	4.52	5.05	5.80	7.14	V
	80.4m	88.6m	99.0m	114m	140m	A
56R	4.30	4.73	5.29	6.08	7.48	V
	76.8m	84.5m	94.5m	109m	134m	A
62R	4.52	4.98	5.57	6.40	7.87	V
	73.0m	80.3m	89.8m	103m	127m	A

Table 7.2 Maximum power for resistors between 0.33 and 1 W   115

R	0.33 W	0.4 W	0.5 W	0.66 W	1 W	
68R	4.74	5.22	5.83	6.70	8.25	V
	69.7m	76.7m	85.7m	98.5m	121m	A
75R	4.97	5.48	6.12	7.04	8.66	V
	66.3m	73.0m	81.6m	93.8m	115m	A
82R	5.20	5.73	6.40	7.36	9.06	V
	63.4m	69.8m	78.1m	89.7m	110m	A
91R	5.48	6.03	6.75	7.75	9.54	V
	60.2m	66.3m	74.1m	85.2m	105m	A
100R	5.74	6.32	7.07	8.12	10.0	V
	57.4m	63.2m	70.7m	81.2m	100m	A
110R	6.02	6.63	7.42	8.52	10.5	V
	54.8m	60.3m	67.4m	77.5m	95.3m	A
120R	6.29	6.93	7.75	8.90	11.0	V
	52.4m	57.7m	64.5m	74.2m	91.3m	A
130R	6.55	7.21	8.06	9.26	11.4	V
	50.4m	55.5m	62.0m	71.3m	87.7m	A
150R	7.04	7.75	8.66	9.95	12.2	V
	46.9m	51.6m	57.7m	66.3m	81.6m	A
160R	7.27	8.00	8.94	10.3	12.6	V
	45.4m	50.0m	55.9m	64.2m	79.1m	A
180R	7.71	8.49	9.49	10.9	13.4	V
	42.8m	47.1m	52.7m	60.6m	74.5m	A
200R	8.12	8.94	10.0	11.5	14.1	V
	40.6m	44.7m	50.0m	57.4m	70.7m	A
220R	8.52	9.38	10.5	12.0	14.8	V
	38.7m	42.6m	47.7m	54.8m	67.4m	A
240R	8.90	9.80	11.0	12.6	15.5	V
	37.1m	40.8m	45.6m	52.4m	64.5m	A
270R	9.44	10.4	11.6	13.3	16.4	V
	35.0m	38.5m	43.0m	49.4m	60.9m	A
300R	9.95	11.0	12.2	14.1	17.3	V
	33.2m	36.5m	40.8m	46.9m	57.7m	A
330R	10.4	11.5	12.8	14.8	18.2	V
	31.6m	34.8m	38.9m	44.7m	55.0m	A
360R	10.9	12.0	13.4	15.4	19.0	V
	30.3m	33.3m	37.3m	42.8m	52.7m	A
390R	11.3	12.5	14.0	16.0	19.7	V
	29.1m	32.0m	35.8m	41.1m	50.6m	A

R	0.33 W	0.4 W	0.5 W	0.66 W	1 W	
430R	11.9	13.1	14.7	16.8	20.7	V
	27.7m	30.5m	34.1m	39.2m	48.2m	A
470R	12.5	13.7	15.3	17.6	21.7	V
	26.5m	29.2m	32.6m	37.5m	46.1m	A
510R	13.0	14.3	16.0	18.3	22.6	V
	25.4m	28.0m	31.3m	36.0m	44.3m	A
560R	13.6	15.0	16.7	19.2	23.7	V
	24.3m	26.7m	29.9m	34.3m	42.3m	A
620R	14.3	15.7	17.6	20.2	24.9	V
	23.1m	25.4m	28.4m	32.6m	40.2m	A
680R	15.0	16.5	18.4	21.2	26.1	V
	22.0m	24.3m	27.1m	31.2m	38.3m	A
750R	15.7	17.3	19.4	22.2	27.4	V
	21.0m	23.1m	25.8m	29.7m	36.5m	A
820R	16.4	18.1	20.2	23.3	28.6	V
	20.1m	22.1m	24.7m	28.4m	34.9m	A
910R	17.3	19.1	21.3	24.5	30.2	V
	19.0m	21.0m	23.4m	26.9m	33.1m	A
1k	18.2	20.0	22.4	25.7	31.6	V
	18.2m	20.0m	22.4m	25.7m	31.6m	A
1k1	19.1	21.0	23.5	26.9	33.2	V
	17.3m	19.1m	21.3m	24.5m	30.2m	A
1k2	19.9	21.9	24.5	28.1	34.6	V
	16.6m	18.3m	20.4m	23.5m	28.9m	A
1k3	20.7	22.8	25.5	29.3	36.1	V
	15.9m	17.5m	19.6m	22.5m	27.7m	A
1k5	22.2	24.5	27.4	31.5	38.7	V
	14.8m	16.3m	18.3m	21.0m	25.8m	A
1k6	23.0	25.3	28.3	32.5	40.0	V
	14.4m	15.8m	17.7m	20.3m	25.0m	A
1k8	24.4	26.8	30.0	34.5	42.4	V
	13.5m	14.9m	16.7m	19.1m	23.6m	A
2k	25.7	28.3	31.6	36.3	44.7	V
	12.8m	14.1m	15.8m	18.2m	22.4m	A
2k2	26.9	29.7	33.2	38.1	46.9	V
	12.2m	13.5m	15.1m	17.3m	21.3m	A
2k4	28.1	31.0	34.6	39.8	49.0	V
	11.7m	12.9m	14.4m	16.6m	20.4m	A

Table 7.2 Maximum power for resistors between 0.33 and 1 W 117

R	0.33 W	0.4 W	0.5 W	0.66 W	1 W	
2k7	29.8	32.9	36.7	42.2	52.0	V
	11.1m	12.2m	13.6m	15.6m	19.2m	A
3k	31.5	34.6	38.7	44.5	54.8	V
	10.5m	11.5m	12.9m	14.8m	18.3m	A
3k3	33.0	36.3	40.6	46.7	57.4	V
	10.0m	11.0m	12.3m	14.1m	17.4m	A
3k6	34.5	37.9	42.4	48.7	60.0	V
	9.57m	10.5m	11.8m	13.5m	16.7m	A
3k9	35.9	39.5	44.2	50.7	62.4	V
	9.20m	10.1m	11.3m	13.0m	16.0m	A
4k3	37.7	41.5	46.4	53.3	65.6	V
	8.76m	9.64m	10.8m	12.4m	15.2m	A
4k7	39.4	43.4	48.5	55.7	68.6	V
	8.38m	9.23m	10.3m	11.9m	14.6m	A
5k1	41.0	45.2	50.5	58.0	71.4	V
	8.04m	8.86m	9.90m	11.4m	14.0m	A
5k6	43.0	47.3	52.9	60.8	74.8	V
	7.68m	8.45m	9.45m	10.9m	13.4m	A
6k2	45.2	49.8	55.7	64.0	78.7	V
	7.30m	8.03m	8.98m	10.3m	12.7m	A
6k8	47.4	52.2	58.3	67.0	82.5	V
	6.97m	7.67m	8.57m	9.85m	12.1m	A
7k5	49.7	54.8	61.2	70.4	86.6	V
	6.63m	7.30m	8.16m	9.38m	11.5m	A
8k2	52.0	57.3	64.0	73.6	90.6	V
	6.34m	6.98m	7.81m	8.97m	11.0m	A
9k1	54.8	60.3	67.5	77.5	95.4	V
	6.02m	6.63m	7.41m	8.52m	10.5m	A
10k	57.4	63.2	70.7	81.2	100	V
	5.74m	6.32m	7.07m	8.12m	10.0m	A
11k	60.2	66.3	74.2	85.2	105	V
	5.48m	6.03m	6.74m	7.75m	9.53m	A
12k	62.9	69.3	77.5	89.0	110	V
	5.24m	5.77m	6.45m	7.42m	9.13m	A
13k	65.5	72.1	80.6	92.6	114	V
	5.04m	5.55m	6.20m	7.13m	8.77m	A
15k	70.4	77.5	86.6	99.5	122	V
	4.69m	5.16m	5.77m	6.63m	8.16m	A

R	0.33 W	0.4 W	0.5 W	0.66 W	1 W	
16k	72.7	80.0	89.4	103	126	V
	4.54m	5.00m	5.59m	6.42m	7.91m	A
18k	77.1	84.9	94.9	109	134	V
	4.28m	4.71m	5.27m	6.06m	7.45m	A
20k	81.2	89.4	100	115	141	V
	4.06m	4.47m	5.00m	5.74m	7.07m	A
22k	85.2	93.8	105	120	148	V
	3.87m	4.26m	4.77m	5.48m	6.74m	A
24k	89.0	98.0	110	126	155	V
	3.71m	4.08m	4.56m	5.24m	6.45m	A
27k	94.4	104	116	133	164	V
	3.50m	3.85m	4.30m	4.94m	6.09m	A
30k	99.5	110	122	141	173	V
	3.32m	3.65m	4.08m	4.69m	5.77m	A
33k	104	115	128	148	182	V
	3.16m	3.48m	3.89m	4.47m	5.50m	A
36k	109	120	134	154	190	V
	3.03m	3.33m	3.73m	4.28m	5.27m	A
39k	113	125	140	160	197	V
	2.91m	3.20m	3.58m	4.11m	5.06m	A
43k	119	131	147	168	207	V
	2.77m	3.05m	3.41m	3.92m	4.82m	A
47k	125	137	153	176	217	V
	2.65m	2.92m	3.26m	3.75m	4.61m	A
51k	130	143	160	183	226	V
	2.54m	2.80m	3.13m	3.60m	4.43m	A
56k	136	150	167	192	237	V
	2.43m	2.67m	2.99m	3.43m	4.23m	A
62k	143	157	176	202	249	V
	2.31m	2.54m	2.84m	3.26m	4.02m	A
68k	150	165	184	212	261	V
	2.20m	2.43m	2.71m	3.12m	3.83m	A
75k	157	173	194	222	274	V
	2.10m	2.31m	2.58m	2.97m	3.65m	A
82k	164	181	202	233	286	V
	2.01m	2.21m	2.47m	2.84m	3.49m	A
91k	173	191	213	245	302	V
	1.90m	2.10m	2.34m	2.69m	3.31m	A

Table 7.2 Maximum power for resistors between 0.33 and 1 W 119

R	0.33 W	0.4 W	0.5 W	0.66 W	1 W	
100k	182	200	224	257	316	V
	1.82m	2.00m	2.24m	2.57m	3.16m	A
110k	191	210	235	269	332	V
	1.73m	1.91m	2.13m	2.45m	3.02m	A
120k	199	219	245	281	346	V
	1.66m	1.83m	2.04m	2.35m	2.89m	A
130k	207	228	255	293	361	V
	1.59m	1.75m	1.96m	2.25m	2.77m	A
150k	222	245	274	315	387	V
	1.48m	1.63m	1.83m	2.10m	2.58m	A
160k	230	253	283	325	400	V
	1.44m	1.58m	1.77m	2.03m	2.50m	A
180k	244	268	300	345	424	V
	1.35m	1.49m	1.67m	1.91m	2.36m	A
200k	257	283	316	363	447	V
	1.28m	1.41m	1.58m	1.82m	2.24m	A
220k	269	297	332	381	469	V
	1.22m	1.35m	1.51m	1.73m	2.13m	A
240k	281	310	346	398	490	V
	1.17m	1.29m	1.44m	1.66m	2.04m	A
270k	298	329	367	422	520	V
	1.11m	1.22m	1.36m	1.56m	1.92m	A
300k	315	346	387	445	548	V
	1.05m	1.15m	1.29m	1.48m	1.83m	A
330k	330	363	406	467	574	V
	1.00m	1.10m	1.23m	1.41m	1.74m	A
360k	345	379	424	487	600	V
	957u	1.05m	1.18m	1.35m	1.67m	A
390k	359	395	442	507	624	V
	920u	1.01m	1.13m	1.30m	1.60m	A
430k	377	415	464	533	656	V
	876u	964u	1.08m	1.24m	1.52m	A
470k	394	434	485	557	686	V
	838u	923u	1.03m	1.19m	1.46m	A
510k	410	452	505	580	714	V
	804u	886u	990u	1.14m	1.40m	A
560k	430	473	529	608	748	V
	768u	845u	945u	1.09m	1.34m	A

R	0.33 W	0.4 W	0.5 W	0.66 W	1 W	
620k	452	498	557	640	787	V
	730u	803u	898u	1.03m	1.27m	A
680k	474	522	583	670	825	V
	697u	767u	857u	985u	1.21m	A
750k	497	548	612	704	866	V
	663u	730u	816u	938u	1.15m	A
820k	520	573	640	736	906	V
	634u	698u	781u	897u	1.10m	A
910k	548	603	675	775	954	V
	602u	663u	741u	852u	1.05m	A
1M	574	632	707	812	1.00k	V
	574u	632u	707u	812u	1.00m	A
1M2	629	693	775	890	1.10k	V
	524u	577u	645u	742u	913u	A
1M5	704	775	866	995	1.22k	V
	469u	516u	577u	663u	816u	A
1M8	771	849	949	1.09k	1.34k	V
	428u	471u	527u	606u	745u	A
2M2	852	938	1.05k	1.20k	1.48k	V
	387u	426u	477u	548u	674u	A
2M7	944	1.04k	1.16k	1.33k	1.64k	V
	350u	385u	430u	494u	609u	A
3M3	1.04k	1.15k	1.28k	1.48k	1.82k	V
	316u	348u	389u	447u	550u	A
3M9	1.13k	1.25k	1.40k	1.60k	1.97k	V
	291u	320u	358u	411u	506u	A
4M7	1.25k	1.37k	1.53k	1.76k	2.17k	V
	265u	292u	326u	375u	461u	A
5M6	1.36k	1.50k	1.67k	1.92k	2.37k	V
	243u	267u	299u	343u	423u	A
6M8	1.50k	1.65k	1.84k	2.12k	2.61k	V
	220u	243u	271u	312u	383u	A
8M2	1.64k	1.81k	2.02k	2.33k	2.86k	V
	201u	221u	247u	284u	349u	A
10M	1.82k	2.00k	2.24k	2.57k	3.16k	V
	182u	200u	224u	257u	316u	A

Table 7.3 *Maximum power for resistors between 2 and 5 W*

R	2 W	2.5 W	3 W	4 W	5 W	
0R01	141m	158m	173m	200m	224m	V
	14.1	15.8	17.3	20.0	22.4	A
0R02	200m	224m	245m	283m	316m	V
	10.0	11.2	12.2	14.1	15.8	A
0R03	245m	274m	300m	346m	387m	V
	8.16	9.13	10.0	11.5	12.9	A
0R033	257m	287m	315m	363m	406m	V
	7.78	8.70	9.53	11.0	12.3	A
0R047	307m	343m	375m	434m	485m	V
	6.52	7.29	7.99	9.23	10.3	A
0R05	316m	354m	387m	447m	500m	V
	6.32	7.07	7.75	8.94	10.0	A
0R056	335m	374m	410m	473m	529m	V
	5.98	6.68	7.32	8.45	9.45	A
0R068	369m	412m	452m	522m	583m	V
	5.42	6.06	6.64	7.67	8.57	A
0R07	374m	418m	458m	529m	592m	V
	5.35	5.98	6.55	7.56	8.45	A
0R10	447m	500m	548m	632m	707m	V
	4.47	5.00	5.48	6.32	7.07	A
0R12	490m	548m	600m	693m	775m	V
	4.08	4.56	5.00	5.77	6.45	A
0R15	548m	612m	671m	775m	866m	V
	3.65	4.08	4.47	5.16	5.77	A
0R18	600m	671m	735m	849m	949m	V
	3.33	3.73	4.08	4.71	5.27	A
0R22	663m	742m	812m	938m	1.05	V
	3.02	3.37	3.69	4.26	4.77	A
0R27	735m	822m	900m	1.04	1.16	V
	2.72	3.04	3.33	3.85	4.30	A
0R33	812m	908m	995m	1.15	1.28	V
	2.46	2.75	3.02	3.48	3.89	A
0R39	883m	987m	1.08	1.25	1.40	V
	2.26	2.53	2.77	3.20	3.58	A
0R47	970m	1.08	1.19	1.37	1.53	V
	2.06	2.31	2.53	2.92	3.26	A

R	2 W	2.5 W	3 W	4 W	5 W	
0R56	1.06	1.18	1.30	1.50	1.67	V
	1.89	2.11	2.31	2.67	2.99	A
0R68	1.17	1.30	1.43	1.65	1.84	V
	1.71	1.92	2.10	2.43	2.71	A
0R82	1.28	1.43	1.57	1.81	2.02	V
	1.56	1.75	1.91	2.21	2.47	A
1R	1.41	1.58	1.73	2.00	2.24	V
	1.41	1.58	1.73	2.00	2.24	A
1R2	1.55	1.73	1.90	2.19	2.45	V
	1.29	1.44	1.58	1.83	2.04	A
1R5	1.73	1.94	2.12	2.45	2.74	V
	1.15	1.29	1.41	1.63	1.83	A
1R8	1.90	2.12	2.32	2.68	3.00	V
	1.05	1.18	1.29	1.49	1.67	A
2R2	2.10	2.35	2.57	2.97	3.32	V
	953m	1.07	1.17	1.35	1.51	A
2R7	2.32	2.60	2.85	3.29	3.67	V
	861m	962m	1.05	1.22	1.36	A
3R3	2.57	2.87	3.15	3.63	4.06	V
	778m	870m	953m	1.10	1.23	A
3R9	2.79	3.12	3.42	3.95	4.42	V
	716m	801m	877m	1.01	1.13	A
4R7	3.07	3.43	3.75	4.34	4.85	V
	652m	729m	799m	923m	1.03	A
5R6	3.35	3.74	4.10	4.73	5.29	V
	598m	668m	732m	845m	945m	A
6R8	3.69	4.12	4.52	5.22	5.83	V
	542m	606m	664m	767m	857m	A
8R2	4.05	4.53	4.96	5.73	6.40	V
	494m	552m	605m	698m	781m	A
10R	4.47	5.00	5.48	6.32	7.07	V
	447m	500m	548m	632m	707m	A
12R	4.90	5.48	6.00	6.93	7.75	V
	408m	456m	500m	577m	645m	A
15R	5.48	6.12	6.71	7.75	8.66	V
	365m	408m	447m	516m	577m	A
18R	6.00	6.71	7.35	8.49	9.49	V

Table 7.3 Maximum power for resistors between 2 and 5 W 123

R	2 W	2.5 W	3 W	4 W	5 W	
	333m	373m	408m	471m	527m	A
22R	6.63	7.42	8.12	9.38	10.5	V
	302m	337m	369m	426m	477m	A
27R	7.35	8.22	9.00	10.4	11.6	V
	272m	304m	333m	385m	430m	A
33R	8.12	9.08	9.95	11.5	12.8	V
	246m	275m	302m	348m	389m	A
39R	8.83	9.87	10.8	12.5	14.0	V
	226m	253m	277m	320m	358m	A
47R	9.70	10.8	11.9	13.7	15.3	V
	206m	231m	253m	292m	326m	A
56R	10.6	11.8	13.0	15.0	16.7	V
	189m	211m	231m	267m	299m	A
68R	11.7	13.0	14.3	16.5	18.4	V
	171m	192m	210m	243m	271m	A
82R	12.8	14.3	15.7	18.1	20.2	V
	156m	175m	191m	221m	247m	A
100R	14.1	15.8	17.3	20.0	22.4	V
	141m	158m	173m	200m	224m	A
120R	15.5	17.3	19.0	21.9	24.5	V
	129m	144m	158m	183m	204m	A
150R	17.3	19.4	21.2	24.5	27.4	V
	115m	129m	141m	163m	183m	A
180R	19.0	21.2	23.2	26.8	30.0	V
	105m	118m	129m	149m	167m	A
220R	21.0	23.5	25.7	29.7	33.2	V
	95.3m	107m	117m	135m	151m	A
270R	23.2	26.0	28.5	32.9	36.7	V
	86.1m	96.2m	105m	122m	136m	A
330R	25.7	28.7	31.5	36.3	40.6	V
	77.8m	87.0m	95.3m	110m	123m	A
390R	27.9	31.2	34.2	39.5	44.2	V
	71.6m	80.1m	87.7m	101m	113m	A
470R	30.7	34.3	37.5	43.4	48.5	V
	65.2m	72.9m	79.9m	92.3m	103m	A
560R	33.5	37.4	41.0	47.3	52.9	V
	59.8m	66.8m	73.2m	84.5m	94.5m	A

R	2 W	2.5 W	3 W	4 W	5 W	
680R	36.9	41.2	45.2	52.2	58.3	V
	54.2m	60.6m	66.4m	76.7m	85.7m	A
820R	40.5	45.3	49.6	57.3	64.0	V
	49.4m	55.2m	60.5m	69.8m	78.1m	A
1k	44.7	50.0	54.8	63.2	70.7	V
	44.7m	50.0m	54.8m	63.2m	70.7m	A
1k2	49.0	54.8	60.0	69.3	77.5	V
	40.8m	45.6m	50.0m	57.7m	64.5m	A
1k5	54.8	61.2	67.1	77.5	86.6	V
	36.5m	40.8m	44.7m	51.6m	57.7m	A
1k8	60.0	67.1	73.5	84.9	94.9	V
	33.3m	37.3m	40.8m	47.1m	52.7m	A
2k2	66.3	74.2	81.2	93.8	105	V
	30.2m	33.7m	36.9m	42.6m	47.7m	A
2k7	73.5	82.2	90.0	104	116	V
	27.2m	30.4m	33.3m	38.5m	43.0m	A
3k3	81.2	90.8	99.5	115	128	V
	24.6m	27.5m	30.2m	34.8m	38.9m	A
3k9	88.3	98.7	108	125	140	V
	22.6m	25.3m	27.7m	32.0m	35.8m	A
4k7	97.0	108	119	137	153	V
	20.6m	23.1m	25.3m	29.2m	32.6m	A
5k6	106	118	130	150	167	V
	18.9m	21.1m	23.1m	26.7m	29.9m	A
6k8	117	130	143	165	184	V
	17.1m	19.2m	21.0m	24.3m	27.1m	A
8k2	128	143	157	181	202	V
	15.6m	17.5m	19.1m	22.1m	24.7m	A
10k	141	158	173	200	224	V
	14.1m	15.8m	17.3m	20.0m	22.4m	A
12k	155	173	190	219	245	V
	12.9m	14.4m	15.8m	18.3m	20.4m	A
15k	173	194	212	245	274	V
	11.5m	12.9m	14.1m	16.3m	18.3m	A
18k	190	212	232	268	300	V
	10.5m	11.8m	12.9m	14.9m	16.7m	A
22k	210	235	257	297	332	V

Table 7.3 Maximum power for resistors between 2 and 5 W   125

R	2 W	2.5 W	3 W	4 W	5 W	
27k	9.53m	10.7m	11.7m	13.5m	15.1m	A
	232	260	285	329	367	V
33k	8.61m	9.62m	10.5m	12.2m	13.6m	A
	257	287	315	363	406	V
39k	7.78m	8.70m	9.53m	11.0m	12.3m	A
	279	312	342	395	442	V
47k	7.16m	8.01m	8.77m	10.1m	11.3m	A
	307	343	375	434	485	V
56k	6.52m	7.29m	7.99m	9.23m	10.3m	A
	335	374	410	473	529	V
68k	5.98m	6.68m	7.32m	8.45m	9.45m	A
	369	412	452	522	583	V
82k	5.42m	6.06m	6.64m	7.67m	8.57m	A
	405	453	496	573	640	V
100k	4.94m	5.52m	6.05m	6.98m	7.81m	A
	447	500	548	632	707	V
120k	4.47m	5.00m	5.48m	6.32m	7.07m	A
	490	548	600	693	775	V
150k	4.08m	4.56m	5.00m	5.77m	6.45m	A
	548	612	671	775	866	V
180k	3.65m	4.08m	4.47m	5.16m	5.77m	A
	600	671	735	849	949	V
220k	3.33m	3.73m	4.08m	4.71m	5.27m	A
	663	742	812	938	1.05k	V
270k	3.02m	3.37m	3.69m	4.26m	4.77m	A
	735	822	900	1.04k	1.16k	V
330k	2.72m	3.04m	3.33m	3.85m	4.30m	A
	812	908	995	1.15k	1.28k	V
390k	2.46m	2.75m	3.02m	3.48m	3.89m	A
	883	987	1.08k	1.25k	1.40k	V
470k	2.26m	2.53m	2.77m	3.20m	3.58m	A
	970	1.08k	1.19k	1.37k	1.53k	V
560k	2.06m	2.31m	2.53m	2.92m	3.26m	A
	1.06k	1.18k	1.30k	1.50k	1.67k	V
680k	1.89m	2.11m	2.31m	2.67m	2.99m	A
	1.17k	1.30k	1.43k	1.65k	1.84k	V
	1.71m	1.92m	2.10m	2.43m	2.71m	A

R	2 W	2.5 W	3 W	4 W	5 W	
820k	1.28k	1.43k	1.57k	1.81k	2.02k	V
	1.56m	1.75m	1.91m	2.21m	2.47m	A
1M	1.41k	1.58k	1.73k	2.00k	2.24k	V
	1.41m	1.58m	1.73m	2.00m	2.24m	A

Table 7.4 *Maximum power for resistors between 6 and 10 W*

R	6 W	7 W	7.5 W	9 W	10 W	
0R01	245m	265m	274m	300m	316m	V
	24.5	26.5	27.4	30.0	31.6	A
0R02	346m	374m	387m	424m	447m	V
	17.3	18.7	19.4	21.2	22.4	A
0R03	424m	458m	474m	520m	548m	V
	14.1	15.3	15.8	17.3	18.3	A
0R033	445m	481m	497m	545m	574m	V
	13.5	14.6	15.1	16.5	17.4	A
0R047	531m	574m	594m	650m	686m	V
	11.3	12.2	12.6	13.8	14.6	A
0R05	548m	592m	612m	671m	707m	V
	11.0	11.8	12.2	13.4	14.1	A
0R056	580m	626m	648m	710m	748m	V
	10.4	11.2	11.6	12.7	13.4	A
0R068	639m	690m	714m	782m	825m	V
	9.39	10.1	10.5	11.5	12.1	A
0R07	648m	700m	725m	794m	837m	V
	9.26	10.0	10.4	11.3	12.0	A
0R10	775m	837m	866m	949m	1.00	V
	7.75	8.37	8.66	9.49	10.0	A
0R12	849m	917m	949m	1.04	1.10	V
	7.07	7.64	7.91	8.66	9.13	A
0R15	949m	1.02	1.06	1.16	1.22	V
	6.32	6.83	7.07	7.75	8.16	A
0R18	1.04	1.12	1.16	1.27	1.34	V
	5.77	6.24	6.45	7.07	7.45	A
0R22	1.15	1.24	1.28	1.41	1.48	V
	5.22	5.64	5.84	6.40	6.74	A
0R27	1.27	1.37	1.42	1.56	1.64	V
	4.71	5.09	5.27	5.77	6.09	A
0R33	1.41	1.52	1.57	1.72	1.82	V
	4.26	4.61	4.77	5.22	5.50	A
0R39	1.53	1.65	1.71	1.87	1.97	V
	3.92	4.24	4.39	4.80	5.06	A
0R47	1.68	1.81	1.88	2.06	2.17	V
	3.57	3.86	3.99	4.38	4.61	A

R	6 W	7 W	7.5 W	9 W	10 W	
0R56	1.83	1.98	2.05	2.24	2.37	V
	3.27	3.54	3.66	4.01	4.23	A
0R68	2.02	2.18	2.26	2.47	2.61	V
	2.97	3.21	3.32	3.64	3.83	A
0R82	2.22	2.40	2.48	2.72	2.86	V
	2.71	2.92	3.02	3.31	3.49	A
1R	2.45	2.65	2.74	3.00	3.16	V
	2.45	2.65	2.74	3.00	3.16	A
1R2	2.68	2.90	3.00	3.29	3.46	V
	2.24	2.42	2.50	2.74	2.89	A
1R5	3.00	3.24	3.35	3.67	3.87	V
	2.00	2.16	2.24	2.45	2.58	A
1R8	3.29	3.55	3.67	4.02	4.24	V
	1.83	1.97	2.04	2.24	2.36	A
2R2	3.63	3.92	4.06	4.45	4.69	V
	1.65	1.78	1.85	2.02	2.13	A
2R7	4.02	4.35	4.50	4.93	5.20	V
	1.49	1.61	1.67	1.83	1.92	A
3R3	4.45	4.81	4.97	5.45	5.74	V
	1.35	1.46	1.51	1.65	1.74	A
3R9	4.84	5.22	5.41	5.92	6.24	V
	1.24	1.34	1.39	1.52	1.60	A
4R7	5.31	5.74	5.94	6.50	6.86	V
	1.13	1.22	1.26	1.38	1.46	A
5R6	5.80	6.26	6.48	7.10	7.48	V
	1.04	1.12	1.16	1.27	1.34	A
6R8	6.39	6.90	7.14	7.82	8.25	V
	939m	1.01	1.05	1.15	1.21	A
8R2	7.01	7.58	7.84	8.59	9.06	V
	855m	924m	956m	1.05	1.10	A
10R	7.75	8.37	8.66	9.49	10.0	V
	775m	837m	866m	949m	1.00	A
12R	8.49	9.17	9.49	10.4	11.0	V
	707m	764m	791m	866m	913m	A
15R	9.49	10.2	10.6	11.6	12.2	V
	632m	683m	707m	775m	816m	A

Table 7.4 Maximum power for resistors between 6 and 10 W 129

R	6 W	7 W	7.5 W	9 W	10 W	
18R	10.4	11.2	11.6	12.7	13.4	V
	577m	624m	645m	707m	745m	A
22R	11.5	12.4	12.8	14.1	14.8	V
	522m	564m	584m	640m	674m	A
27R	12.7	13.7	14.2	15.6	16.4	V
	471m	509m	527m	577m	609m	A
33R	14.1	15.2	15.7	17.2	18.2	V
	426m	461m	477m	522m	550m	A
39R	15.3	16.5	17.1	18.7	19.7	V
	392m	424m	439m	480m	506m	A
47R	16.8	18.1	18.8	20.6	21.7	V
	357m	386m	399m	438m	461m	A
56R	18.3	19.8	20.5	22.4	23.7	V
	327m	354m	366m	401m	423m	A
68R	20.2	21.8	22.6	24.7	26.1	V
	297m	321m	332m	364m	383m	A
82R	22.2	24.0	24.8	27.2	28.6	V
	271m	292m	302m	331m	349m	A
100R	24.5	26.5	27.4	30.0	31.6	V
	245m	265m	274m	300m	316m	A
120R	26.8	29.0	30.0	32.9	34.6	V
	224m	242m	250m	274m	289m	A
150R	30.0	32.4	33.5	36.7	38.7	V
	200m	216m	224m	245m	258m	A
180R	32.9	35.5	36.7	40.2	42.4	V
	183m	197m	204m	224m	236m	A
220R	36.3	39.2	40.6	44.5	46.9	V
	165m	178m	185m	202m	213m	A
270R	40.2	43.5	45.0	49.3	52.0	V
	149m	161m	167m	183m	192m	A
330R	44.5	48.1	49.7	54.5	57.4	V
	135m	146m	151m	165m	174m	A
390R	48.4	52.2	54.1	59.2	62.4	V
	124m	134m	139m	152m	160m	A
470R	53.1	57.4	59.4	65.0	68.6	V
	113m	122m	126m	138m	146m	A

R	6 W	7 W	7.5 W	9 W	10 W	
560R	58.0	62.6	64.8	71.0	74.8	V
	104m	112m	116m	127m	134m	A
680R	63.9	69.0	71.4	78.2	82.5	V
	93.9m	101m	105m	115m	121m	A
820R	70.1	75.8	78.4	85.9	90.6	V
	85.5m	92.4m	95.6m	105m	110m	A
1k	77.5	83.7	86.6	94.9	100	V
	77.5m	83.7m	86.6m	94.9m	100m	A
1k2	84.9	91.7	94.9	104	110	V
	70.7m	76.4m	79.1m	86.6m	91.3m	A
1k5	94.9	102	106	116	122	V
	63.2m	68.3m	70.7m	77.5m	81.6m	A
1k8	104	112	116	127	134	V
	57.7m	62.4m	64.5m	70.7m	74.5m	A
2k2	115	124	128	141	148	V
	52.2m	56.4m	58.4m	64.0m	67.4m	A
2k7	127	137	142	156	164	V
	47.1m	50.9m	52.7m	57.7m	60.9m	A
3k3	141	152	157	172	182	V
	42.6m	46.1m	47.7m	52.2m	55.0m	A
3k9	153	165	171	187	197	V
	39.2m	42.4m	43.9m	48.0m	50.6m	A
4k7	168	181	188	206	217	V
	35.7m	38.6m	39.9m	43.8m	46.1m	A
5k6	183	198	205	224	237	V
	32.7m	35.4m	36.6m	40.1m	42.3m	A
6k8	202	218	226	247	261	V
	29.7m	32.1m	33.2m	36.4m	38.3m	A
8k2	222	240	248	272	286	V
	27.1m	29.2m	30.2m	33.1m	34.9m	A
10k	245	265	274	300	316	V
	24.5m	26.5m	27.4m	30.0m	31.6m	A
12k	268	290	300	329	346	V
	22.4m	24.2m	25.0m	27.4m	28.9m	A
15k	300	324	335	367	387	V
	20.0m	21.6m	22.4m	24.5m	25.8m	A

Table 7.4 Maximum power for resistors between 6 and 10 W   131

R	6 W	7 W	7.5 W	9 W	10 W	
18k	329	355	367	402	424	V
	18.3m	19.7m	20.4m	22.4m	23.6m	A
22k	363	392	406	445	469	V
	16.5m	17.8m	18.5m	20.2m	21.3m	A
27k	402	435	450	493	520	V
	14.9m	16.1m	16.7m	18.3m	19.2m	A
33k	445	481	497	545	574	V
	13.5m	14.6m	15.1m	16.5m	17.4m	A
39k	484	522	541	592	624	V
	12.4m	13.4m	13.9m	15.2m	16.0m	A
47k	531	574	594	650	686	V
	11.3m	12.2m	12.6m	13.8m	14.6m	A
56k	580	626	648	710	748	V
	10.4m	11.2m	11.6m	12.7m	13.4m	A
68k	639	690	714	782	825	V
	9.39m	10.1m	10.5m	11.5m	12.1m	A
82k	701	758	784	859	906	V
	8.55m	9.24m	9.56m	10.5m	11.0m	A
100k	775	837	866	949	1.00k	V
	7.75m	8.37m	8.66m	9.49m	10.0m	A

Table 7.5 *Maximum power for resistors between 11 and 25 W*

R	11 W	12 W	15 W	17 W	25 W	
0R01	332m	346m	387m	412m	500m	V
	33.2	34.6	38.7	41.2	50.0	A
0R02	469m	490m	548m	583m	707m	V
	23.5	24.5	27.4	29.2	35.4	A
0R03	574m	600m	671m	714m	866m	V
	19.1	20.0	22.4	23.8	28.9	A
0R033	602m	629m	704m	749m	908m	V
	18.3	19.1	21.3	22.7	27.5	A
0R047	719m	751m	840m	894m	1.08	V
	15.3	16.0	17.9	19.0	23.1	A
0R05	742m	775m	866m	922m	1.12	V
	14.8	15.5	17.3	18.4	22.4	A
0R056	785m	820m	917m	976m	1.18	V
	14.0	14.6	16.4	17.4	21.1	A
0R068	865m	903m	1.01	1.08	1.30	V
	12.7	13.3	14.9	15.8	19.2	A
0R07	877m	917m	1.02	1.09	1.32	V
	12.5	13.1	14.6	15.6	18.9	A
0R10	1.05	1.10	1.22	1.30	1.58	V
	10.5	11.0	12.2	13.0	15.8	A
0R12	1.15	1.20	1.34	1.43	1.73	V
	9.57	10.0	11.2	11.9	14.4	A
0R15	1.28	1.34	1.50	1.60	1.94	V
	8.56	8.94	10.0	10.6	12.9	A
0R18	1.41	1.47	1.64	1.75	2.12	V
	7.82	8.16	9.13	9.72	11.8	A
0R22	1.56	1.62	1.82	1.93	2.35	V
	7.07	7.39	8.26	8.79	10.7	A
0R27	1.72	1.80	2.01	2.14	2.60	V
	6.38	6.67	7.45	7.93	9.62	A
0R33	1.91	1.99	2.22	2.37	2.87	V
	5.77	6.03	6.74	7.18	8.70	A
0R39	2.07	2.16	2.42	2.57	3.12	V
	5.31	5.55	6.20	6.60	8.01	A
0R47	2.27	2.37	2.66	2.83	3.43	V
	4.84	5.05	5.65	6.01	7.29	A

Table 7.5 Maximum power for resistors between 11 and 25 W   133

R	11 W	12 W	15 W	17 W	25 W	
0R56	2.48	2.59	2.90	3.09	3.74	V
	4.43	4.63	5.18	5.51	6.68	A
0R68	2.73	2.86	3.19	3.40	4.12	V
	4.02	4.20	4.70	5.00	6.06	A
0R82	3.00	3.14	3.51	3.73	4.53	V
	3.66	3.83	4.28	4.55	5.52	A
1R	3.32	3.46	3.87	4.12	5.00	V
	3.32	3.46	3.87	4.12	5.00	A
1R2	3.63	3.79	4.24	4.52	5.48	V
	3.03	3.16	3.54	3.76	4.56	A
1R5	4.06	4.24	4.74	5.05	6.12	V
	2.71	2.83	3.16	3.37	4.08	A
1R8	4.45	4.65	5.20	5.53	6.71	V
	2.47	2.58	2.89	3.07	3.73	A
2R2	4.92	5.14	5.74	6.12	7.42	V
	2.24	2.34	2.61	2.78	3.37	A
2R7	5.45	5.69	6.36	6.77	8.22	V
	2.02	2.11	2.36	2.51	3.04	A
3R3	6.02	6.29	7.04	7.49	9.08	V
	1.83	1.91	2.13	2.27	2.75	A
3R9	6.55	6.84	7.65	8.14	9.87	V
	1.68	1.75	1.96	2.09	2.53	A
4R7	7.19	7.51	8.40	8.94	10.8	V
	1.53	1.60	1.79	1.90	2.31	A
5R6	7.85	8.20	9.17	9.76	11.8	V
	1.40	1.46	1.64	1.74	2.11	A
6R8	8.65	9.03	10.1	10.8	13.0	V
	1.27	1.33	1.49	1.58	1.92	A
8R2	9.50	9.92	11.1	11.8	14.3	V
	1.16	1.21	1.35	1.44	1.75	A
10R	10.5	11.0	12.2	13.0	15.8	V
	1.05	1.10	1.22	1.30	1.58	A
12R	11.5	12.0	13.4	14.3	17.3	V
	957m	1.00	1.12	1.19	1.44	A
15R	12.8	13.4	15.0	16.0	19.4	V
	856m	894m	1.00	1.06	1.29	A

R	11 W	12 W	15 W	17 W	25 W	
18R	14.1	14.7	16.4	17.5	21.2	V
	782m	816m	913m	972m	1.18	A
22R	15.6	16.2	18.2	19.3	23.5	V
	707m	739m	826m	879m	1.07	A
27R	17.2	18.0	20.1	21.4	26.0	V
	638m	667m	745m	793m	962m	A
33R	19.1	19.9	22.2	23.7	28.7	V
	577m	603m	674m	718m	870m	A
39R	20.7	21.6	24.2	25.7	31.2	V
	531m	555m	620m	660m	801m	A
47R	22.7	23.7	26.6	28.3	34.3	V
	484m	505m	565m	601m	729m	A
56R	24.8	25.9	29.0	30.9	37.4	V
	443m	463m	518m	551m	668m	A
68R	27.3	28.6	31.9	34.0	41.2	V
	402m	420m	470m	500m	606m	A
82R	30.0	31.4	35.1	37.3	45.3	V
	366m	383m	428m	455m	552m	A
100R	33.2	34.6	38.7	41.2	50.0	V
	332m	346m	387m	412m	500m	A
120R	36.3	37.9	42.4	45.2	54.8	V
	303m	316m	354m	376m	456m	A
150R	40.6	42.4	47.4	50.5	61.2	V
	271m	283m	316m	337m	408m	A
180R	44.5	46.5	52.0	55.3	67.1	V
	247m	258m	289m	307m	373m	A
220R	49.2	51.4	57.4	61.2	74.2	V
	224m	234m	261m	278m	337m	A
270R	54.5	56.9	63.6	67.7	82.2	V
	202m	211m	236m	251m	304m	A
330R	60.2	62.9	70.4	74.9	90.8	V
	183m	191m	213m	227m	275m	A
390R	65.5	68.4	76.5	81.4	98.7	V
	168m	175m	196m	209m	253m	A
470R	71.9	75.1	84.0	89.4	108	V
	153m	160m	179m	190m	231m	A

Table 7.5 Maximum power for resistors between 11 and 25 W   135

R	11 W	12 W	15 W	17 W	25 W	
560R	78.5	82.0	91.7	97.6	118	V
	140m	146m	164m	174m	211m	A
680R	86.5	90.3	101	108	130	V
	127m	133m	149m	158m	192m	A
820R	95.0	99.2	111	118	143	V
	116m	121m	135m	144m	175m	A
1k	105	110	122	130	158	V
	105m	110m	122m	130m	158m	A
1k2	115	120	134	143	173	V
	95.7m	100m	112m	119m	144m	A
1k5	128	134	150	160	194	V
	85.6m	89.4m	100m	106m	129m	A
1k8	141	147	164	175	212	V
	78.2m	81.6m	91.3m	97.2m	118m	A
2k2	156	162	182	193	235	V
	70.7m	73.9m	82.6m	87.9m	107m	A
2k7	172	180	201	214	260	V
	63.8m	66.7m	74.5m	79.3m	96.2m	A
3k3	191	199	222	237	287	V
	57.7m	60.3m	67.4m	71.8m	87.0m	A
3k9	207	216	242	257	312	V
	53.1m	55.5m	62.0m	66.0m	80.1m	A
4k7	227	237	266	283	343	V
	48.4m	50.5m	56.5m	60.1m	72.9m	A
5k6	248	259	290	309	374	V
	44.3m	46.3m	51.8m	55.1m	66.8m	A
6k8	273	286	319	340	412	V
	40.2m	42.0m	47.0m	50.0m	60.6m	A
8k2	300	314	351	373	453	V
	36.6m	38.3m	42.8m	45.5m	55.2m	A
10k	332	346	387	412	500	V
	33.2m	34.6m	38.7m	41.2m	50.0m	A
12k	363	379	424	452	548	V
	30.3m	31.6m	35.4m	37.6m	45.6m	A
15k	406	424	474	505	612	V
	27.1m	28.3m	31.6m	33.7m	40.8m	A

R	11 W	12 W	15 W	17 W	25 W	
18k	445	465	520	553	671	V
	24.7m	25.8m	28.9m	30.7m	37.3m	A
22k	492	514	574	612	742	V
	22.4m	23.4m	26.1m	27.8m	33.7m	A
27k	545	569	636	677	822	V
	20.2m	21.1m	23.6m	25.1m	30.4m	A
33k	602	629	704	749	908	V
	18.3m	19.1m	21.3m	22.7m	27.5m	A
39k	655	684	765	814	987	V
	16.8m	17.5m	19.6m	20.9m	25.3m	A
47k	719	751	840	894	1.08k	V
	15.3m	16.0m	17.9m	19.0m	23.1m	A
56k	785	820	917	976	1.18k	V
	14.0m	14.6m	16.4m	17.4m	21.1m	A
68k	865	903	1.01k	1.08k	1.30k	V
	12.7m	13.3m	14.9m	15.8m	19.2m	A
82k	950	992	1.11k	1.18k	1.43k	V
	11.6m	12.1m	13.5m	14.4m	17.5m	A
100k	1.05k	1.10k	1.22k	1.30k	1.58k	V
	10.5m	11.0m	12.2m	13.0m	15.8m	A

Table 7.6 *Maximum power for resistors between 50 and 300 W*

R	50 W	100 W	200 W	250 W	300 W	
0R01	707m	1.00	1.41	1.58	1.73	V
	70.7	100	141	158	173	A
0R02	1.00	1.41	2.00	2.24	2.45	V
	50.0	70.7	100	112	122	A
0R03	1.22	1.73	2.45	2.74	3.00	V
	40.8	57.7	81.6	91.3	100	A
0R033	1.28	1.82	2.57	2.87	3.15	V
	38.9	55.0	77.8	87.0	95.3	A
0R047	1.53	2.17	3.07	3.43	3.75	V
	32.6	46.1	65.2	72.9	79.9	A
0R05	1.58	2.24	3.16	3.54	3.87	V
	31.6	44.7	63.2	70.7	77.5	A
0R056	1.67	2.37	3.35	3.74	4.10	V
	29.9	42.3	59.8	66.8	73.2	A
0R068	1.84	2.61	3.69	4.12	4.52	V
	27.1	38.3	54.2	60.6	66.4	A
0R07	1.87	2.65	3.74	4.18	4.58	V
	26.7	37.8	53.5	59.8	65.5	A
0R10	2.24	3.16	4.47	5.00	5.48	V
	22.4	31.6	44.7	50.0	54.8	A
0R12	2.45	3.46	4.90	5.48	6.00	V
	20.4	28.9	40.8	45.6	50.0	A
0R15	2.74	3.87	5.48	6.12	6.71	V
	18.3	25.8	36.5	40.8	44.7	A
0R18	3.00	4.24	6.00	6.71	7.35	V
	16.7	23.6	33.3	37.3	40.8	A
0R22	3.32	4.69	6.63	7.42	8.12	V
	15.1	21.3	30.2	33.7	36.9	A
0R27	3.67	5.20	7.35	8.22	9.00	V
	13.6	19.2	27.2	30.4	33.3	A
0R33	4.06	5.74	8.12	9.08	9.95	V
	12.3	17.4	24.6	27.5	30.2	A
0R39	4.42	6.24	8.83	9.87	10.8	V
	11.3	16.0	22.6	25.3	27.7	A
0R47	4.85	6.86	9.70	10.8	11.9	V
	10.3	14.6	20.6	23.1	25.3	A

R	50 W	100 W	200 W	250 W	300 W	
0R56	5.29	7.48	10.6	11.8	13.0	V
	9.45	13.4	18.9	21.1	23.1	A
0R68	5.83	8.25	11.7	13.0	14.3	V
	8.57	12.1	17.1	19.2	21.0	A
0R82	6.40	9.06	12.8	14.3	15.7	V
	7.81	11.0	15.6	17.5	19.1	A
1R	7.07	10.0	14.1	15.8	17.3	V
	7.07	10.0	14.1	15.8	17.3	A
1R2	7.75	11.0	15.5	17.3	19.0	V
	6.45	9.13	12.9	14.4	15.8	A
1R5	8.66	12.2	17.3	19.4	21.2	V
	5.77	8.16	11.5	12.9	14.1	A
1R8	9.49	13.4	19.0	21.2	23.2	V
	5.27	7.45	10.5	11.8	12.9	A
2R2	10.5	14.8	21.0	23.5	25.7	V
	4.77	6.74	9.53	10.7	11.7	A
2R7	11.6	16.4	23.2	26.0	28.5	V
	4.30	6.09	8.61	9.62	10.5	A
3R3	12.8	18.2	25.7	28.7	31.5	V
	3.89	5.50	7.78	8.70	9.53	A
3R9	14.0	19.7	27.9	31.2	34.2	V
	3.58	5.06	7.16	8.01	8.77	A
4R7	15.3	21.7	30.7	34.3	37.5	V
	3.26	4.61	6.52	7.29	7.99	A
5R6	16.7	23.7	33.5	37.4	41.0	V
	2.99	4.23	5.98	6.68	7.32	A
6R8	18.4	26.1	36.9	41.2	45.2	V
	2.71	3.83	5.42	6.06	6.64	A
8R2	20.2	28.6	40.5	45.3	49.6	V
	2.47	3.49	4.94	5.52	6.05	A
10R	22.4	31.6	44.7	50.0	54.8	V
	2.24	3.16	4.47	5.00	5.48	A
12R	24.5	34.6	49.0	54.8	60.0	V
	2.04	2.89	4.08	4.56	5.00	A
15R	27.4	38.7	54.8	61.2	67.1	V
	1.83	2.58	3.65	4.08	4.47	A

Table 7.6 Maximum power for resistors between 50 and 300 W   139

R	50 W	100 W	200 W	250 W	300 W	
18R	30.0	42.4	60.0	67.1	73.5	V
	1.67	2.36	3.33	3.73	4.08	A
22R	33.2	46.9	66.3	74.2	81.2	V
	1.51	2.13	3.02	3.37	3.69	A
27R	36.7	52.0	73.5	82.2	90.0	V
	1.36	1.92	2.72	3.04	3.33	A
33R	40.6	57.4	81.2	90.8	99.5	V
	1.23	1.74	2.46	2.75	3.02	A
39R	44.2	62.4	88.3	98.7	108	V
	1.13	1.60	2.26	2.53	2.77	A
47R	48.5	68.6	97.0	108	119	V
	1.03	1.46	2.06	2.31	2.53	A
56R	52.9	74.8	106	118	130	V
	945m	1.34	1.89	2.11	2.31	A
68R	58.3	82.5	117	130	143	V
	857m	1.21	1.71	1.92	2.10	A
82R	64.0	90.6	128	143	157	V
	781m	1.10	1.56	1.75	1.91	A
100R	70.7	100	141	158	173	V
	707m	1.00	1.41	1.58	1.73	A
120R	77.5	110	155	173	190	V
	645m	913m	1.29	1.44	1.58	A
150R	86.6	122	173	194	212	V
	577m	816m	1.15	1.29	1.41	A
180R	94.9	134	190	212	232	V
	527m	745m	1.05	1.18	1.29	A
220R	105	148	210	235	257	V
	477m	674m	953m	1.07	1.17	A
270R	116	164	232	260	285	V
	430m	609m	861m	962m	1.05	A
330R	128	182	257	287	315	V
	389m	550m	778m	870m	953m	A
390R	140	197	279	312	342	V
	358m	506m	716m	801m	877m	A
470R	153	217	307	343	375	V
	326m	461m	652m	729m	799m	A

R	50 W	100 W	200 W	250 W	300 W	
560R	167	237	335	374	410	V
	299m	423m	598m	668m	732m	A
680R	184	261	369	412	452	V
	271m	383m	542m	606m	664m	A
820R	202	286	405	453	496	V
	247m	349m	494m	552m	605m	A
1k	224	316	447	500	548	V
	224m	316m	447m	500m	548m	A
1k2	245	346	490	548	600	V
	204m	289m	408m	456m	500m	A
1k5	274	387	548	612	671	V
	183m	258m	365m	408m	447m	A
1k8	300	424	600	671	735	V
	167m	236m	333m	373m	408m	A
2k2	332	469	663	742	812	V
	151m	213m	302m	337m	369m	A
2k7	367	520	735	822	900	V
	136m	192m	272m	304m	333m	A
3k3	406	574	812	908	995	V
	123m	174m	246m	275m	302m	A
3k9	442	624	883	987	1.08k	V
	113m	160m	226m	253m	277m	A
4k7	485	686	970	1.08k	1.19k	V
	103m	146m	206m	231m	253m	A
5k6	529	748	1.06k	1.18k	1.30k	V
	94.5m	134m	189m	211m	231m	A
6k8	583	825	1.17k	1.30k	1.43k	V
	85.7m	121m	171m	192m	210m	A
8k2	640	906	1.28k	1.43k	1.57k	V
	78.1m	110m	156m	175m	191m	A
10k	707	1.00k	1.41k	1.58k	1.73k	V
	70.7m	100m	141m	158m	173m	A
12k	775	1.10k	1.55k	1.73k	1.90k	V
	64.5m	91.3m	129m	144m	158m	A
15k	866	1.22k	1.73k	1.94k	2.12k	V
	57.7m	81.6m	115m	129m	141m	A

Table 7.6 Maximum power for resistors between 50 and 300 W   141

R	50 W	100 W	200 W	250 W	300 W	
18k	949	1.34k	1.90k	2.12k	2.32k	V
	52.7m	74.5m	105m	118m	129m	A
22k	1.05k	1.48k	2.10k	2.35k	2.57k	V
	47.7m	67.4m	95.3m	107m	117m	A
27k	1.16k	1.64k	2.32k	2.60k	2.85k	V
	43.0m	60.9m	86.1m	96.2m	105m	A
33k	1.28k	1.82k	2.57k	2.87k	3.15k	V
	38.9m	55.0m	77.8m	87.0m	95.3m	A
39k	1.40k	1.97k	2.79k	3.12k	3.42k	V
	35.8m	50.6m	71.6m	80.1m	87.7m	A
47k	1.53k	2.17k	3.07k	3.43k	3.75k	V
	32.6m	46.1m	65.2m	72.9m	79.9m	A
56k	1.67k	2.37k	3.35k	3.74k	4.10k	V
	29.9m	42.3m	59.8m	66.8m	73.2m	A
68k	1.84k	2.61k	3.69k	4.12k	4.52k	V
	27.1m	38.3m	54.2m	60.6m	66.4m	A
82k	2.02k	2.86k	4.05k	4.53k	4.96k	V
	24.7m	34.9m	49.4m	55.2m	60.5m	A
100k	2.24k	3.16k	4.47k	5.00k	5.48k	V
	22.4m	31.6m	44.7m	50.0m	54.8m	A

8 Voltage dividers

Introduction

Figure 8.1 shows a voltage divider. It is used everywhere in electronics to reduce the level of AC and/or DC voltages. The attenuation, input resistance and output resistance are as follows:

$$A = \frac{R_2}{R_1 + R_2} \tag{8.1}$$

$$dB = -20 \, \log_{10}(A) \tag{8.2}$$

$$R_I = R_1 + R_2 \tag{8.3}$$

$$R_O = \frac{R_1 R_2}{R_1 + R_2} \tag{8.4}$$

All of these parameters are given in Table 8.1 for various combinations of E24 resistor, and sorted in order of increasing attenuation. The Table starts at 1 dB and ends at 60 dB.

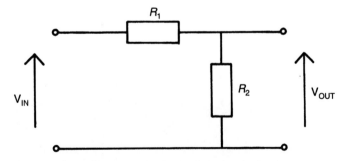

Figure 8.1 A voltage divider

The equations assume that the circuit is driven by a perfect voltage source, i.e. a circuit with zero output impedance, and drives a circuit of infinite input impedance. If the source is the output pin of an op-amp, and the load a non-inverting op-amp circuit, this is as true as makes no difference. If source resistance is a tenth of R_I, or load impedance ten times R_O, then the error is just under a dB; if the ratios are 100:1, then it is less than 0.1 dB.

Using Table 8.1 in cases where source and load impedances can be ignored

We can simply look up the desired attenuation in column A or dB as appropriate, and choose a pair.

> Example 8.1: A voltage divider of 4 dB could be formed by resistors $R_1, R_2 = 3k6, 6k2$ and $3k, 5k1$ respectively, amongst many.

When source and load impedances cannot be ignored

Selecting values for a voltage divider with ideal source and load impedances is pretty straightforward, as we saw above, but what happens when source and load begin affecting the nominal attenuation noticeably? Let us consider what happens.

We'll call source impedance R_S, input and output impedances of the attenuator R_I and R_O, as before, and load impedance R_L. Ideally, we want R_I to be much greater than R_S, and R_L much greater than R_O. This implies that (as R_O cannot be greater than R_I) R_L should be much greater than R_S.

(Of course, the situation can arise where the loss due to R_S and R_L is actually greater than the desired attenuation. If we are in doubt, we can check this by calculating a value for $\frac{R_L}{(R_S+R_L)}$; if this gives a greater attenuation than that desired, we need a buffer amplifier, rather than an attenuator.)

In Table 8.1, the columns R_{SMAX} and R_{LMIN} give suggested values for, as their names suggest, a maximum value of R_S and a minimum of R_L. This is calculated so that, with both in place, they cause the circuit to have an attenuation of 1 dB more than the value given in the dB column. This is true so long as R_S is still less than or equal to R_{SMAX}, and R_L greater than or equal to R_{LMIN}. The column $R_{LS}(1 \text{ dB})$ is the ratio of them.

(If we wish, we can scale both R_1 and R_2 by factors of 10 in the same sense. Then R_I, R_O, R_{SMAX} and R_{LMIN} will also be scaled. A, dB and $R_{LS}(1 \text{ dB})$ will be unaffected.)

If we set out to design an attenuator to work in a circuit where R_S and R_L are significant, we can start out by calculating their ratio, $\frac{R_L}{R_S}$. I'm going to call this R_{LS} (real). Two possible situations now arise:

1. If R_{LS} (real) is greater than or equal to $R_{LS}(1 \text{ dB})$ for the desired value of dB in

Table 8.1, the attenuation with R_L and R_S in place should be within 1 dB of nominal. (Provided that we can find a pair where $R_S \leq R_{SMAX}$ and $R_L \geq R_{LMIN}$.)

2. If R_{LS} (real) is less than $R_{LS}(1\ dB)$ the error will be more than 1 dB. In this situation the problem could be approached by other methods, or the values in the table used as a starting point.

Having selected a pair of values, we can check them numerically, if we are concerned about the accuracy. First we calculate the parallel resistance of R_L and R_2:

$$R_P = \frac{R_L R_2}{R_L + R_2} \tag{8.5}$$

Then:

$$\text{attenuation} = \frac{R_P}{R_S + R_1 + R_P} \tag{8.6}$$

The best way to see how things work is to look at some examples.

Example 8.2: Form a 10 dB attenuator to go between a source of 100 Ω and a load of 10k.

Our ratio of load to source is 100. We look down the dB column at values around 10 dB. $R_{LS}(1\ dB)$ is given as 63 or so here, so our error is less than 1 dB. Now look through the available pairs, to find one where we will be able to make R_S and R_L in range. We can see that we will need to divide by 10 to make $R_{SMAX} \geq 100\ \Omega$ and $R_{LMIN} \leq 10k$. We might choose:

R_1	R_2	A	dB	R_I	R_O	R_{SMAX}	R_{LMIN}	R_{LS} (1 dB)
12k	5k6	318m	09.95	17.6k	3.82k	1.03k	65.1k	63.1

This would yield $R_{SMAX} = 103\ \Omega$, $R_{LMIN} = 6.51k$. We conclude that the error will be around 0.7 dB. (R_S is about equal to R_{SMAX}, so say 0.5 dB, R_L is about 1.5 times R_{LMIN}, we could guess at 0.2 dB. Or we might calculate the actual attenuation using Eqns(3.5) and (3.6). So we aim up the table to lines where dB = 9.3 or thereabouts. We find (again selecting for $R_{SMAX} \geq 100\ \Omega$ and $R_{LMIN} \leq 10k$):

R_1	R_2	A	dB	R_I	R_O	R_{SMAX}	R_{LMIN}	R_{LS} (1 dB)
13k	6k8	343m	09.28	19.8k	4.46k	1.16k	76.1k	65.6

and decide to try $R_1 = 1k3$, $R_2 = 680\ \Omega$, as $R_{SMAX} = 116\ \Omega$, $R_{LMIN} = 7.6k$ then. To check, we work backwards and calculate that $R_P = 680\ \Omega$ in parallel with 10k, or 634 Ω, and we get:

$$\text{attenuation} = \frac{634}{100 + 1300 + 634} = 0.312 \text{ or } 10.1\ dB$$

which is pretty good.

Here's another example, where the values of R_L and R_S make things tougher.

Example 8.3: An attenuator of 3 dB, with a source of 400 Ω, and a load of 3k.

Here we see that our ratio is a lot lower, $\frac{3000}{400} = 7.5$. If we go to the part of the table where dB = 3, we see that the R_{LS} (1 dB) column gives figures of about 67, telling us that we will lose a lot more than 1 dB due to R_S and R_L. Obviously we need less nominal attenuation; we want the error due to R_S and R_L and the nominal attenuation of the attenuator chosen to total 3 dB.

A possible solution is to use the load resistance itself as R_2. The load is 3k; looking up the table, we see:

R_1	R_2	A	dB	R_I	R_O	R_{SMAX}	R_{LMIN}	R_{LS} (1 dB)
1k2	3k	714m	02.92	4.20k	857	235	15.3k	65.4

which is, by good luck, ideal. R_1 would ideally be 1k2, but we already have a source of 400 Ω, so we want the nearest value to 800 Ω, which is 820 Ω. If we wanted to check, we would calculate $\frac{3000}{3000+400+820} = 0.711 = 2.96$ dB, very close.

Example 8.4: $R_S = $ 1k, $R_L = $ 5k and the attenuation desired is 8 dB.

Here our ratio is again low, 5. Our loss will be too great again, so we look for values of R_2 near to 5k. We find:

R_1	R_2	A	dB	R_I	R_O	R_{SMAX}	R_{LMIN}	R_{LS} (1 dB)
7k5	5k1	405m	07.86	12.6k	3.04k	732	52.3k	71.5

suggesting that we need a series resistor between source and load of (7k5 − 1k), or 6k5. The nearest preferred value is 6k8, giving an attenuation of $\frac{5}{5+6.8+1} = 0.391$ or 8.16 dB, very close.

Note that in cases where the load and source resistances form a part of the attenuator itself, we should check that they have a suitable power rating.

Table 8.1 *Voltage dividers*

R_1	R_2	A	dB	R_I	R_O	R_{SMAX}	R_{LMIN}	$R_{LS}(1\,dB)$
6k8	56k	892m	01.00	62.8k	6.06k	3.13k	122k	38.9
6k2	51k	892m	01.00	57.2k	5.53k	2.85k	111k	38.9
1k	8k2	891m	01.00	9.20k	891	459	17.9k	39.0
10k	82k	891m	01.00	92.0k	8.91k	4.59k	179k	39.0
2k2	18k	891m	01.00	20.2k	1.96k	1.01k	39.3k	39.1
3k3	27k	891m	01.00	30.3k	2.94k	1.51k	59.0k	39.1
2k7	22k	891m	01.01	24.7k	2.40k	1.23k	48.2k	39.2
1k6	13k	890m	01.01	14.6k	1.42k	728	28.6k	39.3
2k	16k	889m	01.02	18.0k	1.78k	895	35.7k	39.9
1k5	12k	889m	01.02	13.5k	1.33k	672	26.8k	39.9
3k	24k	889m	01.02	27.0k	2.67k	1.34k	53.6k	39.9
1k3	10k	885m	01.06	11.3k	1.15k	568	22.9k	40.3
3k9	30k	885m	01.06	33.9k	3.45k	1.70k	68.7k	40.3
5k6	43k	885m	01.06	48.6k	4.95k	2.44k	98.6k	40.3
4k3	33k	885m	01.06	37.3k	3.80k	1.87k	75.7k	40.4
4k7	36k	885m	01.07	40.7k	4.16k	2.05k	82.7k	40.4
5k1	39k	884m	01.07	44.1k	4.51k	2.22k	89.7k	40.5
1k2	9k1	883m	01.08	10.3k	1.06k	517	21.1k	40.9
6k2	47k	883m	01.08	53.2k	5.48k	2.67k	109k	40.9
8k2	62k	883m	01.08	70.2k	7.24k	3.52k	144k	41.0
1k	7k5	882m	01.09	8.50k	882	426	17.6k	41.2
1k6	12k	882m	01.09	13.6k	1.41k	682	28.1k	41.2
2k	15k	882m	01.09	17.0k	1.76k	853	35.2k	41.2
2k4	18k	882m	01.09	20.4k	2.12k	1.02k	42.2k	41.2
3k6	27k	882m	01.09	30.6k	3.18k	1.54k	63.3k	41.2
6k8	51k	882m	01.09	57.8k	6.00k	2.90k	120k	41.2
10k	75k	882m	01.09	85.0k	8.82k	4.26k	176k	41.2
9k1	68k	882m	01.09	77.1k	8.03k	3.87k	160k	41.3
7k5	56k	882m	01.09	63.5k	6.61k	3.19k	132k	41.4
1k1	8k2	882m	01.09	9.30k	970	467	19.3k	41.4
2k7	20k	881m	01.10	22.7k	2.38k	1.14k	47.4k	41.6
1k5	11k	880m	01.11	12.5k	1.32k	627	26.3k	41.9
3k	22k	880m	01.11	25.0k	2.64k	1.25k	52.6k	41.9
2k2	16k	879m	01.12	18.2k	1.93k	923	38.1k	41.3
3k3	24k	879m	01.12	27.3k	2.90k	1.38k	57.2k	41.3
1k8	13k	878m	01.13	14.8k	1.58k	749	31.2k	41.7

Table 8.1 Voltage dividers 147

R_1	R_2	A	dB	R_I	R_O	R_{SMAX}	R_{LMIN}	$R_{LS}(1\,dB)$
5k1	36k	876m	01.15	41.1k	4.47k	2.08k	88.3k	42.5
4k7	33k	875m	01.16	37.7k	4.11k	1.91k	81.3k	42.6
1k3	9k1	875m	01.16	10.4k	1.14k	526	22.5k	42.7
4k3	30k	875m	01.16	34.3k	3.76k	1.74k	74.3k	42.8
5k6	39k	874m	01.17	44.6k	4.90k	2.26k	96.7k	42.9
6k2	43k	874m	01.17	49.2k	5.42k	2.49k	107k	43.0
3k9	27k	874m	01.17	30.9k	3.41k	1.56k	67.3k	43.0
6k8	47k	874m	01.17	53.8k	5.94k	2.72k	117k	43.1
1k6	11k	873m	01.18	12.6k	1.40k	643	27.4k	42.6
1k2	8k2	872m	01.19	9.40k	1.05k	480	20.5k	42.7
8k2	56k	872m	01.19	64.2k	7.15k	3.28k	140k	42.8
1k1	7k5	872m	01.19	8.60k	959	439	18.8k	42.8
2k2	15k	872m	01.19	17.2k	1.92k	878	37.6k	42.8
9k1	62k	872m	01.19	71.1k	7.94k	3.63k	155k	42.8
1k	6k8	872m	01.19	7.80k	872	398	17.1k	42.9
7k5	51k	872m	01.19	58.5k	6.54k	2.99k	128k	42.9
10k	68k	872m	01.19	78.0k	8.72k	3.98k	171k	42.9
2k4	16k	870m	01.21	18.4k	2.09k	939	40.9k	43.5
1k5	10k	870m	01.21	11.5k	1.30k	587	25.6k	43.5
1k8	12k	870m	01.21	13.8k	1.57k	704	30.7k	43.5
2k7	18k	870m	01.21	20.7k	2.35k	1.06k	46.0k	43.5
3k	20k	870m	01.21	23.0k	2.61k	1.17k	51.1k	43.5
3k3	22k	870m	01.21	25.3k	2.87k	1.29k	56.2k	43.5
3k6	24k	870m	01.21	27.6k	3.13k	1.41k	61.3k	43.5
2k	13k	867m	01.24	15.0k	1.73k	764	34.0k	44.5
5k1	33k	866m	01.25	38.1k	4.42k	1.94k	86.6k	44.6
5k6	36k	865m	01.26	41.6k	4.85k	2.14k	94.1k	43.9
4k7	30k	865m	01.26	34.7k	4.06k	1.79k	78.9k	44.2
6k8	43k	863m	01.28	49.8k	5.87k	2.56k	114k	44.5
1k3	8k2	863m	01.28	9.50k	1.12k	489	21.8k	44.6
6k2	39k	863m	01.28	45.2k	5.35k	2.33k	104k	44.6
4k3	27k	863m	01.28	31.3k	3.71k	1.61k	72.0k	44.7
7k5	47k	862m	01.29	54.5k	6.47k	2.81k	126k	44.8
1k2	7k5	862m	01.29	8.70k	1.03k	448	20.1k	44.9
2k4	15k	862m	01.29	17.4k	2.07k	896	40.2k	44.9
1k6	10k	862m	01.29	11.6k	1.38k	597	26.8k	44.9
8k2	51k	861m	01.30	59.2k	7.06k	3.05k	137k	45.0

R_1	R_2	A	dB	R_I	R_O	R_{SMAX}	R_{LMIN}	$R_{LS}(1\,dB)$
1k	6k2	861m	01.30	7.20k	861	371	16.7k	45.1
10k	62k	861m	01.30	72.0k	8.61k	3.71k	167k	45.1
1k1	6k8	861m	01.30	7.90k	947	407	18.4k	45.2
3k9	24k	860m	01.31	27.9k	3.35k	1.44k	65.2k	45.4
9k1	56k	860m	01.31	65.1k	7.83k	3.35k	152k	45.4
1k8	11k	859m	01.32	12.8k	1.55k	658	30.1k	45.7
3k6	22k	859m	01.32	25.6k	3.09k	1.32k	60.2k	45.7
1k5	9k1	858m	01.33	10.6k	1.29k	545	25.0k	46.0
3k3	20k	858m	01.33	23.3k	2.83k	1.20k	55.1k	46.0
2k	12k	857m	01.34	14.0k	1.71k	720	33.3k	46.3
3k	18k	857m	01.34	21.0k	2.57k	1.08k	50.0k	46.3
2k7	16k	856m	01.35	18.7k	2.31k	971	44.5k	45.8
2k2	13k	855m	01.36	15.2k	1.88k	790	36.2k	45.9
5k6	33k	855m	01.36	38.6k	4.79k	2.00k	92.2k	46.0
5k1	30k	855m	01.36	35.1k	4.36k	1.82k	83.9k	46.0
6k2	36k	853m	01.38	42.2k	5.29k	2.19k	102k	46.5
1k3	7k5	852m	01.39	8.80k	1.11k	457	21.3k	46.7
4k7	27k	852m	01.39	31.7k	4.00k	1.64k	77.2k	47.0
6k8	39k	852m	01.40	45.8k	5.79k	2.38k	112k	47.0
7k5	43k	851m	01.40	50.5k	6.39k	2.62k	123k	47.0
8k2	47k	851m	01.40	55.2k	6.98k	2.86k	135k	47.0
1k6	9k1	850m	01.41	10.7k	1.36k	555	26.2k	47.3
1k2	6k8	850m	01.41	8.00k	1.02k	415	19.7k	47.4
3k9	22k	849m	01.42	25.9k	3.31k	1.34k	63.9k	47.6
1k1	6k2	849m	01.42	7.30k	934	379	18.0k	47.6
9k1	51k	849m	01.43	60.1k	7.72k	3.12k	149k	47.8
1k	5k6	848m	01.43	6.60k	848	342	16.4k	47.8
10k	56k	848m	01.43	66.0k	8.48k	3.42k	164k	47.8
4k3	24k	848m	01.43	28.3k	3.65k	1.47k	70.3k	47.9
1k8	10k	847m	01.44	11.8k	1.53k	612	29.4k	48.0
2k7	15k	847m	01.44	17.7k	2.29k	918	44.1k	48.0
3k6	20k	847m	01.44	23.6k	3.05k	1.22k	58.8k	48.0
2k	11k	846m	01.45	13.0k	1.69k	681	32.3k	47.4
1k5	8k2	845m	01.46	9.70k	1.27k	508	24.2k	47.6
2k2	12k	845m	01.46	14.2k	1.86k	744	35.5k	47.7
3k3	18k	845m	01.46	21.3k	2.79k	1.12k	53.2k	47.7
2k4	13k	844m	01.47	15.4k	2.03k	807	38.7k	47.9

Table 8.1 Voltage dividers   149

R_1	R_2	A	dB	R_I	R_O	R_{SMAX}	R_{LMIN}	$R_{LS}(1\,dB)$
5k6	30k	843m	01.49	35.6k	4.72k	1.86k	90.2k	48.4
3k	16k	842m	01.49	19.0k	2.53k	994	48.3k	48.6
6k2	33k	842m	01.50	39.2k	5.22k	2.05k	99.7k	48.6
5k1	27k	841m	01.50	32.1k	4.29k	1.68k	82.0k	48.8
6k8	36k	841m	01.50	42.8k	5.72k	2.24k	109k	48.8
8k2	43k	840m	01.52	51.2k	6.89k	2.68k	132k	49.1
1k3	6k8	840m	01.52	8.10k	1.09k	424	20.9k	49.2
7k5	39k	839m	01.53	46.5k	6.29k	2.43k	120k	49.4
1k2	6k2	838m	01.54	7.40k	1.01k	387	19.2k	49.6
9k1	47k	838m	01.54	56.1k	7.62k	2.94k	146k	49.6
3k9	20k	837m	01.55	23.9k	3.26k	1.25k	62.4k	49.8
1k6	8k2	837m	01.55	9.80k	1.34k	513	25.6k	49.9
4k3	22k	837m	01.55	26.3k	3.60k	1.38k	68.7k	49.9
4k7	24k	836m	01.55	28.7k	3.93k	1.50k	75.1k	50.0
1k	5k1	836m	01.56	6.10k	836	319	16.0k	50.0
10k	51k	836m	01.56	61.0k	8.36k	3.19k	160k	50.0
1k1	5k6	836m	01.56	6.70k	919	351	17.6k	50.1
1k8	9k1	835m	01.57	10.9k	1.50k	571	28.7k	50.2
1k5	7k5	833m	01.58	9.00k	1.25k	475	23.7k	49.8
2k	10k	833m	01.58	12.0k	1.67k	634	31.6k	49.8
2k2	11k	833m	01.58	13.2k	1.83k	697	34.7k	49.8
2k4	12k	833m	01.58	14.4k	2.00k	760	37.9k	49.8
3k	15k	833m	01.58	18.0k	2.50k	950	47.3k	49.8
3k6	18k	833m	01.58	21.6k	3.00k	1.14k	56.8k	49.8
6k8	33k	829m	01.63	39.8k	5.64k	2.10k	107k	50.8
3k3	16k	829m	01.63	19.3k	2.74k	1.02k	51.8k	50.8
6k2	30k	829m	01.63	36.2k	5.14k	1.91k	97.3k	50.9
5k6	27k	828m	01.64	32.6k	4.64k	1.72k	87.8k	51.0
2k7	13k	828m	01.64	15.7k	2.24k	829	42.3k	51.1
7k5	36k	828m	01.64	43.5k	6.21k	2.30k	118k	51.2
1k3	6k2	827m	01.65	7.50k	1.07k	396	20.4k	51.4
8k2	39k	826m	01.66	47.2k	6.78k	2.49k	128k	51.5
9k1	43k	825m	01.67	52.1k	7.51k	2.75k	142k	51.7
5k1	24k	825m	01.67	29.1k	4.21k	1.54k	79.7k	51.8
1k	4k7	825m	01.68	5.70k	825	301	15.6k	51.9
10k	47k	825m	01.68	57.0k	8.25k	3.01k	156k	51.9
1k6	7k5	824m	01.68	9.10k	1.32k	481	25.0k	52.0

R_1	R_2	A	dB	R_I	R_O	R_{SMAX}	R_{LMIN}	$R_{LS}(1\,dB)$
4k7	22k	824m	01.68	26.7k	3.87k	1.41k	73.3k	52.0
1k2	5k6	824m	01.69	6.80k	988	359	18.7k	52.1
4k3	20k	823m	01.69	24.3k	3.54k	1.28k	67.0k	52.2
1kl	5kl	823m	01.70	6.20k	905	328	17.1k	52.3
3k9	18k	822m	01.70	21.9k	3.21k	1.17k	60.2k	51.5
2k4	11k	821m	01.71	13.4k	1.97k	714	37.0k	51.8
1k8	8k2	820m	01.72	10.0k	1.48k	533	27.7k	52.0
2k	9kl	820m	01.73	11.1k	1.64k	591	30.8k	52.0
3k3	15k	820m	01.73	18.3k	2.70k	975	50.8k	52.1
2k2	10k	820m	01.73	12.2k	1.80k	650	33.8k	52.1
1k5	6k8	819m	01.73	8.30k	1.23k	442	23.1k	52.1
2k7	12k	816m	01.76	14.7k	2.20k	783	41.4k	52.8
3k6	16k	816m	01.76	19.6k	2.94k	1.04k	55.1k	52.8
6k8	30k	815m	01.77	36.8k	5.54k	1.96k	104k	53.0
7k5	33k	815m	01.78	40.5k	6.11k	2.16k	115k	53.1
8k2	36k	814m	01.78	44.2k	6.68k	2.36k	125k	53.2
6k2	27k	813m	01.80	33.2k	5.04k	1.77k	94.6k	53.5
3k	13k	813m	01.80	16.0k	2.44k	853	45.7k	53.7
5kl	22k	812m	01.81	27.1k	4.14k	1.44k	77.7k	53.8
1k3	5k6	812m	01.81	6.90k	1.06k	368	19.8k	53.8
1k	4k3	811m	01.82	5.30k	811	282	15.2k	53.9
10k	43k	811m	01.82	53.0k	8.11k	2.82k	152k	53.9
5k6	24k	811m	01.82	29.6k	4.54k	1.58k	85.2k	54.0
9kl	39k	811m	01.82	48.1k	7.38k	2.56k	138k	54.0
1kl	4k7	810m	01.83	5.80k	891	309	16.7k	54.1
4k7	20k	810m	01.83	24.7k	3.81k	1.32k	71.5k	54.4
1k6	6k8	810m	01.84	8.40k	1.30k	447	24.3k	54.4
1k2	5kl	810m	01.84	6.30k	971	335	18.2k	54.4
4k3	18k	807m	01.86	22.3k	3.47k	1.20k	64.5k	53.8
1k8	7k5	806m	01.87	9.30k	1.45k	500	27.0k	54.0
2k4	10k	806m	01.87	12.4k	1.94k	667	36.0k	54.0
3k6	15k	806m	01.87	18.6k	2.90k	1.00k	54.0k	54.0
2k2	9kl	805m	01.88	11.3k	1.77k	608	32.9k	54.2
1k5	6k2	805m	01.88	7.70k	1.21k	414	22.5k	54.2
3k9	16k	804m	01.89	19.9k	3.14k	1.07k	58.3k	54.5
2k	8k2	804m	01.90	10.2k	1.61k	549	29.9k	54.5
2k7	11k	803m	01.91	13.7k	2.17k	737	40.3k	54.7

Table 8.1 Voltage dividers 151

R_1	R_2	A	dB	R_I	R_O	R_{SMAX}	R_{LMIN}	$R_{LS}(1\,dB)$
8k2	33k	801m	01.93	41.2k	6.57k	2.22k	122k	55.1
7k5	30k	800m	01.94	37.5k	6.00k	2.02k	112k	55.3
3k	12k	800m	01.94	15.0k	2.40k	807	44.6k	55.3
6k8	27k	799m	01.95	33.8k	5.43k	1.82k	101k	55.6
9k1	36k	798m	01.96	45.1k	7.26k	2.43k	135k	55.7
3k3	13k	798m	01.96	16.3k	2.63k	877	48.9k	55.8
5k6	22k	797m	01.97	27.6k	4.46k	1.48k	83.0k	55.9
1k3	5k1	797m	01.97	6.40k	1.04k	344	19.3k	56.0
5k1	20k	797m	01.97	25.1k	4.06k	1.35k	75.6k	56.0
1k2	4k7	797m	01.98	5.90k	956	317	17.8k	56.0
1k1	4k3	796m	01.98	5.40k	876	290	16.3k	56.1
1k	3k9	796m	01.98	4.90k	796	264	14.8k	56.2
10k	39k	796m	01.98	49.0k	7.96k	2.64k	148k	56.2
1k6	6k2	795m	01.99	7.80k	1.27k	419	23.7k	56.5
6k2	24k	795m	02.00	30.2k	4.93k	1.62k	91.7k	56.5
3k9	15k	794m	02.01	18.9k	3.10k	1.02k	57.6k	56.7
4k7	18k	793m	02.02	22.7k	3.73k	1.22k	69.4k	56.9
2k4	9k1	791m	02.03	11.5k	1.90k	618	35.3k	57.1
1k8	6k8	791m	02.04	8.60k	1.42k	467	26.2k	56.2
2k	7k5	789m	02.05	9.50k	1.58k	516	29.1k	56.4
1k5	5k6	789m	02.06	7.10k	1.18k	385	21.8k	56.6
2k2	8k2	788m	02.06	10.4k	1.73k	565	32.0k	56.6
4k3	16k	788m	02.07	20.3k	3.39k	1.10k	62.4k	56.7
2k7	10k	787m	02.08	12.7k	2.13k	689	39.2k	56.8
3k	11k	786m	02.09	14.0k	2.36k	760	43.4k	57.1
8k2	30k	785m	02.10	38.2k	6.44k	2.07k	119k	57.2
3k3	12k	784m	02.11	15.3k	2.59k	831	47.7k	57.4
9k1	33k	784m	02.12	42.1k	7.13k	2.29k	131k	57.5
1k3	4k7	783m	02.12	6.00k	1.02k	326	18.8k	57.6
3k6	13k	783m	02.12	16.6k	2.82k	901	51.9k	57.6
1k	3k6	783m	02.13	4.60k	783	250	14.4k	57.7
7k5	27k	783m	02.13	34.5k	5.87k	1.87k	108k	57.7
10k	36k	783m	02.13	46.0k	7.83k	2.50k	144k	57.7
1k2	4k3	782m	02.14	5.50k	938	299	17.3k	57.9
5k6	20k	781m	02.14	25.6k	4.38k	1.39k	80.6k	58.0
6k2	22k	780m	02.16	28.2k	4.84k	1.53k	89.1k	58.2
1k1	3k9	780m	02.16	5.00k	858	271	15.8k	58.2

R_1	R_2	A	dB	R_I	R_O	R_{SMAX}	R_{LMIN}	$R_{LS}(1\,dB)$
5k1	18k	779m	02.17	23.1k	3.97k	1.25k	73.2k	58.4
6k8	24k	779m	02.17	30.8k	5.30k	1.67k	97.6k	58.4
1k6	5k6	778m	02.18	7.20k	1.24k	390	22.9k	58.8
4k3	15k	777m	02.19	19.3k	3.34k	1.05k	61.6k	58.9
1k8	6k2	775m	02.21	8.00k	1.40k	434	25.7k	59.3
2k4	8k2	774m	02.23	10.6k	1.86k	575	34.2k	59.6
2k2	7k5	773m	02.23	9.70k	1.70k	526	31.4k	59.6
4k7	16k	773m	02.24	20.7k	3.63k	1.12k	67.0k	59.7
2k	6k8	773m	02.24	8.80k	1.55k	477	28.5k	59.7
1k5	5k1	773m	02.24	6.60k	1.16k	358	21.4k	59.7
2k7	9k1	771m	02.26	11.8k	2.08k	640	38.4k	59.9
3k9	13k	769m	02.28	16.9k	3.00k	926	54.8k	59.1
3k	10k	769m	02.28	13.0k	2.31k	712	42.1k	59.1
3k3	11k	769m	02.28	14.3k	2.54k	784	46.3k	59.1
3k6	12k	769m	02.28	15.6k	2.77k	855	50.5k	59.1
1k3	4k3	768m	02.29	5.60k	998	307	18.2k	59.4
1k	3k3	767m	02.30	4.30k	767	236	14.0k	59.4
10k	33k	767m	02.30	43.0k	7.67k	2.36k	140k	59.4
9k1	30k	767m	02.30	39.1k	6.98k	2.14k	127k	59.5
8k2	27k	767m	02.30	35.2k	6.29k	1.93k	115k	59.5
1k1	3k6	766m	02.32	4.70k	843	258	15.4k	59.7
1k2	3k9	765m	02.33	5.10k	918	279	16.7k	59.9
6k8	22k	764m	02.34	28.8k	5.19k	1.58k	94.8k	60.1
6k2	20k	763m	02.35	26.2k	4.73k	1.44k	86.4k	60.2
5k6	18k	763m	02.35	23.6k	4.27k	1.29k	78.0k	60.3
7k5	24k	762m	02.36	31.5k	5.71k	1.73k	104k	60.4
4k7	15k	761m	02.37	19.7k	3.58k	1.08k	65.3k	60.5
1k6	5k1	761m	02.37	6.70k	1.22k	367	22.2k	60.5
5k1	16k	758m	02.40	21.1k	3.87k	1.16k	70.6k	61.2
1k5	4k7	758m	02.41	6.20k	1.14k	339	20.8k	61.2
2k4	7k5	758m	02.41	9.90k	1.82k	542	33.2k	61.3
1k8	5k6	757m	02.42	7.40k	1.36k	405	24.9k	61.4
2k	6k2	756m	02.43	8.20k	1.51k	449	27.6k	61.5
2k2	6k8	756m	02.43	9.00k	1.66k	493	30.4k	61.6
3k9	12k	755m	02.44	15.9k	2.94k	870	53.8k	61.8
3k6	11k	753m	02.46	14.6k	2.71k	799	49.5k	62.0
2k7	8k2	752m	02.47	10.9k	2.03k	597	37.1k	62.2

Table 8.1 Voltage dividers 153

R_1	R_2	A	dB	R_I	R_O	R_{SMAX}	R_{LMIN}	R_{LS} (1 dB)
3k	9k1	752m	02.47	12.1k	2.26k	662	41.2k	62.2
3k3	10k	752m	02.48	13.3k	2.48k	728	45.3k	62.2
4k3	13k	751m	02.48	17.3k	3.23k	947	59.0k	62.3
1k	3k	750m	02.50	4.00k	750	219	13.7k	62.6
1k1	3k3	750m	02.50	4.40k	825	241	15.1k	62.6
1k2	3k6	750m	02.50	4.80k	900	263	16.4k	62.6
1k3	3k9	750m	02.50	5.20k	975	285	17.8k	62.6
10k	30k	750m	02.50	40.0k	7.50k	2.19k	137k	62.6
9k1	27k	748m	02.52	36.1k	6.81k	1.98k	124k	62.9
5k1	15k	746m	02.54	20.1k	3.81k	1.10k	69.5k	63.1
6k8	20k	746m	02.54	26.8k	5.07k	1.47k	92.6k	63.1
1k6	4k7	746m	02.54	6.30k	1.19k	345	21.8k	63.1
7k5	22k	746m	02.55	29.5k	5.59k	1.63k	101k	62.0
8k2	24k	745m	02.55	32.2k	6.11k	1.78k	110k	62.0
6k2	18k	744m	02.57	24.2k	4.61k	1.34k	83.4k	62.3
1k5	4k3	741m	02.60	5.80k	1.11k	321	20.1k	62.7
5k6	16k	741m	02.61	21.6k	4.15k	1.19k	75.0k	62.8
2k4	6k8	739m	02.63	9.20k	1.77k	509	32.1k	63.0
1k8	5k1	739m	02.63	6.90k	1.33k	382	24.1k	63.0
3k9	11k	738m	02.64	14.9k	2.88k	824	52.1k	63.2
2k2	6k2	738m	02.64	8.40k	1.62k	465	29.4k	63.2
2k	5k6	737m	02.65	7.60k	1.47k	420	26.6k	63.4
4k3	12k	736m	02.66	16.3k	3.17k	902	57.2k	63.5
2k7	7k5	735m	02.67	10.2k	1.99k	564	35.9k	63.6
3k6	10k	735m	02.67	13.6k	2.65k	752	47.9k	63.6
1k3	3k6	735m	02.68	4.90k	955	271	17.3k	63.8
4k7	13k	734m	02.68	17.7k	3.45k	978	62.5k	63.8
3k3	9k1	734m	02.69	12.4k	2.42k	685	43.8k	63.9
1k2	3k3	733m	02.69	4.50k	880	249	15.9k	64.0
3k	8k2	732m	02.71	11.2k	2.20k	619	39.7k	64.2
1k1	3k	732m	02.71	4.10k	805	227	14.6k	64.3
10k	27k	730m	02.74	37.0k	7.30k	2.05k	132k	64.6
1k	2k7	730m	02.74	3.70k	730	205	13.2k	64.6
1k6	4k3	729m	02.75	5.90k	1.17k	326	21.1k	64.7
8k2	22k	728m	02.75	30.2k	5.97k	1.67k	108k	64.7
5k6	15k	728m	02.76	20.6k	4.08k	1.14k	73.8k	64.8
7k5	20k	727m	02.77	27.5k	5.45k	1.52k	98.7k	64.9

R_1	R_2	A	dB	R_I	R_O	R_{SMAX}	R_{LMIN}	$R_{LS}(1\,dB)$
6k8	18k	726m	02.78	24.8k	4.94k	1.37k	89.3k	65.1
9k1	24k	725m	02.79	33.1k	6.60k	1.83k	119k	65.3
1k8	4k7	723m	02.82	6.50k	1.30k	359	23.5k	65.5
1k5	3k9	722m	02.83	5.40k	1.08k	298	19.6k	65.7
2k4	6k2	721m	02.84	8.60k	1.73k	475	31.3k	65.9
6k2	16k	721m	02.84	22.2k	4.47k	1.23k	80.8k	65.9
3k9	10k	719m	02.86	13.9k	2.81k	768	50.8k	66.1
4k3	11k	719m	02.87	15.3k	3.09k	846	55.9k	66.1
4k7	12k	719m	02.87	16.7k	3.38k	923	61.1k	66.2
2k	5k1	718m	02.87	7.10k	1.44k	392	26.0k	66.2
5k1	13k	718m	02.87	18.1k	3.66k	1.00k	66.3k	66.2
2k2	5k6	718m	02.88	7.80k	1.58k	431	28.6k	66.3
1k3	3k3	717m	02.88	4.60k	933	254	16.9k	66.3
3k6	9k1	717m	02.90	12.7k	2.58k	702	46.7k	66.4
2k7	6k8	716m	02.90	9.50k	1.93k	525	34.9k	66.5
1k2	3k	714m	02.92	4.20k	857	235	15.3k	65.4
3k	7k5	714m	02.92	10.5k	2.14k	586	38.4k	65.4
3k3	8k2	713m	02.94	11.5k	2.35k	642	42.1k	65.6
1k1	2k7	711m	02.97	3.80k	782	212	14.0k	65.9
8k2	20k	709m	02.98	28.2k	5.82k	1.57k	104k	66.1
1k6	3k9	709m	02.99	5.50k	1.13k	307	20.3k	66.1
6k2	15k	708m	03.00	21.2k	4.39k	1.18k	78.6k	66.3
9k1	22k	707m	03.01	31.1k	6.44k	1.74k	115k	66.4
1k5	3k6	706m	03.03	5.10k	1.06k	285	19.0k	66.7
10k	24k	706m	03.03	34.0k	7.06k	1.90k	126k	66.7
1k	2k4	706m	03.03	3.40k	706	190	12.6k	66.7
7k5	18k	706m	03.03	25.5k	5.29k	1.42k	94.9k	66.7
1k8	4k3	705m	03.04	6.10k	1.27k	340	22.7k	66.8
5k1	12k	702m	03.08	17.1k	3.58k	954	64.1k	67.2
6k8	16k	702m	03.08	22.8k	4.77k	1.27k	85.5k	67.2
2k	4k7	701m	03.08	6.70k	1.40k	374	25.1k	67.2
4k7	11k	701m	03.09	15.7k	3.29k	876	59.0k	67.3
2k4	5k6	700m	03.10	8.00k	1.68k	446	30.1k	67.4
3k9	9k1	700m	03.10	13.0k	2.73k	726	48.9k	67.4
4k3	10k	699m	03.11	14.3k	3.01k	798	53.9k	67.5
5k6	13k	699m	03.11	18.6k	3.91k	1.04k	70.1k	67.6
2k2	5k1	699m	03.12	7.30k	1.54k	407	27.5k	67.6

Table 8.1 Voltage dividers 155

R_1	R_2	A	dB	R_I	R_O	R_{SMAX}	R_{LMIN}	$R_{LS}(1\,dB)$
1k3	3k	698m	03.13	4.30k	907	240	16.3k	67.7
2k7	6k2	697m	03.14	8.90k	1.88k	497	33.7k	67.9
3k6	8k2	695m	03.16	11.8k	2.50k	659	44.8k	68.1
3k3	7k5	694m	03.17	10.8k	2.29k	603	41.1k	68.1
3k	6k8	694m	03.17	9.80k	2.08k	547	37.3k	68.2
1k6	3k6	692m	03.19	5.20k	1.11k	290	19.8k	68.4
1k2	2k7	692m	03.19	3.90k	831	218	14.9k	68.4
6k8	15k	688m	03.25	21.8k	4.68k	1.22k	83.8k	68.9
1k	2k2	688m	03.25	3.20k	688	179	12.3k	69.0
1k5	3k3	688m	03.25	4.80k	1.03k	268	18.5k	69.0
10k	22k	688m	03.25	32.0k	6.88k	1.79k	123k	69.0
9k1	20k	687m	03.26	29.1k	6.25k	1.62k	112k	69.0
8k2	18k	687m	03.26	26.2k	5.63k	1.46k	101k	69.0
1k1	2k4	686m	03.28	3.50k	754	195	13.5k	69.2
1k8	3k9	684m	03.30	5.70k	1.23k	318	22.1k	69.4
5k1	11k	683m	03.31	16.1k	3.48k	899	62.4k	69.5
2k	4k3	683m	03.32	6.30k	1.37k	352	24.5k	69.6
5k6	12k	682m	03.33	17.6k	3.82k	982	68.4k	69.7
2k2	4k7	681m	03.34	6.90k	1.50k	385	26.9k	69.7
7k5	16k	681m	03.34	23.5k	5.11k	1.31k	91.5k	69.8
4k7	10k	680m	03.35	14.7k	3.20k	821	57.3k	69.8
2k4	5k1	680m	03.35	7.50k	1.63k	419	29.2k	69.8
4k3	9k1	679m	03.36	13.4k	2.92k	748	52.3k	69.9
3k9	8k2	678m	03.38	12.1k	2.64k	682	46.9k	68.7
6k2	13k	677m	03.39	19.2k	4.20k	1.08k	74.4k	68.8
3k6	7k5	676m	03.41	11.1k	2.43k	626	43.1k	68.9
1k3	2k7	675m	03.41	4.00k	878	226	15.6k	69.0
2k7	5k6	675m	03.42	8.30k	1.82k	468	32.3k	69.0
3k	6k2	674m	03.43	9.20k	2.02k	519	35.9k	69.1
1k6	3k3	673m	03.43	4.90k	1.08k	276	19.1k	69.2
3k3	6k8	673m	03.44	10.1k	2.22k	570	39.4k	69.2
1k	2k	667m	03.52	3.00k	667	169	11.8k	70.0
1k5	3k	667m	03.52	4.50k	1.00k	254	17.7k	70.0
1k8	3k6	667m	03.52	5.40k	1.20k	304	21.3k	70.0
7k5	15k	667m	03.52	22.5k	5.00k	1.27k	88.7k	70.0
1k1	2k2	667m	03.52	3.30k	733	186	13.0k	70.0
1k2	2k4	667m	03.52	3.60k	800	203	14.2k	70.0

R_1	R_2	A	dB	R_I	R_O	R_{SMAX}	R_{LMIN}	$R_{LS}(1\,dB)$
10k	20k	667m	03.52	30.0k	6.67k	1.69k	118k	70.0
9k1	18k	664m	03.55	27.1k	6.04k	1.53k	107k	70.2
5k6	11k	663m	03.57	16.6k	3.71k	935	65.8k	70.4
5k1	10k	662m	03.58	15.1k	3.38k	851	59.9k	70.4
2k4	4k7	662m	03.58	7.10k	1.59k	400	28.2k	70.5
2k2	4k3	662m	03.59	6.50k	1.46k	366	25.8k	70.5
8k2	16k	661m	03.59	24.2k	5.42k	1.36k	96.2k	70.5
2k	3k9	661m	03.60	5.90k	1.32k	332	23.5k	70.6
4k7	9k1	659m	03.62	13.8k	3.10k	778	55.0k	70.7
6k2	12k	659m	03.62	18.2k	4.09k	1.03k	72.5k	70.7
3k9	7k5	658m	03.64	11.4k	2.57k	642	45.5k	70.9
6k8	13k	657m	03.65	19.8k	4.46k	1.12k	79.2k	71.0
4k3	8k2	656m	03.66	12.5k	2.82k	704	50.1k	71.1
2k7	5k1	654m	03.69	7.80k	1.77k	440	31.3k	71.3
3k6	6k8	654m	03.69	10.4k	2.35k	586	41.8k	71.3
3k3	6k2	653m	03.71	9.50k	2.15k	535	38.2k	71.4
1k6	3k	652m	03.71	4.60k	1.04k	259	18.5k	71.4
3k	5k6	651m	03.73	8.60k	1.95k	485	34.7k	71.5
1k3	2k4	649m	03.76	3.70k	843	209	15.0k	71.8
1k8	3k3	647m	03.78	5.10k	1.16k	287	20.7k	71.9
1k2	2k2	647m	03.78	3.40k	776	192	13.8k	71.9
8k2	15k	647m	03.79	23.2k	5.30k	1.31k	94.1k	72.0
1k1	2k	645m	03.81	3.10k	710	175	12.6k	72.1
1k	1k8	643m	03.84	2.80k	643	158	11.4k	72.3
1k5	2k7	643m	03.84	4.20k	964	237	17.1k	72.3
2k	3k6	643m	03.84	5.60k	1.29k	316	22.8k	72.3
10k	18k	643m	03.84	28.0k	6.43k	1.58k	114k	72.3
2k4	4k3	642m	03.85	6.70k	1.54k	378	27.3k	72.4
5k6	10k	641m	03.86	15.6k	3.59k	879	63.7k	72.5
5k1	9k1	641m	03.86	14.2k	3.27k	800	58.0k	72.5
6k2	11k	640m	03.88	17.2k	3.97k	969	70.4k	72.6
2k2	3k9	639m	03.89	6.10k	1.41k	344	25.0k	72.6
6k8	12k	638m	03.90	18.8k	4.34k	1.06k	77.0k	72.7
9k1	16k	637m	03.91	25.1k	5.80k	1.41k	103k	72.8
4k7	8k2	636m	03.94	12.9k	2.99k	727	53.0k	72.9
4k3	7k5	636m	03.94	11.8k	2.73k	665	48.5k	72.9
3k9	6k8	636m	03.94	10.7k	2.48k	603	44.0k	72.9

Table 8.1 Voltage dividers   157

R_1	R_2	A	dB	R_I	R_O	R_{SMAX}	R_{LMIN}	$R_{LS}(1\,dB)$
2k7	4k7	635m	03.94	7.40k	1.71k	417	30.4k	73.0
7k5	13k	634m	03.96	20.5k	4.76k	1.16k	84.4k	73.1
3k6	6k2	633m	03.98	9.80k	2.28k	552	40.4k	73.1
3k	5k1	630m	04.02	8.10k	1.89k	461	33.2k	71.9
3k3	5k6	629m	04.02	8.90k	2.08k	507	36.5k	72.0
1k3	2k2	629m	04.03	3.50k	817	199	14.4k	72.0
1k6	2k7	628m	04.04	4.30k	1.00k	245	17.6k	72.1
1k2	2k	625m	04.08	3.20k	750	182	13.2k	72.3
1k8	3k	625m	04.08	4.80k	1.13k	273	19.8k	72.3
2k	3k3	623m	04.12	5.30k	1.25k	302	21.9k	72.5
9k1	15k	622m	04.12	24.1k	5.66k	1.37k	99.5k	72.6
1k1	1k8	621m	04.14	2.90k	683	165	12.0k	72.7
2k2	3k6	621m	04.14	5.80k	1.37k	330	24.0k	72.7
2k4	3k9	619m	04.17	6.30k	1.49k	359	26.1k	72.8
5k6	9k1	619m	04.17	14.7k	3.47k	837	60.9k	72.8
6k8	11k	618m	04.18	17.8k	4.20k	1.01k	73.8k	72.9
6k2	10k	617m	04.19	16.2k	3.83k	922	67.2k	72.9
5k1	8k2	617m	04.20	13.3k	3.14k	757	55.3k	73.0
1k	1k6	615m	04.22	2.60k	615	148	10.8k	73.1
7k5	12k	615m	04.22	19.5k	4.62k	1.11k	81.1k	73.1
10k	16k	615m	04.22	26.0k	6.15k	1.48k	108k	73.1
1k5	2k4	615m	04.22	3.90k	923	222	16.2k	73.1
4k7	7k5	615m	04.23	12.2k	2.89k	694	50.8k	73.1
2k7	4k3	614m	04.23	7.00k	1.66k	398	29.1k	73.2
3k9	6k2	614m	04.24	10.1k	2.39k	575	42.1k	73.2
8k2	13k	613m	04.25	21.2k	5.03k	1.21k	88.4k	73.2
4k3	6k8	613m	04.26	11.1k	2.63k	632	46.3k	73.3
3k	4k7	610m	04.29	7.70k	1.83k	438	32.2k	73.4
3k6	5k6	609m	04.31	9.20k	2.19k	524	38.5k	73.5
3k3	5k1	607m	04.33	8.40k	2.00k	478	35.2k	73.6
1k3	2k	606m	04.35	3.30k	788	188	13.8k	73.7
1k	1k5	600m	04.44	2.50k	600	142	10.5k	74.1
1k2	1k8	600m	04.44	3.00k	720	171	12.7k	74.1
1k8	2k7	600m	04.44	4.50k	1.08k	256	19.0k	74.1
2k	3k	600m	04.44	5.00k	1.20k	285	21.1k	74.1
2k2	3k3	600m	04.44	5.50k	1.32k	313	23.2k	74.1
2k4	3k6	600m	04.44	6.00k	1.44k	341	25.3k	74.1

R_1	R_2	A	dB	R_I	R_O	R_{SMAX}	R_{LMIN}	$R_{LS}(1\,dB)$
10k	15k	600m	04.44	25.0k	6.00k	1.42k	105k	74.1
1k6	2k4	600m	04.44	4.00k	960	228	16.9k	74.1
5k1	7k5	595m	04.51	12.6k	3.04k	717	53.3k	74.4
6k8	10k	595m	04.51	16.8k	4.05k	956	71.1k	74.4
6k2	9k1	595m	04.51	15.3k	3.69k	871	64.8k	74.4
1k5	2k2	595m	04.52	3.70k	892	211	15.7k	74.4
7k5	11k	595m	04.52	18.5k	4.46k	1.05k	78.4k	74.4
5k6	8k2	594m	04.52	13.8k	3.33k	785	58.5k	74.4
8k2	12k	594m	04.52	20.2k	4.87k	1.15k	85.6k	74.5
1k1	1k6	593m	04.54	2.70k	652	154	11.5k	74.5
4k7	6k8	591m	04.56	11.5k	2.78k	654	48.8k	74.6
2k7	3k9	591m	04.57	6.60k	1.60k	376	28.0k	74.6
4k3	6k2	590m	04.58	10.5k	2.54k	598	44.6k	74.7
3k9	5k6	589m	04.59	9.50k	2.30k	541	40.4k	74.7
3k	4k3	589m	04.60	7.30k	1.77k	415	31.1k	74.7
9k1	13k	588m	04.61	22.1k	5.35k	1.26k	94.1k	74.8
3k3	4k7	588m	04.62	8.00k	1.94k	455	34.1k	74.8
3k6	5k1	586m	04.64	8.70k	2.11k	495	37.1k	74.9
1k3	1k8	581m	04.72	3.10k	755	176	13.3k	75.2
1k6	2k2	579m	04.75	3.80k	926	216	16.3k	75.3
2k4	3k3	579m	04.75	5.70k	1.39k	324	24.4k	75.3
1k1	1k5	577m	04.78	2.60k	635	148	11.2k	75.4
2k2	3k	577m	04.78	5.20k	1.27k	296	22.3k	75.4
2k	2k7	574m	04.81	4.70k	1.15k	267	20.2k	75.5
8k2	11k	573m	04.84	19.2k	4.70k	1.09k	82.5k	75.5
5k6	7k5	573m	04.84	13.1k	3.21k	746	56.3k	75.6
6k8	9k1	572m	04.85	15.9k	3.89k	905	68.4k	75.6
1k2	1k6	571m	04.86	2.80k	686	159	12.0k	75.6
1k5	2k	571m	04.86	3.50k	857	199	15.1k	75.6
1k8	2k4	571m	04.86	4.20k	1.03k	239	18.1k	75.6
2k7	3k6	571m	04.86	6.30k	1.54k	359	27.1k	75.6
5k1	6k8	571m	04.86	11.9k	2.91k	677	51.2k	75.6
7k5	10k	571m	04.86	17.5k	4.29k	996	75.3k	75.6
6k2	8k2	569m	04.89	14.4k	3.53k	820	62.0k	75.7
4k7	6k2	569m	04.90	10.9k	2.67k	621	47.0k	75.7
9k1	12k	569m	04.90	21.1k	5.18k	1.20k	90.9k	75.7
3k9	5k1	567m	04.93	9.00k	2.21k	518	38.4k	74.3

Table 8.1 Voltage dividers 159

R_1	R_2	A	dB	R_I	R_O	R_{SMAX}	R_{LMIN}	$R_{LS}(1\,dB)$
3k6	4k7	566m	04.94	8.30k	2.04k	477	35.5k	74.3
3k3	4k3	566m	04.95	7.60k	1.87k	437	32.5k	74.3
4k3	5k6	566m	04.95	9.90k	2.43k	569	42.3k	74.3
1k	1k3	565m	04.96	2.30k	565	132	9.83k	74.3
3k	3k9	565m	04.96	6.90k	1.70k	397	29.5k	74.3
10k	13k	565m	04.96	23.0k	5.65k	1.32k	98.3k	74.3
1k2	1k5	556m	05.11	2.70k	667	155	11.6k	74.7
1k6	2k	556m	05.11	3.60k	889	207	15.5k	74.7
2k4	3k	556m	05.11	5.40k	1.33k	310	23.2k	74.7
1k3	1k6	552m	05.17	2.90k	717	167	12.5k	74.9
2k2	2k7	551m	05.18	4.90k	1.21k	282	21.1k	74.9
2k7	3k3	550m	05.19	6.00k	1.49k	345	25.8k	74.9
1k8	2k2	550m	05.19	4.00k	990	230	17.2k	74.9
8k2	10k	549m	05.20	18.2k	4.51k	1.05k	78.4k	74.9
5k1	6k2	549m	05.21	11.3k	2.80k	650	48.7k	75.0
5k6	6k8	548m	05.22	12.4k	3.07k	713	53.4k	75.0
7k5	9k1	548m	05.22	16.6k	4.11k	954	71.5k	75.0
6k2	7k5	547m	05.23	13.7k	3.39k	787	59.1k	75.0
9k1	11k	547m	05.24	20.1k	4.98k	1.16k	86.6k	75.0
6k8	8k2	547m	05.25	15.0k	3.72k	862	64.7k	75.0
3k9	4k7	547m	05.25	8.60k	2.13k	494	37.1k	75.0
1k	1k2	545m	05.26	2.20k	545	126	9.49k	75.0
1k5	1k8	545m	05.26	3.30k	818	190	14.2k	75.0
2k	2k4	545m	05.26	4.40k	1.09k	253	19.0k	75.0
3k	3k6	545m	05.26	6.60k	1.64k	379	28.5k	75.0
10k	12k	545m	05.26	22.0k	5.45k	1.26k	94.9k	75.0
3k6	4k3	544m	05.28	7.90k	1.96k	454	34.1k	75.1
4k7	5k6	544m	05.29	10.3k	2.56k	592	44.5k	75.1
4k3	5k1	543m	05.31	9.40k	2.33k	540	40.6k	75.1
1k1	1k3	542m	05.33	2.40k	596	138	10.4k	75.1
3k3	3k9	542m	05.33	7.20k	1.79k	414	31.1k	75.1
1k3	1k5	536m	05.42	2.80k	696	161	12.1k	75.3
1k6	1k8	529m	05.52	3.40k	847	195	14.7k	75.4
2k4	2k7	529m	05.52	5.10k	1.27k	293	22.1k	75.4
1k8	2k	526m	05.58	3.80k	947	218	16.5k	75.5
2k7	3k	526m	05.58	5.70k	1.42k	328	24.7k	75.5
8k2	9k1	526m	05.58	17.3k	4.31k	994	75.0k	75.5

R_1	R_2	A	dB	R_I	R_O	R_{SMAX}	R_{LMIN}	$R_{LS}(1\,dB)$
5k6	6k2	525m	05.59	11.8k	2.94k	678	51.2k	75.5
6k8	7k5	524m	05.61	14.3k	3.57k	822	62.0k	75.5
3k9	4k3	524m	05.61	8.20k	2.05k	471	35.6k	75.5
1k	1k1	524m	05.62	2.10k	524	121	9.11k	75.5
2k	2k2	524m	05.62	4.20k	1.05k	241	18.2k	75.5
3k	3k3	524m	05.62	6.30k	1.57k	362	27.3k	75.5
10k	11k	524m	05.62	21.0k	5.24k	1.21k	91.1k	75.5
9k1	10k	524m	05.62	19.1k	4.76k	1.10k	82.9k	75.5
5k1	5k6	523m	05.62	10.7k	2.67k	615	46.4k	75.5
6k2	6k8	523m	05.63	13.0k	3.24k	747	56.4k	75.5
7k5	8k2	522m	05.64	15.7k	3.92k	902	68.1k	75.5
4k3	4k7	522m	05.64	9.00k	2.25k	517	39.1k	75.5
1k1	1k2	522m	05.65	2.30k	574	132	9.98k	75.5
2k2	2k4	522m	05.65	4.60k	1.15k	264	20.0k	75.5
3k3	3k6	522m	05.65	6.90k	1.72k	397	30.0k	75.5
4k7	5k1	520m	05.67	9.80k	2.45k	563	42.6k	75.5
1k2	1k3	520m	05.68	2.50k	624	144	10.9k	75.5
3k6	3k9	520m	05.68	7.50k	1.87k	431	32.6k	75.5
1k5	1k6	516m	05.74	3.10k	774	178	13.5k	75.6
1k	1k	500m	06.02	2.00k	500	115	8.70k	75.7
1k1	1k1	500m	06.02	2.20k	550	126	9.57k	75.7
1k2	1k2	500m	06.02	2.40k	600	138	10.4k	75.7
1k3	1k3	500m	06.02	2.60k	650	149	11.3k	75.7
1k5	1k5	500m	06.02	3.00k	750	172	13.0k	75.7
1k6	1k6	500m	06.02	3.20k	800	184	13.9k	75.7
1k8	1k8	500m	06.02	3.60k	900	207	15.7k	75.7
2k	2k	500m	06.02	4.00k	1.00k	230	17.4k	75.7
2k2	2k2	500m	06.02	4.40k	1.10k	253	19.1k	75.7
2k4	2k4	500m	06.02	4.80k	1.20k	276	20.9k	75.7
2k7	2k7	500m	06.02	5.40k	1.35k	310	23.5k	75.7
3k	3k	500m	06.02	6.00k	1.50k	345	26.1k	75.7
3k3	3k3	500m	06.02	6.60k	1.65k	379	28.7k	75.7
3k6	3k6	500m	06.02	7.20k	1.80k	414	31.3k	75.7
3k9	3k9	500m	06.02	7.80k	1.95k	448	33.9k	75.7
4k3	4k3	500m	06.02	8.60k	2.15k	494	37.4k	75.7
4k7	4k7	500m	06.02	9.40k	2.35k	540	40.9k	75.7
5k1	5k1	500m	06.02	10.2k	2.55k	586	44.4k	75.7

Table 8.1 Voltage dividers   161

R_1	R_2	A	dB	R_I	R_O	R_{SMAX}	R_{LMIN}	$R_{LS}(1\,dB)$
5k6	5k6	500m	06.02	11.2k	2.80k	644	48.7k	75.7
6k2	6k2	500m	06.02	12.4k	3.10k	713	53.9k	75.7
6k8	6k8	500m	06.02	13.6k	3.40k	782	59.2k	75.7
7k5	7k5	500m	06.02	15.0k	3.75k	862	65.2k	75.7
8k2	8k2	500m	06.02	16.4k	4.10k	943	71.3k	75.7
9k1	9k1	500m	06.02	18.2k	4.55k	1.05k	79.2k	75.7
10k	10k	500m	06.02	20.0k	5.00k	1.15k	87.0k	75.7
1k6	1k5	484m	06.31	3.10k	774	178	13.5k	75.6
1k3	1k2	480m	06.38	2.50k	624	145	10.7k	74.0
3k9	3k6	480m	06.38	7.50k	1.87k	436	32.2k	74.0
5k1	4k7	480m	06.38	9.80k	2.45k	569	42.1k	74.0
1k2	1k1	478m	06.41	2.30k	574	134	9.89k	74.0
2k4	2k2	478m	06.41	4.60k	1.15k	267	19.8k	74.0
3k6	3k3	478m	06.41	6.90k	1.72k	401	29.7k	74.0
4k7	4k3	478m	06.42	9.00k	2.25k	523	38.7k	74.0
8k2	7k5	478m	06.42	15.7k	3.92k	912	67.5k	74.0
6k8	6k2	477m	06.43	13.0k	3.24k	755	55.9k	74.0
5k6	5k1	477m	06.44	10.7k	2.67k	621	46.0k	74.0
10k	9k1	476m	06.44	19.1k	4.76k	1.11k	82.1k	74.0
1k1	1k	476m	06.44	2.10k	524	122	9.02k	74.0
2k2	2k	476m	06.44	4.20k	1.05k	244	18.0k	74.0
11k	10k	476m	06.44	21.0k	5.24k	1.22k	90.2k	74.0
3k3	3k	476m	06.44	6.30k	1.57k	366	27.1k	74.0
4k3	3k9	476m	06.45	8.20k	2.05k	476	35.2k	74.0
7k5	6k8	476m	06.46	14.3k	3.57k	830	61.4k	74.0
6k2	5k6	475m	06.47	11.8k	2.94k	685	50.7k	74.0
9k1	8k2	474m	06.48	17.3k	4.31k	1.00k	74.3k	74.0
2k	1k8	474m	06.49	3.80k	947	221	16.3k	74.0
3k	2k7	474m	06.49	5.70k	1.42k	331	24.5k	74.0
1k8	1k6	471m	06.55	3.40k	847	197	14.6k	73.9
2k7	2k4	471m	06.55	5.10k	1.27k	296	21.9k	73.9
1k5	1k3	464m	06.66	2.80k	696	163	12.0k	73.8
1k3	1k1	458m	06.78	2.40k	596	139	10.3k	73.7
3k9	3k3	458m	06.78	7.20k	1.79k	418	30.8k	73.7
5k1	4k3	457m	06.79	9.40k	2.33k	546	40.2k	73.6
5k6	4k7	456m	06.81	10.3k	2.56k	598	44.0k	73.6
4k3	3k6	456m	06.83	7.90k	1.96k	459	33.8k	73.6

R_1	R_2	A	dB	R_I	R_O	R_{SMAX}	R_{LMIN}	$R_{LS}(1\,dB)$
1k2	1k	455m	06.85	2.20k	545	128	9.39k	73.6
2k4	2k	455m	06.85	4.40k	1.09k	255	18.8k	73.6
12k	10k	455m	06.85	22.0k	5.45k	1.28k	93.9k	73.6
1k8	1k5	455m	06.85	3.30k	818	192	14.1k	73.6
3k6	3k	455m	06.85	6.60k	1.64k	383	28.2k	73.6
4k7	3k9	453m	06.87	8.60k	2.13k	499	36.7k	73.5
8k2	6k8	453m	06.87	15.0k	3.72k	871	64.0k	73.5
11k	9k1	453m	06.88	20.1k	4.98k	1.17k	85.8k	73.5
7k5	6k2	453m	06.89	13.7k	3.39k	795	58.5k	73.5
9k1	7k5	452m	06.90	16.6k	4.11k	964	70.8k	73.5
6k8	5k6	452m	06.90	12.4k	3.07k	720	52.9k	73.5
6k2	5k1	451m	06.91	11.3k	2.80k	656	48.2k	73.5
10k	8k2	451m	06.93	18.2k	4.51k	1.06k	77.6k	73.4
2k2	1k8	450m	06.94	4.00k	990	232	17.1k	73.4
3k3	2k7	450m	06.94	6.00k	1.49k	348	25.6k	73.4
2k7	2k2	449m	06.96	4.90k	1.21k	284	20.9k	73.4
1k6	1k3	448m	06.97	2.90k	717	168	12.4k	73.4
1k5	1k2	444m	07.04	2.70k	667	157	11.5k	73.3
2k	1k6	444m	07.04	3.60k	889	209	15.3k	73.3
3k	2k4	444m	07.04	5.40k	1.33k	314	23.0k	73.3
1k3	1k	435m	07.23	2.30k	565	134	9.74k	72.9
13k	10k	435m	07.23	23.0k	5.65k	1.34k	97.4k	72.9
3k9	3k	435m	07.23	6.90k	1.70k	401	29.2k	72.9
5k6	4k3	434m	07.24	9.90k	2.43k	575	41.9k	72.9
4k3	3k3	434m	07.25	7.60k	1.87k	441	32.2k	72.9
4k7	3k6	434m	07.26	8.30k	2.04k	482	35.1k	72.9
5k1	3k9	433m	07.26	9.00k	2.21k	523	38.1k	72.8
12k	9k1	431m	07.30	21.1k	5.18k	1.23k	89.1k	72.8
6k2	4k7	431m	07.31	10.9k	2.67k	633	46.0k	72.8
8k2	6k2	431m	07.32	14.4k	3.53k	836	60.8k	72.7
2k4	1k8	429m	07.36	4.20k	1.03k	244	17.7k	72.7
10k	7k5	429m	07.36	17.5k	4.29k	1.02k	73.8k	72.7
1k6	1k2	429m	07.36	2.80k	686	163	11.8k	72.7
2k	1k5	429m	07.36	3.50k	857	203	14.8k	72.7
3k6	2k7	429m	07.36	6.30k	1.54k	366	26.6k	72.7
6k8	5k1	429m	07.36	11.9k	2.91k	691	50.2k	72.7
9k1	6k8	428m	07.38	15.9k	3.89k	923	67.0k	72.6

Table 8.1 Voltage dividers 163

R_1	R_2	A	dB	R_I	R_O	R_{SMAX}	R_{LMIN}	$R_{LS}(1\,dB)$
7k5	5k6	427m	07.38	13.1k	3.21k	761	55.2k	72.6
11k	8k2	427m	07.39	19.2k	4.70k	1.11k	80.9k	72.6
2k7	2k	426m	07.42	4.70k	1.15k	273	19.8k	72.5
1k5	1k1	423m	07.47	2.60k	635	151	10.9k	72.4
3k	2k2	423m	07.47	5.20k	1.27k	302	21.9k	72.4
2k2	1k6	421m	07.51	3.80k	926	221	16.0k	72.3
3k3	2k4	421m	07.51	5.70k	1.39k	331	23.9k	72.3
1k8	1k3	419m	07.55	3.10k	755	180	13.0k	72.2
5k1	3k6	414m	07.66	8.70k	2.11k	505	36.3k	72.0
4k7	3k3	413m	07.69	8.00k	1.94k	464	33.4k	71.9
13k	9k1	412m	07.71	22.1k	5.35k	1.28k	92.2k	71.9
4k3	3k	411m	07.72	7.30k	1.77k	424	30.4k	71.8
5k6	3k9	411m	07.73	9.50k	2.30k	552	39.6k	71.8
6k2	4k3	410m	07.75	10.5k	2.54k	610	43.7k	71.7
3k9	2k7	409m	07.76	6.60k	1.60k	383	27.5k	71.7
6k8	4k7	409m	07.77	11.5k	2.78k	668	47.9k	71.7
1k6	1k1	407m	07.80	2.70k	652	157	11.2k	71.6
12k	8k2	406m	07.83	20.2k	4.87k	1.17k	83.9k	71.5
8k2	5k6	406m	07.83	13.8k	3.33k	801	57.3k	71.5
11k	7k5	405m	07.84	18.5k	4.46k	1.07k	76.8k	71.5
2k2	1k5	405m	07.84	3.70k	892	215	15.4k	71.5
9k1	6k2	405m	07.85	15.3k	3.69k	888	63.5k	71.5
10k	6k8	405m	07.86	16.8k	4.05k	975	69.7k	71.5
7k5	5k1	405m	07.86	12.6k	3.04k	732	52.3k	71.5
15k	10k	400m	07.96	25.0k	6.00k	1.45k	103k	71.2
1k5	1k	400m	07.96	2.50k	600	145	10.3k	71.2
1k8	1k2	400m	07.96	3.00k	720	174	12.4k	71.2
2k4	1k6	400m	07.96	4.00k	960	232	16.5k	71.2
2k7	1k8	400m	07.96	4.50k	1.08k	261	18.6k	71.2
3k	2k	400m	07.96	5.00k	1.20k	290	20.7k	71.2
3k3	2k2	400m	07.96	5.50k	1.32k	319	22.7k	71.2
3k6	2k4	400m	07.96	6.00k	1.44k	348	24.8k	71.2
2k	1k3	394m	08.09	3.30k	788	192	13.6k	70.8
5k1	3k3	393m	08.12	8.40k	2.00k	488	34.5k	70.8
5k6	3k6	391m	08.15	9.20k	2.19k	534	37.7k	70.7
4k7	3k	390m	08.19	7.70k	1.83k	447	31.5k	70.6
6k8	4k3	387m	08.24	11.1k	2.63k	644	45.4k	70.4

R_1	R_2	A	dB	R_I	R_O	R_{SMAX}	R_{LMIN}	$R_{LS}(1\,dB)$
13k	8k2	387m	08.25	21.2k	5.03k	1.23k	86.6k	70.4
6k2	3k9	386m	08.27	10.1k	2.39k	586	41.2k	70.3
4k3	2k7	386m	08.27	7.00k	1.66k	406	28.6k	70.3
7k5	4k7	385m	08.29	12.2k	2.89k	708	49.8k	70.3
12k	7k5	385m	08.30	19.5k	4.62k	1.13k	79.5k	70.2
1k6	1k	385m	08.30	2.60k	615	151	10.6k	70.2
16k	10k	385m	08.30	26.0k	6.15k	1.51k	106k	70.2
2k4	1k5	385m	08.30	3.90k	923	226	15.9k	70.2
8k2	5k1	383m	08.33	13.3k	3.14k	772	54.2k	70.1
10k	6k2	383m	08.34	16.2k	3.83k	941	65.9k	70.1
11k	6k8	382m	08.36	17.8k	4.20k	1.03k	72.4k	70.0
3k9	2k4	381m	08.38	6.30k	1.49k	366	25.6k	70.0
9k1	5k6	381m	08.38	14.7k	3.47k	853	59.7k	70.0
1k8	1k1	379m	08.42	2.90k	683	168	11.8k	69.8
3k6	2k2	379m	08.42	5.80k	1.37k	337	23.5k	69.8
15k	9k1	378m	08.46	24.1k	5.66k	1.40k	97.6k	69.7
3k3	2k	377m	08.46	5.30k	1.25k	308	21.4k	69.7
2k	1k2	375m	08.52	3.20k	750	186	12.9k	69.5
3k	1k8	375m	08.52	4.80k	1.13k	279	19.4k	69.5
2k7	1k6	372m	08.59	4.30k	1.00k	250	17.3k	69.3
2k2	1k3	371m	08.60	3.50k	817	203	14.1k	69.3
5k6	3k3	371m	08.62	8.90k	2.08k	517	35.8k	69.2
5k1	3k	370m	08.63	8.10k	1.89k	470	32.5k	69.2
6k2	3k6	367m	08.70	9.80k	2.28k	569	39.2k	68.9
13k	7k5	366m	08.73	20.5k	4.76k	1.19k	81.9k	68.8
4k7	2k7	365m	08.76	7.40k	1.71k	430	29.5k	68.7
6k8	3k9	364m	08.77	10.7k	2.48k	621	42.7k	68.7
7k5	4k3	364m	08.77	11.8k	2.73k	685	47.1k	68.7
8k2	4k7	364m	08.77	12.9k	2.99k	749	51.5k	68.7
16k	9k1	363m	08.81	25.1k	5.80k	1.46k	99.9k	68.6
12k	6k8	362m	08.83	18.8k	4.34k	1.09k	74.8k	68.5
3k9	2k2	361m	08.86	6.10k	1.41k	354	24.2k	68.4
11k	6k2	360m	08.86	17.2k	3.97k	999	68.3k	68.4
9k1	5k1	359m	08.89	14.2k	3.27k	824	56.3k	68.3
10k	5k6	359m	08.90	15.6k	3.59k	906	61.8k	68.3
4k3	2k4	358m	08.92	6.70k	1.54k	389	26.5k	68.2
18k	10k	357m	08.94	28.0k	6.43k	1.63k	111k	68.1

Table 8.1 Voltage dividers 165

R_1	R_2	A	dB	R_I	R_O	R_{SMAX}	R_{LMIN}	$R_{LS}(1\,dB)$
1k8	1k	357m	08.94	2.80k	643	163	11.1k	68.1
2k7	1k5	357m	08.94	4.20k	964	244	16.6k	68.1
3k6	2k	357m	08.94	5.60k	1.29k	325	22.1k	68.1
2k	1k1	355m	09.00	3.10k	710	182	12.1k	66.6
15k	8k2	353m	09.03	23.2k	5.30k	1.36k	90.4k	66.4
3k3	1k8	353m	09.05	5.10k	1.16k	299	19.9k	66.4
2k2	1k2	353m	09.05	3.40k	776	199	13.2k	66.4
2k4	1k3	351m	09.09	3.70k	843	217	14.4k	66.3
5k6	3k	349m	09.15	8.60k	1.95k	504	33.3k	66.0
3k	1k6	348m	09.17	4.60k	1.04k	270	17.8k	65.9
6k2	3k3	347m	09.18	9.50k	2.15k	557	36.7k	65.9
6k8	3k6	346m	09.21	10.4k	2.35k	610	40.1k	65.8
5k1	2k7	346m	09.21	7.80k	1.77k	457	30.1k	65.8
8k2	4k3	344m	09.27	12.5k	2.82k	733	48.1k	65.6
13k	6k8	343m	09.28	19.8k	4.46k	1.16k	76.1k	65.6
7k5	3k9	342m	09.32	11.4k	2.57k	669	43.7k	65.4
12k	6k2	341m	09.35	18.2k	4.09k	1.07k	69.7k	65.3
9k1	4k7	341m	09.36	13.8k	3.10k	809	52.8k	65.3
3k9	2k	339m	09.40	5.90k	1.32k	346	22.5k	65.1
16k	8k2	339m	09.40	24.2k	5.42k	1.42k	92.4k	65.1
4k3	2k2	338m	09.41	6.50k	1.46k	381	24.8k	65.1
4k7	2k4	338m	09.42	7.10k	1.59k	416	27.1k	65.1
10k	5k1	338m	09.43	15.1k	3.38k	886	57.6k	65.0
11k	5k6	337m	09.44	16.6k	3.71k	974	63.3k	65.0
18k	9k1	336m	09.48	27.1k	6.04k	1.59k	103k	64.8
15k	7k5	333m	09.54	22.5k	5.00k	1.32k	85.3k	64.6
2k	1k	333m	09.54	3.00k	667	176	11.4k	64.6
3k	1k5	333m	09.54	4.50k	1.00k	264	17.1k	64.6
3k6	1k8	333m	09.54	5.40k	1.20k	317	20.5k	64.6
20k	10k	333m	09.54	30.0k	6.67k	1.76k	114k	64.6
2k2	1k1	333m	09.54	3.30k	733	194	12.5k	64.6
2k4	1k2	333m	09.54	3.60k	800	211	13.6k	64.6
6k8	3k3	327m	09.72	10.1k	2.22k	592	37.9k	64.0
3k3	1k6	327m	09.72	4.90k	1.08k	287	18.4k	63.9
6k2	3k	326m	09.73	9.20k	2.02k	540	34.5k	63.9
5k6	2k7	325m	09.75	8.30k	1.82k	487	31.1k	63.8
2k7	1k3	325m	09.76	4.00k	878	235	15.0k	63.8

R_1	R_2	A	dB	R_I	R_O	R_{SMAX}	R_{LMIN}	$R_{LS}(1\,dB)$
7k5	3k6	324m	09.78	11.1k	2.43k	651	41.5k	63.7
13k	6k2	323m	09.82	19.2k	4.20k	1.13k	71.6k	63.6
8k2	3k9	322m	09.83	12.1k	2.64k	710	45.1k	63.5
9k1	4k3	321m	09.87	13.4k	2.92k	786	49.8k	63.4
5k1	2k4	320m	09.90	7.50k	1.63k	440	27.8k	63.3
10k	4k7	320m	09.90	14.7k	3.20k	862	54.5k	63.2
16k	7k5	319m	09.92	23.5k	5.11k	1.38k	87.1k	63.2
4k7	2k2	319m	09.93	6.90k	1.50k	405	25.6k	63.1
12k	5k6	318m	09.95	17.6k	3.82k	1.03k	65.1k	63.1
4k3	2k	317m	09.97	6.30k	1.37k	369	23.3k	63.0
11k	5k1	317m	09.99	16.1k	3.48k	944	59.4k	62.9
3k9	1k8	316m	10.01	5.70k	1.23k	334	21.0k	62.8
2k4	1k1	314m	10.05	3.50k	754	205	12.9k	62.7
18k	8k2	313m	10.09	26.2k	5.63k	1.54k	96.1k	62.5
20k	9k1	313m	10.10	29.1k	6.25k	1.71k	107k	62.5
22k	10k	313m	10.10	32.0k	6.88k	1.88k	117k	62.5
2k2	1k	313m	10.10	3.20k	688	188	11.7k	62.5
3k3	1k5	313m	10.10	4.80k	1.03k	282	17.6k	62.5
15k	6k8	312m	10.12	21.8k	4.68k	1.28k	79.8k	62.4
2k7	1k2	308m	10.24	3.90k	831	229	14.2k	61.9
3k6	1k6	308m	10.24	5.20k	1.11k	305	18.9k	61.9
6k8	3k	306m	10.28	9.80k	2.08k	575	35.5k	61.8
7k5	3k3	306m	10.30	10.8k	2.29k	633	39.1k	61.7
8k2	3k6	305m	10.31	11.8k	2.50k	692	42.7k	61.6
6k2	2k7	303m	10.36	8.90k	1.88k	522	32.1k	61.4
3k	1k3	302m	10.39	4.30k	907	252	15.5k	61.3
5k1	2k2	301m	10.42	7.30k	1.54k	428	26.2k	61.2
13k	5k6	301m	10.43	18.6k	3.91k	1.09k	66.7k	61.2
10k	4k3	301m	10.44	14.3k	3.01k	839	51.3k	61.1
9k1	3k9	300m	10.46	13.0k	2.73k	762	46.5k	61.0
5k6	2k4	300m	10.46	8.00k	1.68k	469	28.6k	61.0
11k	4k7	299m	10.48	15.7k	3.29k	921	56.1k	61.0
4k7	2k	299m	10.50	6.70k	1.40k	393	23.9k	60.9
12k	5k1	298m	10.51	17.1k	3.58k	1.00k	61.0k	60.8
16k	6k8	298m	10.51	22.8k	4.77k	1.34k	81.4k	60.8
4k3	1k8	295m	10.60	6.10k	1.27k	358	21.6k	60.5
18k	7k5	294m	10.63	25.5k	5.29k	1.50k	90.3k	60.4

Table 8.1 Voltage dividers 167

R_1	R_2	A	dB	R_I	R_O	R_{SMAX}	R_{LMIN}	R_{LS} (1 dB)
24k	10k	294m	10.63	34.0k	7.06k	1.99k	120k	60.4
2k4	1k	294m	10.63	3.40k	706	199	12.0k	60.4
3k6	1k5	294m	10.63	5.10k	1.06k	299	18.1k	60.4
22k	9k1	293m	10.67	31.1k	6.44k	1.82k	110k	60.2
15k	6k2	292m	10.68	21.2k	4.39k	1.24k	74.8k	60.2
3k9	1k6	291m	10.72	5.50k	1.13k	323	19.3k	60.0
20k	8k2	291m	10.73	28.2k	5.82k	1.65k	99.2k	60.0
2k7	1k1	289m	10.77	3.80k	782	223	13.3k	59.8
8k2	3k3	287m	10.84	11.5k	2.35k	674	40.1k	59.5
3k	1k2	286m	10.88	4.20k	857	246	14.6k	59.3
7k5	3k	286m	10.88	10.5k	2.14k	616	36.5k	59.3
6k8	2k7	284m	10.93	9.50k	1.93k	557	33.0k	59.1
9k1	3k6	283m	10.95	12.7k	2.58k	745	44.0k	59.0
3k3	1k3	283m	10.98	4.60k	933	270	15.9k	58.9
5k6	2k2	282m	10.99	7.80k	1.58k	457	26.9k	58.9
13k	5k1	282m	11.00	18.1k	3.66k	1.06k	62.5k	58.8
5k1	2k	282m	11.00	7.10k	1.44k	416	24.5k	58.8
12k	4k7	281m	11.01	16.7k	3.38k	979	57.6k	58.8
11k	4k3	281m	11.02	15.3k	3.09k	897	52.7k	58.7
10k	3k9	281m	11.04	13.9k	2.81k	815	47.8k	58.7
16k	6k2	279m	11.08	22.2k	4.47k	1.30k	76.2k	58.5
6k2	2k4	279m	11.09	8.60k	1.73k	504	29.5k	58.5
3k9	1k5	278m	11.13	5.40k	1.08k	317	18.5k	58.3
4k7	1k8	277m	11.15	6.50k	1.30k	381	22.2k	58.2
24k	9k1	275m	11.22	33.1k	6.60k	1.94k	113k	58.0
18k	6k8	274m	11.24	24.8k	4.94k	1.45k	84.2k	57.9
20k	7k5	273m	11.29	27.5k	5.45k	1.61k	93.0k	57.7
15k	5k6	272m	11.31	20.6k	4.08k	1.21k	69.5k	57.5
22k	8k2	272m	11.32	30.2k	5.97k	1.77k	102k	57.5
4k3	1k6	271m	11.33	5.90k	1.17k	346	19.9k	57.5
27k	10k	270m	11.36	37.0k	7.30k	2.17k	124k	57.3
2k7	1k	270m	11.36	3.70k	730	217	12.4k	57.3
3k	1k1	268m	11.43	4.10k	805	240	13.7k	57.1
8k2	3k	268m	11.44	11.2k	2.20k	657	37.4k	57.0
3k3	1k2	267m	11.48	4.50k	880	264	15.0k	56.9
9k1	3k3	266m	11.50	12.4k	2.42k	727	41.3k	56.8
13k	4k7	266m	11.52	17.7k	3.45k	1.04k	58.9k	56.7

R_1	R_2	A	dB	R_I	R_O	R_{SMAX}	R_{LMIN}	$R_{LS}(1\,dB)$
3k6	1k3	265m	11.53	4.90k	955	287	16.3k	56.7
7k5	2k7	265m	11.54	10.2k	1.99k	598	33.8k	56.6
10k	3k6	265m	11.54	13.6k	2.65k	798	45.1k	56.6
12k	4k3	264m	11.57	16.3k	3.17k	956	54.0k	56.5
5k6	2k	263m	11.60	7.60k	1.47k	446	25.1k	56.4
6k2	2k2	262m	11.64	8.40k	1.62k	493	27.7k	56.2
11k	3k9	262m	11.64	14.9k	2.88k	874	49.1k	56.2
5k1	1k8	261m	11.67	6.90k	1.33k	405	22.7k	56.1
6k8	2k4	261m	11.67	9.20k	1.77k	540	30.2k	56.1
16k	5k6	259m	11.73	21.6k	4.15k	1.27k	70.7k	55.8
4k3	1k5	259m	11.75	5.80k	1.11k	340	19.0k	55.7
18k	6k2	256m	11.83	24.2k	4.61k	1.42k	78.6k	55.4
24k	8k2	255m	11.88	32.2k	6.11k	1.89k	104k	55.2
22k	7k5	254m	11.90	29.5k	5.59k	1.73k	95.4k	55.1
4k7	1k6	254m	11.90	6.30k	1.19k	369	20.4k	55.1
15k	5k1	254m	11.91	20.1k	3.81k	1.18k	64.9k	55.0
20k	6k8	254m	11.91	26.8k	5.07k	1.57k	86.5k	55.0
27k	9k1	252m	11.97	36.1k	6.81k	2.12k	116k	54.8
30k	10k	250m	12.04	40.0k	7.50k	2.35k	128k	54.5
3k	1k	250m	12.04	4.00k	750	235	12.8k	54.5
3k3	1k1	250m	12.04	4.40k	825	258	14.1k	54.5
3k6	1k2	250m	12.04	4.80k	900	282	15.3k	54.5
3k9	1k3	250m	12.04	5.20k	975	305	16.6k	54.5
13k	4k3	249m	12.09	17.3k	3.23k	1.01k	55.1k	54.3
10k	3k3	248m	12.11	13.3k	2.48k	780	42.3k	54.2
9k1	3k	248m	12.11	12.1k	2.26k	710	38.5k	54.2
8k2	2k7	248m	12.12	10.9k	2.03k	639	34.6k	54.2
11k	3k6	247m	12.16	14.6k	2.71k	856	46.2k	54.0
12k	3k9	245m	12.21	15.9k	2.94k	933	50.2k	53.8
6k8	2k2	244m	12.24	9.00k	1.66k	528	28.3k	53.7
6k2	2k	244m	12.26	8.20k	1.51k	481	25.8k	53.6
5k6	1k8	243m	12.28	7.40k	1.36k	434	23.2k	53.5
7k5	2k4	242m	12.31	9.90k	1.82k	581	31.0k	53.4
4k7	1k5	242m	12.33	6.20k	1.14k	364	19.4k	53.3
16k	5k1	242m	12.33	21.1k	3.87k	1.24k	65.9k	53.3
5k1	1k6	239m	12.44	6.70k	1.22k	393	20.8k	52.8
15k	4k7	239m	12.45	19.7k	3.58k	1.16k	61.0k	52.8

Table 8.1 Voltage dividers 169

R_1	R_2	A	dB	R_I	R_O	R_{SMAX}	R_{LMIN}	$R_{LS}(1\,dB)$
24k	7k5	238m	12.46	31.5k	5.71k	1.85k	97.4k	52.7
18k	5k6	237m	12.49	23.6k	4.27k	1.38k	72.8k	52.6
20k	6k2	237m	12.52	26.2k	4.73k	1.54k	80.7k	52.5
22k	6k8	236m	12.54	28.8k	5.19k	1.69k	88.6k	52.4
3k9	1k2	235m	12.57	5.10k	918	299	15.6k	52.3
3k6	1k1	234m	12.61	4.70k	843	276	14.4k	52.1
27k	8k2	233m	12.65	35.2k	6.29k	2.06k	107k	51.9
30k	9k1	233m	12.66	39.1k	6.98k	2.29k	119k	51.9
3k3	1k	233m	12.67	4.30k	767	252	13.1k	51.9
33k	10k	233m	12.67	43.0k	7.67k	2.52k	131k	51.9
4k3	1k3	232m	12.68	5.60k	998	328	17.0k	51.8
11k	3k3	231m	12.74	14.3k	2.54k	839	43.3k	51.6
12k	3k6	231m	12.74	15.6k	2.77k	915	47.2k	51.6
13k	3k9	231m	12.74	16.9k	3.00k	991	51.2k	51.6
10k	3k	231m	12.74	13.0k	2.31k	762	39.3k	51.6
9k1	2k7	229m	12.81	11.8k	2.08k	692	35.5k	51.3
6k8	2k	227m	12.87	8.80k	1.55k	516	26.4k	51.1
5k1	1k5	227m	12.87	6.60k	1.16k	387	19.8k	51.1
16k	4k7	227m	12.88	20.7k	3.63k	1.21k	61.9k	51.0
7k5	2k2	227m	12.89	9.70k	1.70k	569	29.0k	51.0
8k2	2k4	226m	12.90	10.6k	1.86k	622	31.7k	50.9
6k2	1k8	225m	12.96	8.00k	1.40k	469	23.8k	50.7
15k	4k3	223m	13.04	19.3k	3.34k	1.13k	57.0k	50.3
5k6	1k6	222m	13.06	7.20k	1.24k	422	21.2k	50.2
18k	5k1	221m	13.12	23.1k	3.97k	1.35k	67.8k	50.0
24k	6k8	221m	13.12	30.8k	5.30k	1.81k	90.3k	50.0
3k9	1k1	220m	13.15	5.00k	858	293	14.6k	49.9
22k	6k2	220m	13.16	28.2k	4.84k	1.65k	82.5k	49.9
20k	5k6	219m	13.20	25.6k	4.38k	1.50k	74.6k	49.7
4k3	1k2	218m	13.22	5.50k	938	323	16.0k	49.6
27k	7k5	217m	13.26	34.5k	5.87k	2.02k	100k	49.5
36k	10k	217m	13.26	46.0k	7.83k	2.70k	133k	49.5
3k6	1k	217m	13.26	4.60k	783	270	13.3k	49.5
13k	3k6	217m	13.28	16.6k	2.82k	974	48.1k	49.4
4k7	1k3	217m	13.28	6.00k	1.02k	352	17.4k	49.3
33k	9k1	216m	13.30	42.1k	7.13k	2.47k	122k	49.3
12k	3k3	216m	13.32	15.3k	2.59k	897	44.1k	49.2

R_1	R_2	A	dB	R_I	R_O	R_{SMAX}	R_{LMIN}	$R_{LS}(1\,dB)$
30k	8k2	215m	13.36	38.2k	6.44k	2.24k	110k	49.0
11k	3k	214m	13.38	14.0k	2.36k	821	40.2k	48.9
10k	2k7	213m	13.45	12.7k	2.13k	745	36.2k	48.7
16k	4k3	212m	13.48	20.3k	3.39k	1.19k	57.8k	48.5
8k2	2k2	212m	13.49	10.4k	1.73k	610	29.6k	48.5
5k6	1k5	211m	13.50	7.10k	1.18k	416	20.2k	48.4
7k5	2k	211m	13.53	9.50k	1.58k	557	26.9k	48.3
6k8	1k8	209m	13.58	8.60k	1.42k	504	24.3k	48.1
9k1	2k4	209m	13.61	11.5k	1.90k	674	32.4k	48.0
18k	4k7	207m	13.68	22.7k	3.73k	1.33k	63.5k	47.7
15k	3k9	206m	13.71	18.9k	3.10k	1.11k	52.8k	47.6
24k	6k2	205m	13.75	30.2k	4.93k	1.77k	84.0k	47.4
6k2	1k6	205m	13.76	7.80k	1.27k	457	21.7k	47.4
39k	10k	204m	13.80	49.0k	7.96k	2.87k	136k	47.2
3k9	1k	204m	13.80	4.90k	796	287	13.6k	47.2
4k3	1k1	204m	13.82	5.40k	876	317	14.9k	47.2
4k7	1k2	203m	13.83	5.90k	956	346	16.3k	47.1
20k	5k1	203m	13.84	25.1k	4.06k	1.47k	69.3k	47.1
5k1	1k3	203m	13.84	6.40k	1.04k	375	17.7k	47.1
22k	5k6	203m	13.85	27.6k	4.46k	1.62k	76.1k	47.0
13k	3k3	202m	13.87	16.3k	2.63k	956	44.9k	46.9
36k	9k1	202m	13.90	45.1k	7.26k	2.65k	124k	46.8
27k	6k8	201m	13.93	33.8k	5.43k	1.98k	92.6k	46.7
30k	7k5	200m	13.98	37.5k	6.00k	2.20k	102k	46.5
12k	3k	200m	13.98	15.0k	2.40k	880	40.9k	46.5
33k	8k2	199m	14.02	41.2k	6.57k	2.42k	112k	46.3
11k	2k7	197m	14.11	13.7k	2.17k	804	37.0k	46.0
8k2	2k	196m	14.15	10.2k	1.61k	598	27.4k	45.8
16k	3k9	196m	14.16	19.9k	3.14k	1.17k	53.5k	45.8
6k2	1k5	195m	14.21	7.70k	1.21k	452	20.6k	45.6
9k1	2k2	195m	14.21	11.3k	1.77k	663	30.2k	45.6
15k	3k6	194m	14.26	18.6k	2.90k	1.09k	49.5k	45.4
7k5	1k8	194m	14.26	9.30k	1.45k	545	24.8k	45.4
10k	2k4	194m	14.26	12.4k	1.94k	727	33.0k	45.4
18k	4k3	193m	14.30	22.3k	3.47k	1.31k	59.2k	45.2
6k8	1k6	190m	14.40	8.40k	1.30k	493	22.1k	44.8
5k1	1k2	190m	14.40	6.30k	971	369	16.6k	44.8

Table 8.1 Voltage dividers   171

R_1	R_2	A	dB	R_I	R_O	R_{SMAX}	R_{LMIN}	$R_{LS}(1\text{ dB})$
20k	4k7	190m	14.41	24.7k	3.81k	1.45k	64.9k	44.8
4k7	1k1	190m	14.44	5.80k	891	340	15.2k	44.7
24k	5k6	189m	14.46	29.6k	4.54k	1.74k	77.4k	44.6
39k	9k1	189m	14.46	48.1k	7.38k	2.82k	126k	44.6
4k3	1k	189m	14.49	5.30k	811	311	13.8k	44.5
43k	10k	189m	14.49	53.0k	8.11k	3.11k	138k	44.5
5k6	1k3	188m	14.50	6.90k	1.06k	405	18.0k	44.5
22k	5k1	188m	14.51	27.1k	4.14k	1.59k	70.6k	44.4
13k	3k	188m	14.54	16.0k	2.44k	938	41.6k	44.3
27k	6k2	187m	14.57	33.2k	5.04k	1.95k	86.0k	44.2
36k	8k2	186m	14.63	44.2k	6.68k	2.59k	114k	43.9
33k	7k5	185m	14.65	40.5k	6.11k	2.38k	104k	43.9
30k	6k8	185m	14.67	36.8k	5.54k	2.16k	94.5k	43.8
12k	2k7	184m	14.72	14.7k	2.20k	862	37.6k	43.6
16k	3k6	184m	14.72	19.6k	2.94k	1.15k	50.1k	43.6
6k8	1k5	181m	14.86	8.30k	1.23k	487	21.0k	43.0
15k	3k3	180m	14.88	18.3k	2.70k	1.07k	46.1k	43.0
10k	2k2	180m	14.88	12.2k	1.80k	716	30.7k	43.0
9k1	2k	180m	14.89	11.1k	1.64k	651	28.0k	42.9
8k2	1k8	180m	14.89	10.0k	1.48k	587	25.2k	42.9
11k	2k4	179m	14.94	13.4k	1.97k	786	33.6k	42.7
18k	3k9	178m	14.99	21.9k	3.21k	1.28k	54.7k	42.6
5k1	1k1	177m	15.02	6.20k	905	364	15.4k	42.4
20k	4k3	177m	15.04	24.3k	3.54k	1.43k	60.3k	42.3
5k6	1k2	176m	15.07	6.80k	988	399	16.8k	42.2
22k	4k7	176m	15.09	26.7k	3.87k	1.57k	66.0k	42.2
7k5	1k6	176m	15.10	9.10k	1.32k	534	22.5k	42.1
47k	10k	175m	15.12	57.0k	8.25k	3.34k	141k	42.1
4k7	1k	175m	15.12	5.70k	825	334	14.1k	42.1
24k	5k1	175m	15.13	29.1k	4.21k	1.71k	71.7k	42.0
43k	9k1	175m	15.16	52.1k	7.51k	3.06k	128k	41.9
39k	8k2	174m	15.20	47.2k	6.78k	2.77k	116k	41.7
6k2	1k3	173m	15.22	7.50k	1.07k	440	18.3k	41.7
36k	7k5	172m	15.27	43.5k	6.21k	2.55k	106k	41.5
13k	2k7	172m	15.29	15.7k	2.24k	921	38.1k	41.4
27k	5k6	172m	15.30	32.6k	4.64k	1.91k	79.1k	41.4
30k	6k2	171m	15.33	36.2k	5.14k	2.12k	87.6k	41.3

R_1	R_2	A	dB	R_I	R_O	R_{SMAX}	R_{LMIN}	$R_{LS}(1\,dB)$
16k	3k3	171m	15.34	19.3k	2.74k	1.13k	46.6k	41.2
33k	6k8	171m	15.35	39.8k	5.64k	2.33k	96.1k	41.2
11k	2k2	167m	15.56	13.2k	1.83k	774	31.3k	40.4
12k	2k4	167m	15.56	14.4k	2.00k	845	34.1k	40.4
15k	3k	167m	15.56	18.0k	2.50k	1.06k	42.6k	40.4
18k	3k6	167m	15.56	21.6k	3.00k	1.27k	51.2k	40.4
7k5	1k5	167m	15.56	9.00k	1.25k	528	21.3k	40.4
10k	2k	167m	15.56	12.0k	1.67k	704	28.4k	40.4
9k1	1k8	165m	15.64	10.9k	1.50k	639	25.6k	40.1
5k6	1k1	164m	15.69	6.70k	919	393	15.7k	39.9
51k	10k	164m	15.71	61.0k	8.36k	3.58k	143k	39.8
5k1	1k	164m	15.71	6.10k	836	358	14.3k	39.8
24k	4k7	164m	15.72	28.7k	3.93k	1.68k	67.0k	39.8
22k	4k3	163m	15.73	26.3k	3.60k	1.54k	61.3k	39.8
8k2	1k6	163m	15.74	9.80k	1.34k	575	22.8k	39.7
20k	3k9	163m	15.75	23.9k	3.26k	1.40k	55.6k	39.7
47k	9k1	162m	15.80	56.1k	7.62k	3.29k	130k	39.5
6k2	1k2	162m	15.80	7.40k	1.01k	434	17.1k	39.5
39k	7k5	161m	15.85	46.5k	6.29k	2.73k	107k	39.3
6k8	1k3	160m	15.89	8.10k	1.09k	475	18.6k	39.2
43k	8k2	160m	15.91	51.2k	6.89k	3.00k	117k	39.1
27k	5k1	159m	15.98	32.1k	4.29k	1.88k	73.1k	38.8
36k	6k8	159m	15.98	42.8k	5.72k	2.51k	97.5k	38.8
33k	6k2	158m	16.02	39.2k	5.22k	2.30k	89.0k	38.7
16k	3k	158m	16.03	19.0k	2.53k	1.11k	43.1k	38.7
30k	5k6	157m	16.07	35.6k	4.72k	2.09k	80.5k	38.5
13k	2k4	156m	16.15	15.4k	2.03k	903	34.5k	38.2
18k	3k3	155m	16.20	21.3k	2.79k	1.26k	47.1k	37.3
12k	2k2	155m	16.20	14.2k	1.86k	841	31.4k	37.3
8k2	1k5	155m	16.21	9.70k	1.27k	575	21.4k	37.2
11k	2k	154m	16.26	13.0k	1.69k	770	28.6k	37.1
15k	2k7	153m	16.33	17.7k	2.29k	1.05k	38.6k	36.8
20k	3k6	153m	16.33	23.6k	3.05k	1.40k	51.5k	36.8
10k	1k8	153m	16.33	11.8k	1.53k	699	25.7k	36.8
24k	4k3	152m	16.37	28.3k	3.65k	1.68k	61.5k	36.7
56k	10k	152m	16.39	66.0k	8.48k	3.91k	143k	36.6
5k6	1k	152m	16.39	6.60k	848	391	14.3k	36.6

Table 8.1 Voltage dividers 173

R_1	R_2	A	dB	R_I	R_O	R_{SMAX}	R_{LMIN}	$R_{LS}(1\,dB)$
51k	9k1	151m	16.40	60.1k	7.72k	3.56k	130k	36.6
6k2	1k1	151m	16.44	7.30k	934	433	15.8k	36.5
22k	3k9	151m	16.44	25.9k	3.31k	1.53k	55.9k	36.4
6k8	1k2	150m	16.48	8.00k	1.02k	474	17.2k	36.3
9k1	1k6	150m	16.51	10.7k	1.36k	634	23.0k	36.2
47k	8k2	149m	16.56	55.2k	6.98k	3.27k	118k	36.0
43k	7k5	149m	16.56	50.5k	6.39k	2.99k	108k	36.0
39k	6k8	148m	16.57	45.8k	5.79k	2.71k	97.7k	36.0
27k	4k7	148m	16.58	31.7k	4.00k	1.88k	67.6k	36.0
7k5	1k3	148m	16.61	8.80k	1.11k	521	18.7k	35.9
36k	6k2	147m	16.66	42.2k	5.29k	2.50k	89.3k	35.7
30k	5k1	145m	16.75	35.1k	4.36k	2.08k	73.6k	35.4
33k	5k6	145m	16.77	38.6k	4.79k	2.29k	80.8k	35.3
13k	2k2	145m	16.79	15.2k	1.88k	901	31.8k	35.3
16k	2k7	144m	16.81	18.7k	2.31k	1.11k	39.0k	35.2
12k	2k	143m	16.90	14.0k	1.71k	830	28.9k	34.9
18k	3k	143m	16.90	21.0k	2.57k	1.24k	43.4k	34.9
20k	3k3	142m	16.98	23.3k	2.83k	1.38k	47.8k	34.6
9k1	1k5	142m	16.98	10.6k	1.29k	628	21.7k	34.6
11k	1k8	141m	17.04	12.8k	1.55k	758	26.1k	34.4
22k	3k6	141m	17.04	25.6k	3.09k	1.52k	52.2k	34.4
24k	3k9	140m	17.09	27.9k	3.35k	1.65k	56.6k	34.2
56k	9k1	140m	17.09	65.1k	7.83k	3.86k	132k	34.2
6k8	1k1	139m	17.12	7.90k	947	468	16.0k	34.1
62k	10k	139m	17.15	72.0k	8.61k	4.27k	145k	34.1
6k2	1k	139m	17.15	7.20k	861	427	14.5k	34.1
51k	8k2	139m	17.17	59.2k	7.06k	3.51k	119k	34.0
15k	2k4	138m	17.21	17.4k	2.07k	1.03k	34.9k	33.9
7k5	1k2	138m	17.21	8.70k	1.03k	516	17.5k	33.9
10k	1k6	138m	17.21	11.6k	1.38k	687	23.3k	33.9
47k	7k5	138m	17.23	54.5k	6.47k	3.23k	109k	33.8
27k	4k3	137m	17.24	31.3k	3.71k	1.85k	62.6k	33.8
39k	6k2	137m	17.25	45.2k	5.35k	2.68k	90.3k	33.7
8k2	1k3	137m	17.28	9.50k	1.12k	563	18.9k	33.6
43k	6k8	137m	17.29	49.8k	5.87k	2.95k	99.1k	33.6
30k	4k7	135m	17.36	34.7k	4.06k	2.06k	68.6k	33.4
36k	5k6	135m	17.42	41.6k	4.85k	2.46k	81.8k	33.2

R_1	R_2	A	dB	R_I	R_O	R_{SMAX}	R_{LMIN}	$R_{LS}(1\,dB)$
33k	5k1	134m	17.47	38.1k	4.42k	2.26k	74.5k	33.0
13k	2k	133m	17.50	15.0k	1.73k	889	29.3k	32.9
12k	1k8	130m	17.69	13.8k	1.57k	818	26.4k	32.3
16k	2k4	130m	17.69	18.4k	2.09k	1.09k	35.2k	32.3
18k	2k7	130m	17.69	20.7k	2.35k	1.23k	39.6k	32.3
20k	3k	130m	17.69	23.0k	2.61k	1.36k	44.0k	32.3
22k	3k3	130m	17.69	25.3k	2.87k	1.50k	48.4k	32.3
24k	3k6	130m	17.69	27.6k	3.13k	1.64k	52.8k	32.3
10k	1k5	130m	17.69	11.5k	1.30k	681	22.0k	32.3
51k	7k5	128m	17.84	58.5k	6.54k	3.47k	110k	31.8
68k	10k	128m	17.84	78.0k	8.72k	4.62k	147k	31.8
6k8	1k	128m	17.84	7.80k	872	462	14.7k	31.8
62k	9k1	128m	17.86	71.1k	7.94k	4.21k	134k	31.8
15k	2k2	128m	17.86	17.2k	1.92k	1.02k	32.4k	31.8
7k5	1k1	128m	17.86	8.60k	959	510	16.2k	31.8
56k	8k2	128m	17.87	64.2k	7.15k	3.80k	121k	31.7
8k2	1k2	128m	17.88	9.40k	1.05k	557	17.7k	31.7
11k	1k6	127m	17.93	12.6k	1.40k	747	23.6k	31.6
47k	6k8	126m	17.97	53.8k	5.94k	3.19k	100k	31.4
27k	3k9	126m	17.98	30.9k	3.41k	1.83k	57.5k	31.4
43k	6k2	126m	17.99	49.2k	5.42k	2.92k	91.4k	31.4
39k	5k6	126m	18.02	44.6k	4.90k	2.64k	82.6k	31.3
30k	4k3	125m	18.04	34.3k	3.76k	2.03k	63.5k	31.2
9k1	1k3	125m	18.06	10.4k	1.14k	616	19.2k	31.2
33k	4k7	125m	18.08	37.7k	4.11k	2.23k	69.4k	31.1
36k	5k1	124m	18.13	41.1k	4.47k	2.44k	75.4k	31.0
13k	1k8	122m	18.30	14.8k	1.58k	877	26.7k	30.4
16k	2k2	121m	18.35	18.2k	1.93k	1.08k	32.6k	30.3
24k	3k3	121m	18.35	27.3k	2.90k	1.62k	49.0k	30.3
11k	1k5	120m	18.42	12.5k	1.32k	741	22.3k	30.1
22k	3k	120m	18.42	25.0k	2.64k	1.48k	44.6k	30.1
20k	2k7	119m	18.49	22.7k	2.38k	1.35k	40.1k	29.8
8k2	1k1	118m	18.54	9.30k	970	551	16.4k	29.7
56k	7k5	118m	18.55	63.5k	6.61k	3.76k	112k	29.7
68k	9k1	118m	18.56	77.1k	8.03k	4.57k	135k	29.6
15k	2k	118m	18.59	17.0k	1.76k	1.01k	29.8k	29.6
7k5	1k	118m	18.59	8.50k	882	504	14.9k	29.6

Table 8.1 Voltage dividers 175

R_1	R_2	A	dB	R_I	R_O	R_{SMAX}	R_{LMIN}	$R_{LS}(1\,dB)$
12k	1k6	118m	18.59	13.6k	1.41k	806	23.8k	29.6
18k	2k4	118m	18.59	20.4k	2.12k	1.21k	35.7k	29.6
27k	3k6	118m	18.59	30.6k	3.18k	1.81k	53.6k	29.6
51k	6k8	118m	18.59	57.8k	6.00k	3.42k	101k	29.6
75k	10k	118m	18.59	85.0k	8.82k	5.04k	149k	29.6
62k	8k2	117m	18.65	70.2k	7.24k	4.16k	122k	29.4
47k	6k2	117m	18.67	53.2k	5.48k	3.15k	92.4k	29.3
9k1	1k2	117m	18.67	10.3k	1.06k	610	17.9k	29.3
39k	5k1	116m	18.74	44.1k	4.51k	2.61k	76.1k	29.1
36k	4k7	115m	18.75	40.7k	4.16k	2.41k	70.2k	29.1
33k	4k3	115m	18.76	37.3k	3.80k	2.21k	64.2k	29.0
43k	5k6	115m	18.77	48.6k	4.95k	2.88k	83.6k	29.0
30k	3k9	115m	18.78	33.9k	3.45k	2.01k	58.2k	29.0
10k	1k3	115m	18.78	11.3k	1.15k	670	19.4k	29.0
12k	1k5	111m	19.08	13.5k	1.33k	800	22.5k	28.1
16k	2k	111m	19.08	18.0k	1.78k	1.07k	30.0k	28.1
24k	3k	111m	19.08	27.0k	2.67k	1.60k	45.0k	28.1
13k	1k6	110m	19.20	14.6k	1.42k	865	24.0k	27.8
22k	2k7	109m	19.23	24.7k	2.40k	1.46k	40.6k	27.7
18k	2k2	109m	19.26	20.2k	1.96k	1.20k	33.1k	27.6
27k	3k3	109m	19.26	30.3k	2.94k	1.80k	49.6k	27.6
82k	10k	109m	19.28	92.0k	8.91k	5.45k	150k	27.6
8k2	1k	109m	19.28	9.20k	891	545	15.0k	27.6
51k	6k2	108m	19.30	57.2k	5.53k	3.39k	93.3k	27.5
56k	6k8	108m	19.31	62.8k	6.06k	3.72k	102k	27.5
75k	9k1	108m	19.32	84.1k	8.12k	4.98k	137k	27.5
62k	7k5	108m	19.34	69.5k	6.69k	4.12k	113k	27.4
9k1	1k1	108m	19.34	10.2k	981	604	16.6k	27.4
68k	8k2	108m	19.36	76.2k	7.32k	4.52k	123k	27.4
39k	4k7	108m	19.37	43.7k	4.19k	2.59k	70.8k	27.3
15k	1k8	107m	19.40	16.8k	1.61k	995	27.1k	27.2
20k	2k4	107m	19.40	22.4k	2.14k	1.33k	36.2k	27.2
30k	3k6	107m	19.40	33.6k	3.21k	1.99k	54.2k	27.2
10k	1k2	107m	19.40	11.2k	1.07k	664	18.1k	27.2
36k	4k3	107m	19.44	40.3k	3.84k	2.39k	64.8k	27.1
47k	5k6	106m	19.46	52.6k	5.00k	3.12k	84.4k	27.1
43k	5k1	106m	19.49	48.1k	4.56k	2.85k	76.9k	27.0

R_1	R_2	A	dB	R_I	R_O	R_{SMAX}	R_{LMIN}	$R_{LS}(1\,dB)$
11k	1k3	106m	19.52	12.3k	1.16k	729	19.6k	26.9
33k	3k9	106m	19.52	36.9k	3.49k	2.19k	58.9k	26.9
13k	1k5	103m	19.71	14.5k	1.34k	859	22.7k	26.4
16k	1k8	101m	19.90	17.8k	1.62k	1.05k	27.3k	25.9
24k	2k7	101m	19.90	26.7k	2.43k	1.58k	41.0k	25.9
18k	2k	100m	20.00	20.0k	1.80k	1.19k	30.4k	25.6
27k	3k	100m	20.00	30.0k	2.70k	1.78k	45.6k	25.6
82k	9k1	99.9m	20.01	91.1k	8.19k	5.40k	138k	25.6
56k	6k2	99.7m	20.03	62.2k	5.58k	3.69k	94.2k	25.6
68k	7k5	99.3m	20.06	75.5k	6.75k	4.47k	114k	25.5
39k	4k3	99.3m	20.06	43.3k	3.87k	2.57k	65.4k	25.5
20k	2k2	99.1m	20.08	22.2k	1.98k	1.32k	33.4k	25.4
30k	3k3	99.1m	20.08	33.3k	2.97k	1.97k	50.2k	25.4
10k	1k1	99.1m	20.08	11.1k	991	658	16.7k	25.4
9k1	1k	99.0m	20.09	10.1k	901	598	15.2k	25.4
91k	10k	99.0m	20.09	10.1k	9.01k	5.98k	152k	25.4
51k	5k6	98.9m	20.09	56.6k	5.05k	3.35k	85.2k	25.4
62k	6k8	98.8m	20.10	68.8k	6.13k	4.08k	103k	25.4
75k	8k2	98.6m	20.13	83.2k	7.39k	4.93k	125k	25.3
43k	4k7	98.5m	20.13	47.7k	4.24k	2.83k	71.5k	25.3
11k	1k2	98.4m	20.14	12.2k	1.08k	723	18.3k	25.3
22k	2k4	98.4m	20.14	24.4k	2.16k	1.45k	36.5k	25.3
33k	3k6	98.4m	20.14	36.6k	3.25k	2.17k	54.8k	25.3
47k	5k1	97.9m	20.19	52.1k	4.60k	3.09k	77.6k	25.2
12k	1k3	97.7m	20.20	13.3k	1.17k	788	19.8k	25.1
36k	3k9	97.7m	20.20	39.9k	3.52k	2.36k	59.4k	25.1
15k	1k6	96.4m	20.32	16.6k	1.45k	984	24.4k	24.8
11k	1k1	90.9m	20.83	12.1k	1.00k	717	16.9k	23.5
12k	1k2	90.9m	20.83	13.2k	1.09k	782	18.4k	23.5
13k	1k3	90.9m	20.83	14.3k	1.18k	847	19.9k	23.5
15k	1k5	90.9m	20.83	16.5k	1.36k	978	23.0k	23.5
16k	1k6	90.9m	20.83	17.6k	1.45k	1.04k	24.5k	23.5
18k	1k8	90.9m	20.83	19.8k	1.64k	1.17k	27.6k	23.5
20k	2k	90.9m	20.83	22.0k	1.82k	1.30k	30.7k	23.5
22k	2k2	90.9m	20.83	24.2k	2.00k	1.43k	33.8k	23.5
24k	2k4	90.9m	20.83	26.4k	2.18k	1.56k	36.8k	23.5
27k	2k7	90.9m	20.83	29.7k	2.45k	1.76k	41.4k	23.5

Table 8.1 Voltage dividers 177

R_1	R_2	A	dB	R_I	R_O	R_SMAX	R_LMIN	R_LS (1 dB)
30k	3k	90.9m	20.83	33.0k	2.73k	1.96k	46.0k	23.5
33k	3k3	90.9m	20.83	36.3k	3.00k	2.15k	50.6k	23.5
36k	3k6	90.9m	20.83	39.6k	3.27k	2.35k	55.2k	23.5
39k	3k9	90.9m	20.83	42.9k	3.55k	2.54k	59.8k	23.5
43k	4k3	90.9m	20.83	47.3k	3.91k	2.80k	66.0k	23.5
47k	4k7	90.9m	20.83	51.7k	4.27k	3.06k	72.1k	23.5
51k	5k1	90.9m	20.83	56.1k	4.64k	3.32k	78.2k	23.5
56k	5k6	90.9m	20.83	61.6k	5.09k	3.65k	85.9k	23.5
62k	6k2	90.9m	20.83	68.2k	5.64k	4.04k	95.1k	23.5
68k	6k8	90.9m	20.83	74.8k	6.18k	4.43k	104k	23.5
75k	7k5	90.9m	20.83	82.5k	6.82k	4.89k	115k	23.5
82k	8k2	90.9m	20.83	90.2k	7.45k	5.34k	126k	23.5
91k	9k1	90.9m	20.83	100.0k	8.27k	5.93k	140k	23.5
100k	10k	90.9m	20.83	110.0k	9.09k	6.52k	153k	23.5
10k	1k	90.9m	20.83	11.0k	909	652	15.3k	23.5
16k	1k5	85.7m	21.34	17.5k	1.37k	1.04k	23.1k	22.3
13k	1k2	84.5m	21.46	14.2k	1.10k	841	18.5k	22.0
39k	3k6	84.5m	21.46	42.6k	3.30k	2.52k	55.6k	22.0
51k	4k7	84.4m	21.48	55.7k	4.30k	3.30k	72.6k	22.0
12k	1k1	84.0m	21.52	13.1k	1.01k	776	17.0k	21.9
24k	2k2	84.0m	21.52	26.2k	2.02k	1.55k	34.0k	21.9
36k	3k3	84.0m	21.52	39.3k	3.02k	2.33k	51.0k	21.9
47k	4k3	83.8m	21.53	51.3k	3.94k	3.04k	66.5k	21.9
82k	7k5	83.8m	21.54	89.5k	6.87k	5.30k	116k	21.9
68k	6k2	83.6m	21.56	74.2k	5.68k	4.40k	95.9k	21.8
56k	5k1	83.5m	21.57	61.1k	4.67k	3.62k	78.9k	21.8
100k	9k1	83.4m	21.58	109.0k	8.34k	6.46k	141k	21.8
11k	1k	83.3m	21.58	12.0k	917	711	15.5k	21.8
22k	2k	83.3m	21.58	24.0k	1.83k	1.42k	30.9k	21.8
33k	3k	83.3m	21.58	36.0k	2.75k	2.13k	46.4k	21.8
43k	3k9	83.2m	21.60	46.9k	3.58k	2.78k	60.3k	21.7
75k	6k8	83.1m	21.60	81.8k	6.23k	4.85k	105k	21.7
62k	5k6	82.8m	21.64	67.6k	5.14k	4.01k	86.7k	21.6
91k	8k2	82.7m	21.65	99.2k	7.52k	5.88k	127k	21.6
20k	1k8	82.6m	21.66	21.8k	1.65k	1.29k	27.9k	21.6
30k	2k7	82.6m	21.66	32.7k	2.48k	1.94k	41.8k	21.6
18k	1k6	81.6m	21.76	19.6k	1.47k	1.16k	24.8k	21.4

R_1	R_2	A	dB	R_I	R_O	R_{SMAX}	R_{LMIN}	$R_{LS}(1\,dB)$
27k	2k4	81.6m	21.76	29.4k	2.20k	1.74k	37.2k	21.4
15k	1k3	79.8m	21.96	16.3k	1.20k	966	20.2k	20.9
13k	1k1	78.0m	22.16	14.1k	1.01k	835	17.1k	20.5
39k	3k3	78.0m	22.16	42.3k	3.04k	2.51k	51.3k	20.5
51k	4k3	77.8m	22.19	55.3k	3.97k	3.28k	66.9k	20.4
56k	4k7	77.4m	22.22	60.7k	4.34k	3.60k	73.2k	20.3
43k	3k6	77.3m	22.24	46.6k	3.32k	2.76k	56.1k	20.3
12k	1k	76.9m	22.28	13.0k	923	770	15.6k	20.2
18k	1k5	76.9m	22.28	19.5k	1.38k	1.16k	23.4k	20.2
24k	2k	76.9m	22.28	26.0k	1.85k	1.54k	31.2k	20.2
36k	3k	76.9m	22.28	39.0k	2.77k	2.31k	46.7k	20.2
47k	3k9	76.6m	22.31	50.9k	3.60k	3.02k	60.8k	20.2
82k	6k8	76.6m	22.32	88.8k	6.28k	5.26k	106k	20.1
11k	910R	76.4m	22.34	11.9k	840	706	14.2k	20.1
75k	6k2	76.4m	22.34	81.2k	5.73k	4.81k	96.6k	20.1
91k	7k5	76.1m	22.37	98.5k	6.93k	5.84k	117k	20.0
68k	5k6	76.1m	22.37	73.6k	5.17k	4.36k	87.3k	20.0
62k	5k1	76.0m	22.38	67.1k	4.71k	3.98k	79.5k	20.0
100k	8k2	75.8m	22.41	108.0k	7.58k	6.41k	128k	19.9
22k	1k8	75.6m	22.43	23.8k	1.66k	1.41k	28.1k	19.9
33k	2k7	75.6m	22.43	35.7k	2.50k	2.12k	42.1k	19.9
27k	2k2	75.3m	22.46	29.2k	2.03k	1.73k	34.3k	19.8
16k	1k3	75.1m	22.48	17.3k	1.20k	1.03k	20.3k	19.8
15k	1k2	74.1m	22.61	16.2k	1.11k	960	18.8k	19.5
20k	1k6	74.1m	22.61	21.6k	1.48k	1.28k	25.0k	19.5
30k	2k4	74.1m	22.61	32.4k	2.22k	1.92k	37.5k	19.5
13k	1k	71.4m	22.92	14.0k	929	830	15.7k	18.9
39k	3k	71.4m	22.92	42.0k	2.79k	2.49k	47.0k	18.9
56k	4k3	71.3m	22.94	60.3k	3.99k	3.57k	67.4k	18.9
43k	3k3	71.3m	22.94	46.3k	3.06k	2.74k	51.7k	18.9
47k	3k6	71.1m	22.96	50.6k	3.34k	3.00k	56.4k	18.8
51k	3k9	71.0m	22.97	54.9k	3.62k	3.25k	61.1k	18.8
12k	910R	70.5m	23.04	12.9k	846	765	14.3k	18.7
62k	4k7	70.5m	23.04	66.7k	4.37k	3.95k	73.7k	18.7
82k	6k2	70.3m	23.06	88.2k	5.76k	5.23k	97.3k	18.6
16k	1k2	69.8m	23.13	17.2k	1.12k	1.02k	18.8k	18.5
20k	1k5	69.8m	23.13	21.5k	1.40k	1.27k	23.5k	18.5

Table 8.1 Voltage dividers 179

R_1	R_2	A	dB	R_I	R_O	R_{SMAX}	R_{LMIN}	$R_{LS}(1\,dB)$
24k	1k8	69.8m	23.13	25.8k	1.67k	1.53k	28.3k	18.5
36k	2k7	69.8m	23.13	38.7k	2.51k	2.29k	42.4k	18.5
68k	5k1	69.8m	23.13	73.1k	4.74k	4.33k	80.1k	18.5
100k	7k5	69.8m	23.13	108.0k	6.98k	6.37k	118k	18.5
91k	6k8	69.5m	23.16	97.8k	6.33k	5.80k	107k	18.4
75k	5k6	69.5m	23.16	80.6k	5.21k	4.78k	87.9k	18.4
11k	820R	69.4m	23.18	11.8k	763	700	12.9k	18.4
27k	2k	69.0m	23.23	29.0k	1.86k	1.72k	31.4k	18.3
15k	1k1	68.3m	23.31	16.1k	1.02k	954	17.3k	18.1
30k	2k2	68.3m	23.31	32.2k	2.05k	1.91k	34.6k	18.1
22k	1k6	67.8m	23.38	23.6k	1.49k	1.40k	25.2k	18.0
33k	2k4	67.8m	23.38	35.4k	2.24k	2.10k	37.8k	18.0
18k	1k3	67.4m	23.43	19.3k	1.21k	1.14k	20.5k	17.9
51k	3k6	65.9m	23.62	54.6k	3.36k	3.24k	56.7k	17.5
47k	3k3	65.6m	23.66	50.3k	3.08k	2.98k	52.0k	17.5
13k	910R	65.4m	23.69	13.9k	850	824	14.4k	17.4
43k	3k	65.2m	23.71	46.0k	2.80k	2.73k	47.3k	17.4
56k	3k9	65.1m	23.73	59.9k	3.65k	3.55k	61.5k	17.3
62k	4k3	64.9m	23.76	66.3k	4.02k	3.93k	67.9k	17.3
39k	2k7	64.7m	23.78	41.7k	2.53k	2.47k	42.6k	17.2
68k	4k7	64.6m	23.79	72.7k	4.40k	4.31k	74.2k	17.2
16k	1k1	64.3m	23.83	17.1k	1.03k	1.01k	17.4k	17.1
12k	820R	64.0m	23.88	12.8k	768	760	13.0k	17.1
82k	5k6	63.9m	23.89	87.6k	5.24k	5.19k	88.5k	17.0
11k	750R	63.8m	23.90	11.8k	702	696	11.8k	17.0
22k	1k5	63.8m	23.90	23.5k	1.40k	1.39k	23.7k1	17.0
91k	6k2	63.8m	23.91	97.2k	5.80k	5.76k	98.0k	17.0
75k	5k1	63.7m	23.92	80.1k	4.78k	4.75k	80.6k	17.0
100k	6k8	63.7m	23.92	107.0k	6.37k	6.33k	107k	17.0
15k	1k	62.5m	24.08	16.0k	938	948	15.8k	16.7
18k	1k2	62.5m	24.08	19.2k	1.13k	1.14k	19.0k	16.7
24k	1k6	62.5m	24.08	25.6k	1.50k	1.52k	25.3k	16.7
27k	1k8	62.5m	24.08	28.8k	1.69k	1.71k	28.5k	16.7
30k	2k	62.5m	24.08	32.0k	1.88k	1.90k	31.6k	16.7
33k	2k2	62.5m	24.08	35.2k	2.06k	2.09k	34.8k	16.7
36k	2k4	62.5m	24.08	38.4k	2.25k	2.28k	38.0k	16.7
20k	1k3	61.0m	24.29	21.3k	1.22k	1.26k	20.6k	16.3

R_1	R_2	A	dB	R_I	R_O	R_{SMAX}	R_{LMIN}	$R_{LS}(1\,dB)$
51k	3k3	60.8m	24.33	54.3k	3.10k	3.22k	52.3k	16.3
56k	3k6	60.4m	24.38	59.6k	3.38k	3.53k	57.1k	16.2
47k	3k	60.0m	24.44	50.0k	2.82k	2.96k	47.6k	16.1
68k	4k3	59.5m	24.51	72.3k	4.04k	4.28k	68.3k	15.9
13k	820R	59.3m	24.53	13.8k	771	819	13.0k	15.9
62k	3k9	59.2m	24.56	65.9k	3.67k	3.90k	61.9k	15.9
43k	2k7	59.1m	24.57	45.7k	2.54k	2.71k	42.9k	15.8
75k	4k7	59.0m	24.59	79.7k	4.42k	4.72k	74.6k	15.8
16k	1k	58.8m	24.61	17.0k	941	1.01k	15.9k	15.8
12k	750R	58.8m	24.61	12.8k	706	755	11.9k	15.8
24k	1k5	58.8m	24.61	25.5k	1.41k	1.51k	23.8k	15.8
82k	5k1	58.6m	24.65	87.1k	4.80k	5.16k	81.0k	15.7
100k	6k2	58.4m	24.67	106.0k	5.84k	6.29k	98.5k	15.7
11k	680R	58.2m	24.70	11.7k	640	692	10.8k	15.6
39k	2k4	58.0m	24.74	41.4k	2.26k	2.45k	38.2k	15.6
91k	5k6	58.0m	24.74	96.6k	5.28k	5.72k	89.0k	15.6
18k	1k1	57.6m	24.79	19.1k	1.04k	1.13k	17.5k	15.5
36k	2k2	57.6m	24.79	38.2k	2.07k	2.26k	35.0k	15.5
15k	910R	57.2m	24.85	15.9k	858	943	14.5k	15.4
33k	2k	57.1m	24.86	35.0k	1.89k	2.07k	31.8k	15.3
20k	1k2	56.6m	24.94	21.2k	1.13k	1.26k	19.1k	15.2
30k	1k8	56.6m	24.94	31.8k	1.70k	1.88k	28.7k	15.2
27k	1k6	55.9m	25.04	28.6k	1.51k	1.69k	25.5k	15.0
22k	1k3	55.8m	25.07	23.3k	1.23k	1.38k	20.7k	15.0
56k	3k3	55.6m	25.09	59.3k	3.12k	3.51k	52.6k	15.0
51k	3k	55.6m	25.11	54.0k	2.83k	3.20k	47.8k	14.9
62k	3k6	54.9m	25.21	65.6k	3.40k	3.89k	57.4k	14.8
13k	750R	54.5m	25.26	13.8k	709	815	12.0k	14.7
47k	2k7	54.3m	25.30	49.7k	2.55k	2.94k	43.1k	14.6
68k	3k9	54.2m	25.31	71.9k	3.69k	4.26k	62.2k	14.6
75k	4k3	54.2m	25.32	79.3k	4.07k	4.70k	68.6k	14.6
82k	4k7	54.2m	25.32	86.7k	4.45k	5.14k	75.0k	14.6
16k	910R	53.8m	25.38	16.9k	861	1.00k	14.5k	14.5
12k	680R	53.6m	25.41	12.7k	644	751	10.9k	14.5
39k	2k2	53.4m	25.45	41.2k	2.08k	2.44k	35.1k	14.4
11k	620R	53.4m	25.46	11.6k	587	689	9.91k	14.4
91k	5k1	53.1m	25.50	96.1k	4.83k	5.69k	81.5k	14.3

Table 8.1 Voltage dividers   181

R_1	R_2	A	dB	R_I	R_O	R_{SMAX}	R_{LMIN}	$R_{LS}(1\,dB)$
100k	5k6	53.0m	25.51	106.0k	5.30k	6.26k	89.5k	14.3
43k	2k4	52.9m	25.54	45.4k	2.27k	2.69k	38.4k	14.3
18k	1k	52.6m	25.58	19.0k	947	1.13k	16.0k	14.2
27k	1k5	52.6m	25.58	28.5k	1.42k	1.69k	24.0k	14.2
36k	2k	52.6m	25.58	38.0k	1.89k	2.25k	32.0k	14.2
20k	1k1	52.1m	25.66	21.1k	1.04k	1.25k	17.6k	14.1
15k	820R	51.8m	25.71	15.8k	777	937	13.1k	14.0
22k	1k2	51.7m	25.73	23.2k	1.14k	1.37k	19.2k	14.0
33k	1k8	51.7m	25.73	34.8k	1.71k	2.06k	28.8k	14.0
24k	1k3	51.4m	25.78	25.3k	1.23k	1.50k	20.8k	13.9
56k	3k	50.8m	25.87	59.0k	2.85k	3.50k	48.1k	13.7
30k	1k6	50.6m	25.91	31.6k	1.52k	1.87k	25.6k	13.7
62k	3k3	50.5m	25.93	65.3k	3.13k	3.87k	52.9k	13.7
51k	2k7	50.3m	25.97	53.7k	2.56k	3.18k	43.3k	13.6
68k	3k6	50.3m	25.97	71.6k	3.42k	4.24k	57.7k	13.6
82k	4k3	49.8m	26.05	86.3k	4.09k	5.11k	69.0k	13.5
13k	680R	49.7m	26.07	13.7k	646	811	10.9k	13.5
75k	3k9	49.4m	26.12	78.9k	3.71k	4.68k	62.6k	13.4
12k	620R	49.1m	26.17	12.6k	590	748	9.95k	13.3
91k	4k7	49.1m	26.18	95.7k	4.47k	5.67k	75.4k	13.3
39k	2k	48.8m	26.24	41.0k	1.90k	2.43k	32.1k	13.2
16k	820R	48.8m	26.24	16.8k	780	997	13.2k	13.2
43k	2k2	48.7m	26.25	45.2k	2.09k	2.68k	35.3k	13.2
47k	2k4	48.6m	26.27	49.4k	2.28k	2.93k	38.5k	13.2
100k	5k1	48.5m	26.28	105.0k	4.85k	6.23k	81.9k	13.2
11k	560R	48.4m	26.30	11.6k	533	685	8.99k	13.1
18k	910R	48.1m	26.35	18.9k	866	1.12k	14.6k	13.0
15k	750R	47.6m	26.44	15.8k	714	933	12.1k	12.9
20k	1k	47.6m	26.44	21.0k	952	1.24k	16.1k	12.9
22k	1k1	47.6m	26.44	23.1k	1.05k	1.37k	17.7k	12.9
24k	1k2	47.6m	26.44	25.2k	1.14k	1.49k	19.3k	12.9
30k	1k5	47.6m	26.44	31.5k	1.43k	1.87k	24.1k	12.9
36k	1k8	47.6m	26.44	37.8k	1.71k	2.24k	28.9k	12.9
68k	3k3	46.3m	26.69	71.3k	3.15k	4.22k	53.1k	12.6
33k	1k6	46.2m	26.70	34.6k	1.53k	2.05k	25.8k	12.6
62k	3k	46.2m	26.72	65.0k	2.86k	3.85k	48.3k	12.5
56k	2k7	46.0m	26.75	58.7k	2.58k	3.48k	43.5k	12.5

R_1	R_2	A	dB	R_I	R_O	R_{SMAX}	R_{LMIN}	$R_{LS}(1\,dB)$
27k	1k3	45.9m	26.76	28.3k	1.24k	1.68k	20.9k	12.5
75k	3k6	45.8m	26.78	78.6k	3.44k	4.66k	58.0k	12.4
13k	620R	45.5m	26.84	13.6k	592	807	9.99k	12.4
82k	3k9	45.4m	26.86	85.9k	3.72k	5.09k	62.8k	12.3
91k	4k3	45.1m	26.91	95.3k	4.11k	5.65k	69.3k	12.3
51k	2k4	44.9m	26.95	53.4k	2.29k	3.16k	38.7k	12.2
100k	4k7	44.9m	26.96	105.0k	4.49k	6.20k	75.8k	12.2
16k	750R	44.8m	26.98	16.8k	716	992	12.1k	12.2
47k	2k2	44.7m	26.99	49.2k	2.10k	2.92k	35.5k	12.2
12k	560R	44.6m	27.02	12.6k	535	744	9.03k	12.1
43k	2k	44.4m	27.04	45.0k	1.91k	2.67k	32.3k	12.1
11k	510R	44.3m	27.07	11.5k	487	682	8.23k	12.1
39k	1k8	44.1m	27.11	40.8k	1.72k	2.42k	29.0k	12.0
24k	1k1	43.8m	27.17	25.1k	1.05k	1.49k	17.8k	11.9
18k	820R	43.6m	27.22	18.8k	784	1.12k	13.2k	11.9
20k	910R	43.5m	27.23	20.9k	870	1.24k	14.7k	11.9
22k	1k	43.5m	27.23	23.0k	957	1.36k	16.1k	11.8
33k	1k5	43.5m	27.23	34.5k	1.43k	2.04k	24.2k	11.8
15k	680R	43.4m	27.26	15.7k	651	929	11.0k	11.8
36k	1k6	42.6m	27.42	37.6k	1.53k	2.23k	25.9k	11.6
27k	1k2	42.6m	27.42	28.2k	1.15k	1.67k	19.4k	11.6
68k	3k	42.3m	27.48	71.0k	2.87k	4.21k	48.5k	11.5
75k	3k3	42.1m	27.50	78.3k	3.16k	4.64k	53.3k	11.5
82k	3k6	42.1m	27.52	85.6k	3.45k	5.07k	58.2k	11.5
62k	2k7	41.7m	27.59	64.7k	2.59k	3.83k	43.7k	11.4
30k	1k3	41.5m	27.63	31.3k	1.25k	1.85k	21.0k	11.3
51k	2k2	41.4m	27.67	53.2k	2.11k	3.15k	35.6k	11.3
13k	560R	41.3m	27.68	13.6k	537	803	9.06k	11.3
100k	4k3	41.2m	27.70	104.0k	4.12k	6.18k	69.6k	11.3
56k	2k4	41.1m	27.72	58.4k	2.30k	3.46k	38.8k	11.2
91k	3k9	41.1m	27.72	94.9k	3.74k	5.62k	63.1k	11.2
11k	470R	41.0m	27.75	11.5k	451	680	7.61k	11.2
47k	2k	40.8m	27.78	49.0k	1.92k	2.90k	32.4k	11.2
12k	510R	40.8m	27.79	12.5k	489	741	8.26k	11.1
16k	680R	40.8m	27.79	16.7k	652	988	11.0k	11.1
43k	1k8	40.2m	27.92	44.8k	1.73k	2.65k	29.2k	11.0
18k	750R	40.0m	27.96	18.8k	720	1.11k	12.2k	10.9

Table 8.1 Voltage dividers 183

R_1	R_2	A	dB	R_I	R_O	R_{SMAX}	R_{LMIN}	$R_{LS}(1\,\text{dB})$
24k	1k	40.0m	27.96	25.0k	960	1.48k	16.2k	10.9
36k	1k5	40.0m	27.96	37.5k	1.44k	2.22k	24.3k	10.9
22k	910R	39.7m	28.02	22.9k	874	1.36k	14.7k	10.9
15k	620R	39.7m	28.03	15.6k	595	926	10.0k	10.9
39k	1k6	39.4m	28.09	40.6k	1.54k	2.41k	25.9k	10.8
20k	820R	39.4m	28.09	20.8k	788	1.23k	13.3k	10.8
27k	1k1	39.1m	28.15	28.1k	1.06k	1.67k	17.8k	10.7
82k	3k3	38.7m	28.25	85.3k	3.17k	5.05k	53.5k	10.6
30k	1k2	38.5m	28.30	31.2k	1.15k	1.85k	19.5k	10.5
75k	3k	38.5m	28.30	78.0k	2.88k	4.62k	48.7k	10.5
68k	2k7	38.2m	28.36	70.7k	2.60k	4.19k	43.8k	10.5
91k	3k6	38.1m	28.39	94.6k	3.46k	5.61k	58.4k	10.4
33k	1k3	37.9m	28.43	34.3k	1.25k	2.03k	21.1k	10.4
56k	2k2	37.8m	28.45	58.2k	2.12k	3.45k	35.7k	10.4
13k	510R	37.7m	28.46	13.5k	491	801	8.28k	10.3
51k	2k	37.7m	28.46	53.0k	1.92k	3.14k	32.5k	10.3
12k	470R	37.7m	28.48	12.5k	452	739	7.63k	10.3
11k	430R	37.6m	28.49	11.4k	414	677	6.98k	10.3
100k	3k9	37.5m	28.51	104.0k	3.75k	6.16k	63.3k	10.3
16k	620R	37.3m	28.56	16.6k	597	985	10.1k	10.2
62k	2k4	37.3m	28.57	64.4k	2.31k	3.82k	39.0k	10.2
39k	1k5	37.0m	28.63	40.5k	1.44k	2.40k	24.4k	10.2
47k	1k8	36.9m	28.66	48.8k	1.73k	2.89k	29.3k	10.1
24k	910R	36.5m	28.75	24.9k	877	1.48k	14.8k	10.0
18k	680R	36.4m	28.78	18.7k	655	1.11k	11.1k	9.99
20k	750R	36.1m	28.84	20.8k	723	1.23k	12.2k	9.92
15k	560R	36.0m	28.88	15.6k	540	922	9.11k	9.88
22k	820R	35.9m	28.89	22.8k	791	1.35k	13.3k	9.87
43k	1k6	35.9m	28.90	44.6k	1.54k	2.64k	26.0k	9.85
27k	1k	35.7m	28.94	28.0k	964	1.66k	16.3k	9.81
30k	1k1	35.4m	29.03	31.1k	1.06k	1.84k	17.9k	9.72
82k	3k	35.3m	29.05	85.0k	2.89k	5.04k	48.8k	9.70
33k	1k2	35.1m	29.10	34.2k	1.16k	2.03k	19.5k	9.64
91k	3k3	35.0m	29.12	94.3k	3.18k	5.59k	53.7k	9.62
13k	470R	34.9m	29.15	13.5k	454	798	7.66k	9.59
36k	1k3	34.9m	29.16	37.3k	1.25k	2.21k	21.2k	9.58
75k	2k7	34.7m	29.18	77.7k	2.61k	4.60k	44.0k	9.55

R_1	R_2	A	dB	R_I	R_O	R_{SMAX}	R_{LMIN}	$R_{LS}(1\,dB)$
100k	3k6	34.7m	29.18	104.0k	3.47k	6.14k	58.6k	9.55
12k	430R	34.6m	29.22	12.4k	415	737	7.01k	9.51
56k	2k	34.5m	29.25	58.0k	1.93k	3.44k	32.6k	9.48
62k	2k2	34.3m	29.30	64.2k	2.12k	3.80k	35.9k	9.43
11k	390R	34.2m	29.31	11.4k	377	675	6.36k	9.42
51k	1k8	34.1m	29.35	52.8k	1.74k	3.13k	29.3k	9.38
68k	2k4	34.1m	29.35	70.4k	2.32k	4.17k	39.1k	9.38
16k	560R	33.8m	29.42	16.6k	541	981	9.13k	9.31
43k	1k5	33.7m	29.45	44.5k	1.45k	2.64k	24.5k	9.28
18k	620R	33.3m	29.55	18.6k	599	1.10k	10.1k	9.17
24k	820R	33.0m	29.62	24.8k	793	1.47k	13.4k	9.10
22k	750R	33.0m	29.64	22.8k	725	1.35k	12.2k	9.08
47k	1k6	32.9m	29.65	48.6k	1.55k	2.88k	26.1k	9.07
15k	510R	32.9m	29.66	15.5k	493	919	8.32k	9.06
20k	680R	32.9m	29.66	20.7k	658	1.23k	11.1k	9.06
27k	910R	32.6m	29.73	27.9k	880	1.65k	14.9k	8.98
30k	1k	32.3m	29.83	31.0k	968	1.84k	16.3k	8.89
33k	1k1	32.3m	29.83	34.1k	1.06k	2.02k	18.0k	8.89
36k	1k2	32.3m	29.83	37.2k	1.16k	2.20k	19.6k	8.89
39k	1k3	32.3m	29.83	40.3k	1.26k	2.39k	21.2k	8.89
13k	430R	32.0m	29.89	13.4k	416	796	7.02k	8.83
100k	3k3	31.9m	29.91	103.0k	3.19k	6.12k	53.9k	8.81
91k	3k	31.9m	29.92	94.0k	2.90k	5.57k	49.0k	8.80
82k	2k7	31.9m	29.93	84.7k	2.61k	5.02k	44.1k	8.79
11k	360R	31.7m	29.98	11.4k	349	673	5.88k	8.74
12k	390R	31.5m	30.04	12.4k	378	734	6.37k	8.68
68k	2k2	31.3m	30.08	70.2k	2.13k	4.16k	36.0k	8.65
62k	2k	31.3m	30.10	64.0k	1.94k	3.79k	32.7k	8.62
56k	1k8	31.1m	30.13	57.8k	1.74k	3.42k	29.4k	8.59
75k	2k4	31.0m	30.17	77.4k	2.33k	4.59k	39.2k	8.56
47k	1k5	30.9m	30.19	48.5k	1.45k	2.87k	24.5k	8.54
16k	510R	30.9m	30.20	16.5k	494	978	8.34k	8.53
51k	1k6	30.4m	30.34	52.6k	1.55k	3.12k	26.2k	8.40
15k	470R	30.4m	30.35	15.5k	456	917	7.69k	8.39
24k	750R	30.3m	30.37	24.8k	727	1.47k	12.3k	8.37
18k	560R	30.2m	30.41	18.6k	543	1.10k	9.17k	8.33
20k	620R	30.1m	30.44	20.6k	601	1.22k	10.1k	8.31

· Table 8.1 Voltage dividers 185

R_1	R_2	A	dB	R_I	R_O	R_{SMAX}	R_{LMIN}	$R_{LS}(1\,dB)$
22k	680R	30.0m	30.46	22.7k	660	1.34k	11.1k	8.28
39k	1k2	29.9m	30.50	40.2k	1.16k	2.38k	19.6k	8.25
36k	1k1	29.6m	30.56	37.1k	1.07k	2.20k	18.0k	8.19
27k	820R	29.5m	30.61	27.8k	796	1.65k	13.4k	8.15
30k	910R	29.4m	30.62	30.9k	883	1.83k	14.9k	8.14
33k	1k	29.4m	30.63	34.0k	971	2.01k	16.4k	8.13
43k	1k3	29.3m	30.65	44.3k	1.26k	2.62k	21.3k	8.11
11k	330R	29.1m	30.71	11.3k	320	671	5.41k	8.05
12k	360R	29.1m	30.71	12.4k	350	732	5.90k	8.05
13k	390R	29.1m	30.71	13.4k	379	793	6.39k	8.05
100k	3k	29.1m	30.71	103.0k	2.91k	6.10k	49.2k	8.05
91k	2k7	28.8m	30.81	93.7k	2.62k	5.55k	44.3k	7.97
51k	1k5	28.6m	30.88	52.5k	1.46k	3.11k	24.6k	7.91
68k	2k	28.6m	30.88	70.0k	1.94k	4.15k	32.8k	7.91
16k	470R	28.5m	30.89	16.5k	457	976	7.71k	7.90
75k	2k2	28.5m	30.90	77.2k	2.14k	4.57k	36.1k	7.89
82k	2k4	28.4m	30.92	84.4k	2.33k	5.00k	39.4k	7.87
62k	1k8	28.2m	30.99	63.8k	1.75k	3.78k	29.5k	7.81
15k	430R	27.9m	31.10	15.4k	418	914	7.05k	7.72
56k	1k6	27.8m	31.13	57.6k	1.56k	3.41k	26.3k	7.69
18k	510R	27.6m	31.20	18.5k	496	1.10k	8.37k	7.63
24k	680R	27.6m	31.20	24.7k	661	1.46k	11.2k	7.63
39k	1k1	27.4m	31.24	40.1k	1.07k	2.38k	18.1k	7.60
22k	620R	27.4m	31.24	22.6k	603	1.34k	10.2k	7.59
20k	560R	27.2m	31.30	20.6k	545	1.22k	9.19k	7.55
43k	1k2	27.1m	31.32	44.2k	1.17k	2.62k	19.7k	7.52
27k	750R	27.0m	31.36	27.8k	730	1.64k	12.3k	7.49
36k	1k	27.0m	31.36	37.0k	973	2.19k	16.4k	7.49
13k	360R	26.9m	31.39	13.4k	350	792	5.91k	7.47
47k	1k3	26.9m	31.40	48.3k	1.27k	2.86k	21.3k	7.46
33k	910R	26.8m	31.43	33.9k	886	2.01k	14.9k	7.44
12k	330R	26.8m	31.45	12.3k	321	731	5.42k	7.42
30k	820R	26.6m	31.50	30.8k	798	1.83k	13.5k	7.38
11k	300R	26.5m	31.52	11.3k	292	670	4.93k	7.36
100k	2k7	26.3m	31.60	103.0k	2.63k	6.09k	44.4k	7.29
16k	430R	26.2m	31.64	16.4k	419	974	7.07k	7.26
82k	2k2	26.1m	31.66	84.2k	2.14k	4.99k	36.2k	7.25

R_1	R_2	A	dB	R_I	R_O	R_{SMAX}	R_{LMIN}	$R_{LS}(1\,dB)$
56k	1k5	26.1m	31.67	57.5k	1.46k	3.41k	24.7k	7.24
75k	2k	26.0m	31.71	77.0k	1.95k	4.56k	32.9k	7.21
68k	1k8	25.8m	31.77	69.8k	1.75k	4.14k	29.6k	7.16
91k	2k4	25.7m	31.80	93.4k	2.34k	5.53k	39.5k	7.13
18k	470R	25.4m	31.89	18.5k	458	1.09k	7.73k	7.06
15k	390R	25.3m	31.92	15.4k	380	912	6.42k	7.03
24k	620R	25.2m	31.98	24.6k	604	1.46k	10.2k	6.99
62k	1k6	25.2m	31.99	63.6k	1.56k	3.77k	26.3k	6.98
39k	1k	25.0m	32.04	40.0k	975	2.37k	16.5k	6.94
43k	1k1	24.9m	32.06	44.1k	1.07k	2.61k	18.1k	6.93
47k	1k2	24.9m	32.08	48.2k	1.17k	2.86k	19.7k	6.91
20k	510R	24.9m	32.09	20.5k	497	1.22k	8.39k	6.91
51k	1k3	24.9m	32.09	52.3k	1.27k	3.10k	21.4k	6.90
22k	560R	24.8m	32.10	22.6k	546	1.34k	9.22k	6.89
13k	330R	24.8m	32.13	13.3k	322	790	5.43k	6.88
36k	910R	24.7m	32.16	36.9k	888	2.19k	15.0k	6.85
27k	680R	24.6m	32.19	27.7k	663	1.64k	11.2k	6.83
12k	300R	24.4m	32.26	12.3k	293	729	4.94k	6.78
30k	750R	24.4m	32.26	30.8k	732	1.82k	12.3k	6.78
33k	820R	24.2m	32.31	33.8k	800	2.00k	13.5k	6.74
11k	270R	24.0m	32.41	11.3k	264	668	4.45k	6.66
82k	2k	23.8m	32.46	84.0k	1.95k	4.98k	32.9k	6.62
16k	390R	23.8m	32.47	16.4k	381	971	6.43k	6.62
62k	1k5	23.6m	32.53	63.5k	1.46k	3.76k	24.7k	6.57
91k	2k2	23.6m	32.54	93.2k	2.15k	5.52k	36.3k	6.56
15k	360R	23.4m	32.60	15.4k	352	910	5.93k	6.52
75k	1k8	23.4m	32.60	76.8k	1.76k	4.55k	29.7k	6.52
100k	2k4	23.4m	32.60	102.0k	2.34k	6.07k	39.6k	6.52
18k	430R	23.3m	32.64	18.4k	420	1.09k	7.09k	6.49
51k	1k2	23.0m	32.77	52.2k	1.17k	3.09k	19.8k	6.40
68k	1k6	23.0m	32.77	69.6k	1.56k	4.12k	26.4k	6.40
20k	470R	23.0m	32.78	20.5k	459	1.21k	7.75k	6.39
47k	1k1	22.9m	32.82	48.1k	1.07k	2.85k	18.1k	6.36
39k	910R	22.8m	32.84	39.9k	889	2.36k	15.0k	6.35
24k	560R	22.8m	32.84	24.6k	547	1.46k	9.24k	6.35
43k	1k	22.7m	32.87	44.0k	977	2.61k	16.5k	6.33
56k	1k3	22.7m	32.88	57.3k	1.27k	3.40k	21.4k	6.32

Table 8.1 Voltage dividers 187

R_1	R_2	A	dB	R_I	R_O	R_{SMAX}	R_{LMIN}	$R_{\text{LS}}(1\,\text{dB})$
22k	510R	22.7m	32.90	22.5k	498	1.33k	8.41k	6.31
13k	300R	22.6m	32.93	13.3k	293	788	4.95k	6.28
27k	620R	22.4m	32.98	27.6k	606	1.64k	10.2k	6.25
36k	820R	22.3m	33.05	36.8k	802	2.18k	13.5k	6.20
33k	750R	22.2m	33.06	33.8k	733	2.00k	12.4k	6.19
30k	680R	22.2m	33.09	30.7k	665	1.82k	11.2k	6.17
12k	270R	22.0m	33.15	12.3k	264	727	4.46k	6.13
16k	360R	22.0m	33.15	16.4k	352	969	5.94k	6.13
68k	1k5	21.6m	33.32	69.5k	1.47k	4.12k	24.8k	6.01
15k	330R	21.5m	33.34	15.3k	323	908	5.45k	6.00
100k	2k2	21.5m	33.34	102.0k	2.15k	6.06k	36.3k	6.00
91k	2k	21.5m	33.35	93.0k	1.96k	5.51k	33.0k	5.99
82k	1k8	21.5m	33.36	83.8k	1.76k	4.97k	29.7k	5.99
11k	240R	21.4m	33.41	11.2k	235	666	3.96k	5.95
18k	390R	21.2m	33.47	18.4k	382	1.09k	6.44k	5.91
51k	1k1	21.1m	33.51	52.1k	1.08k	3.09k	18.2k	5.89
20k	430R	21.0m	33.54	20.4k	421	1.21k	7.10k	5.87
56k	1k2	21.0m	33.56	57.2k	1.17k	3.39k	19.8k	5.85
22k	470R	20.9m	33.59	22.5k	460	1.33k	7.77k	5.83
75k	1k6	20.9m	33.60	76.6k	1.57k	4.54k	26.4k	5.82
47k	1k	20.8m	33.62	48.0k	979	2.84k	16.5k	5.81
24k	510R	20.8m	33.64	24.5k	499	1.45k	8.43k	5.80
43k	910R	20.7m	33.67	43.9k	891	2.60k	15.0k	5.78
39k	820R	20.6m	33.73	39.8k	803	2.36k	13.6k	5.74
62k	1k3	20.5m	33.75	63.3k	1.27k	3.75k	21.5k	5.73
36k	750R	20.4m	33.80	36.8k	735	2.18k	12.4k	5.69
13k	270R	20.3m	33.83	13.3k	265	786	4.46k	5.68
27k	560R	20.3m	33.84	27.6k	549	1.63k	9.26k	5.67
30k	620R	20.2m	33.87	30.6k	607	1.81k	10.3k	5.65
16k	330R	20.2m	33.89	16.3k	323	968	5.46k	5.64
33k	680R	20.2m	33.90	33.7k	666	2.00k	11.2k	5.63
11k	220R	19.6m	34.15	11.2k	216	665	3.64k	5.48
12k	240R	19.6m	34.15	12.2k	235	725	3.97k	5.48
15k	300R	19.6m	34.15	15.3k	294	907	4.96k	5.48
18k	360R	19.6m	34.15	18.4k	353	1.09k	5.96k	5.48
75k	1k5	19.6m	34.15	76.5k	1.47k	4.53k	24.8k	5.48
100k	2k	19.6m	34.15	102.0k	1.96k	6.04k	33.1k	5.48

R_1	R_2	A	dB	R_I	R_O	R_{SMAX}	R_{LMIN}	$R_{LS}(1\,dB)$
91k	1k8	19.4m	34.25	92.8k	1.77k	5.50k	29.8k	5.42
56k	1k1	19.3m	34.30	57.1k	1.08k	3.38k	18.2k	5.38
51k	1k	19.2m	34.32	52.0k	981	3.08k	16.6k	5.37
24k	470R	19.2m	34.33	24.5k	461	1.45k	7.78k	5.37
22k	430R	19.2m	34.35	22.4k	422	1.33k	7.12k	5.36
82k	1k6	19.1m	34.36	83.6k	1.57k	4.95k	26.5k	5.35
20k	390R	19.1m	34.37	20.4k	383	1.21k	6.46k	5.34
47k	910R	19.0m	34.43	47.9k	893	2.84k	15.1k	5.31
62k	1k2	19.0m	34.43	63.2k	1.18k	3.74k	19.9k	5.31
39k	750R	18.9m	34.49	39.8k	736	2.36k	12.4k	5.27
68k	1k3	18.8m	34.54	69.3k	1.28k	4.11k	21.5k	5.24
43k	820R	18.7m	34.56	43.8k	805	2.60k	13.6k	5.23
27k	510R	18.5m	34.64	27.5k	501	1.63k	8.45k	5.18
36k	680R	18.5m	34.64	36.7k	667	2.17k	11.3k	5.18
33k	620R	18.4m	34.68	33.6k	609	1.99k	10.3k	5.16
16k	300R	18.4m	34.70	16.3k	294	966	4.97k	5.15
30k	560R	18.3m	34.74	30.6k	550	1.81k	9.28k	5.12
13k	240R	18.1m	34.83	13.2k	236	785	3.98k	5.07
12k	220R	18.0m	34.89	12.2k	216	724	3.65k	5.04
18k	330R	18.0m	34.89	18.3k	324	1.09k	5.47k	5.04
82k	1k5	18.0m	34.91	83.5k	1.47k	4.95k	24.9k	5.02
11k	200R	17.9m	34.96	11.2k	196	664	3.32k	5.00
20k	360R	17.7m	35.05	20.4k	354	1.21k	5.97k	4.95
100k	1k8	17.7m	35.05	102.0k	1.77k	6.03k	29.8k	4.95
15k	270R	17.7m	35.05	15.3k	265	905	4.48k	4.95
24k	430R	17.6m	35.09	24.4k	422	1.45k	7.13k	4.92
56k	1k	17.5m	35.12	57.0k	982	3.38k	16.6k	4.91
51k	910R	17.5m	35.12	51.9k	894	3.08k	15.1k	4.91
62k	1k1	17.4m	35.17	63.1k	1.08k	3.74k	18.2k	4.88
22k	390R	17.4m	35.18	22.4k	383	1.33k	6.47k	4.87
68k	1k2	17.3m	35.22	69.2k	1.18k	4.10k	19.9k	4.85
91k	1k6	17.3m	35.25	92.6k	1.57k	5.49k	26.5k	4.84
47k	820R	17.1m	35.32	47.8k	806	2.83k	13.6k	4.80
43k	750R	17.1m	35.32	43.8k	737	2.59k	12.4k	4.80
39k	680R	17.1m	35.32	39.7k	668	2.35k	11.3k	4.80
27k	470R	17.1m	35.34	27.5k	462	1.63k	7.80k	4.79
75k	1k3	17.0m	35.37	76.3k	1.28k	4.52k	21.6k	4.77

Table 8.1 Voltage dividers 189

R_1	R_2	A	dB	R_I	R_O	R_{SMAX}	R_{LMIN}	$R_{LS}(1\,dB)$
36k	620R	16.9m	35.43	36.6k	610	2.17k	10.3k	4.74
30k	510R	16.7m	35.54	30.5k	501	1.81k	8.46k	4.68
33k	560R	16.7m	35.55	33.6k	551	1.99k	9.29k	4.67
13k	220R	16.6m	35.58	13.2k	216	783	3.65k	4.66
16k	270R	16.6m	35.60	16.3k	266	964	4.48k	4.65
12k	200R	16.4m	35.71	12.2k	197	723	3.32k	4.59
18k	300R	16.4m	35.71	18.3k	295	1.08k	4.98k	4.59
20k	330R	16.2m	35.79	20.3k	325	1.20k	5.48k	4.55
91k	1k5	16.2m	35.80	92.5k	1.48k	5.48k	24.9k	4.54
11k	180R	16.1m	35.86	11.2k	177	662	2.99k	4.51
22k	360R	16.1m	35.86	22.4k	354	1.32k	5.98k	4.51
24k	390R	16.0m	35.92	24.4k	384	1.45k	6.48k	4.48
56k	910R	16.0m	35.92	56.9k	895	3.37k	15.1k	4.48
68k	1k1	15.9m	35.96	69.1k	1.08k	4.09k	18.3k	4.46
62k	1k	15.9m	35.99	63.0k	984	3.73k	16.6k	4.45
51k	820R	15.8m	36.01	51.8k	807	3.07k	13.6k	4.44
15k	240R	15.7m	36.06	15.2k	236	903	3.99k	4.41
75k	1k2	15.7m	36.06	76.2k	1.18k	4.52k	19.9k	4.41
100k	1k6	15.7m	36.06	102.0k	1.57k	6.02k	26.6k	4.41
47k	750R	15.7m	36.08	47.8k	738	2.83k	12.5k	4.40
27k	430R	15.7m	36.10	27.4k	423	1.63k	7.14k	4.39
39k	620R	15.6m	36.11	39.6k	610	2.35k	10.3k	4.39
82k	1k3	15.6m	36.13	83.3k	1.28k	4.94k	21.6k	4.38
43k	680R	15.6m	36.16	43.7k	669	2.59k	11.3k	4.36
30k	470R	15.4m	36.24	30.5k	463	1.81k	7.81k	4.33
36k	560R	15.3m	36.30	36.6k	551	2.17k	9.31k	4.30
33k	510R	15.2m	36.35	33.5k	502	1.99k	8.48k	4.27
13k	200R	15.2m	36.39	13.2k	197	782	3.32k	4.25
12k	180R	14.8m	36.61	12.2k	177	722	2.99k	4.15
16k	240R	14.8m	36.61	16.2k	236	962	3.99k	4.15
18k	270R	14.8m	36.61	18.3k	266	1.08k	4.49k	4.15
20k	300R	14.8m	36.61	20.3k	296	1.20k	4.99k	4.15
22k	330R	14.8m	36.61	22.3k	325	1.32k	5.49k	4.15
24k	360R	14.8m	36.61	24.4k	355	1.44k	5.99k	4.15
100k	1k5	14.8m	36.61	102.0k	1.48k	6.01k	24.9k	4.15
51k	750R	14.5m	36.78	51.8k	739	3.07k	12.5k	4.07
68k	1k	14.5m	36.78	69.0k	986	4.09k	16.6k	4.07

R_1	R_2	A	dB	R_I	R_O	R_{SMAX}	R_{LMIN}	$R_{LS}(1\,dB)$
62k	910R	14.5m	36.79	62.9k	897	3.73k	15.1k	4.06
15k	220R	14.5m	36.80	15.2k	217	902	3.66k	4.06
75k	1k1	14.5m	36.80	76.1k	1.08k	4.51k	18.3k	4.06
56k	820R	14.4m	36.81	56.8k	808	3.37k	13.6k	4.05
82k	1k2	14.4m	36.82	83.2k	1.18k	4.93k	20.0k	4.05
11k	160R	14.3m	36.87	11.2k	158	661	2.66k	4.02
47k	680R	14.3m	36.92	47.7k	670	2.83k	11.3k	4.00
27k	390R	14.2m	36.93	27.4k	384	1.62k	6.49k	4.00
43k	620R	14.2m	36.95	43.6k	611	2.58k	10.3k	3.99
39k	560R	14.2m	36.98	39.6k	552	2.34k	9.32k	3.97
30k	430R	14.1m	37.00	30.4k	424	1.80k	7.15k	3.97
91k	1k3	14.1m	37.03	92.3k	1.28k	5.47k	21.6k	3.96
33k	470R	14.0m	37.05	33.5k	463	1.98k	7.82k	3.94
36k	510R	14.0m	37.10	36.5k	503	2.16k	8.49k	3.92
13k	180R	13.7m	37.29	13.2k	178	781	3.00k	3.84
16k	220R	13.6m	37.35	16.2k	217	961	3.66k	3.81
24k	330R	13.6m	37.35	24.3k	326	1.44k	5.49k	3.81
11k	150R	13.5m	37.42	11.2k	148	661	2.50k	3.78
22k	300R	13.5m	37.42	22.3k	296	1.32k	4.99k	3.78
20k	270R	13.3m	37.51	20.3k	266	1.20k	4.50k	3.74
82k	1k1	13.2m	37.56	83.1k	1.09k	4.92k	18.3k	3.72
56k	750R	13.2m	37.58	56.8k	740	3.36k	12.5k	3.71
68k	910R	13.2m	37.58	68.9k	898	4.08k	15.2k	3.71
12k	160R	13.2m	37.62	12.2k	158	721	2.66k	3.70
15k	200R	13.2m	37.62	15.2k	197	901	3.33k	3.70
18k	240R	13.2m	37.62	18.2k	237	1.08k	4.00k	3.70
27k	360R	13.2m	37.62	27.4k	355	1.62k	6.00k	3.70
51k	680R	13.2m	37.62	51.7k	671	3.06k	11.3k	3.70
75k	1k	13.2m	37.62	76.0k	987	4.50k	16.7k	3.70
62k	820R	13.1m	37.69	62.8k	809	3.72k	13.7k	3.67
47k	620R	13.0m	37.71	47.6k	612	2.82k	10.3k	3.66
91k	1k2	13.0m	37.71	92.2k	1.18k	5.46k	20.0k	3.66
39k	510R	12.9m	37.78	39.5k	503	2.34k	8.50k	3.63
36k	470R	12.9m	37.80	36.5k	464	2.16k	7.83k	3.62
33k	430R	12.9m	37.81	33.4k	424	1.98k	7.16k	3.62
43k	560R	12.9m	37.82	43.6k	553	2.58k	9.33k	3.61
30k	390R	12.8m	37.83	30.4k	385	1.80k	6.50k	3.61

Table 8.1 Voltage dividers 191

R_1	R_2	A	dB	R_I	R_O	R_{SMAX}	R_{LMIN}	$R_{LS}(1\,dB)$
100k	1k3	12.8m	37.83	101.0k	1.28k	6.00k	21.7k	3.61
12k	150R	12.3m	38.17	12.2k	148	720	2.50k	3.47
16k	200R	12.3m	38.17	16.2k	198	960	3.33k	3.47
24k	300R	12.3m	38.17	24.3k	296	1.44k	5.00k	3.47
13k	160R	12.2m	38.30	13.2k	158	780	2.67k	3.42
22k	270R	12.1m	38.33	22.3k	267	1.32k	4.50k	3.41
18k	220R	12.1m	38.36	18.2k	217	1.08k	3.67k	3.40
27k	330R	12.1m	38.36	27.3k	326	1.62k	5.50k	3.40
82k	1k	12.0m	38.38	83.0k	988	4.92k	16.7k	3.39
51k	620R	12.0m	38.41	51.6k	613	3.06k	10.3k	3.38
56k	680R	12.0m	38.42	56.7k	672	3.36k	11.3k	3.38
75k	910R	12.0m	38.43	75.9k	899	4.50k	15.2k	3.37
62k	750R	12.0m	38.45	62.8k	741	3.72k	12.5k	3.36
91k	1k1	11.9m	38.46	92.1k	1.09k	5.46k	18.3k	3.36
68k	820R	11.9m	38.48	68.8k	810	4.08k	13.7k	3.35
39k	470R	11.9m	38.48	39.5k	464	2.34k	7.84k	3.35
15k	180R	11.9m	38.52	15.2k	178	899	3.00k	3.34
20k	240R	11.9m	38.52	20.2k	237	1.20k	4.00k	3.34
30k	360R	11.9m	38.52	30.4k	356	1.80k	6.00k	3.34
100k	1k2	11.9m	38.52	101.0k	1.19k	6.00k	20.0k	3.34
36k	430R	11.8m	38.56	36.4k	425	2.16k	7.17k	3.32
47k	560R	11.8m	38.58	47.6k	553	2.82k	9.34k	3.31
43k	510R	11.7m	38.62	43.5k	504	2.58k	8.51k	3.30
11k	130R	11.7m	38.65	11.1k	128	659	2.17k	3.29
33k	390R	11.7m	38.65	33.4k	385	1.98k	6.50k	3.29
13k	150R	11.4m	38.86	13.2k	148	779	2.50k	3.21
16k	180R	11.1m	39.07	16.2k	178	959	3.00k	3.13
24k	270R	11.1m	39.07	24.3k	267	1.44k	4.51k	3.13
18k	200R	11.0m	39.18	18.2k	198	1.08k	3.34k	3.10
27k	300R	11.0m	39.18	27.3k	297	1.62k	5.01k	3.10
82k	910R	11.0m	39.19	82.9k	900	4.91k	15.2k	3.09
56k	620R	11.0m	39.21	56.6k	613	3.35k	10.3k	3.08
68k	750R	10.9m	39.24	68.8k	742	4.07k	12.5k	3.07
39k	430R	10.9m	39.25	39.4k	425	2.34k	7.18k	3.07
20k	220R	10.9m	39.27	20.2k	218	1.20k	3.67k	3.07
30k	330R	10.9m	39.27	30.3k	326	1.80k	5.51k	3.07
100k	1k1	10.9m	39.27	101.0k	1.09k	5.99k	18.4k	3.07

R_1	R_2	A	dB	R_I	R_O	R_{SMAX}	R_{LMIN}	$R_{LS}(1\,dB)$
91k	1k	10.9m	39.28	92.0k	989	5.45k	16.7k	3.06
51k	560R	10.9m	39.28	51.6k	554	3.06k	9.35k	3.06
62k	680R	10.8m	39.29	62.7k	673	3.71k	11.4k	3.06
75k	820R	10.8m	39.32	75.8k	811	4.49k	13.7k	3.05
43k	470R	10.8m	39.32	43.5k	465	2.58k	7.85k	3.05
11k	120R	10.8m	39.34	11.1k	119	659	2.00k	3.04
22k	240R	10.8m	39.34	22.2k	237	1.32k	4.01k	3.04
33k	360R	10.8m	39.34	33.4k	356	1.98k	6.01k	3.04
47k	510R	10.7m	39.38	47.5k	505	2.82k	8.51k	3.02
36k	390R	10.7m	39.40	36.4k	386	2.16k	6.51k	3.02
12k	130R	10.7m	39.40	12.1k	129	719	2.17k	3.02
15k	160R	10.6m	39.53	15.2k	158	898	2.67k	2.97
11k	110R	9.90m	40.09	11.1k	109	658	1.84k	2.79
12k	120R	9.90m	40.09	12.1k	119	718	2.01k	2.79
13k	130R	9.90m	40.09	13.1k	129	778	2.17k	2.79
15k	150R	9.90m	40.09	15.2k	149	898	2.51k	2.79
16k	160R	9.90m	40.09	16.2k	158	958	2.67k	2.79
18k	180R	9.90m	40.09	18.2k	178	1.08k	3.01k	2.79
20k	200R	9.90m	40.09	20.2k	198	1.20k	3.34k	2.79
22k	220R	9.90m	40.09	22.2k	218	1.32k	3.68k	2.79
24k	240R	9.90m	40.09	24.2k	238	1.44k	4.01k	2.79
27k	270R	9.90m	40.09	27.3k	267	1.62k	4.51k	2.79
30k	300R	9.90m	40.09	30.3k	297	1.80k	5.01k	2.79
33k	330R	9.90m	40.09	33.3k	327	1.97k	5.51k	2.79
36k	360R	9.90m	40.09	36.4k	356	2.15k	6.02k	2.79
39k	390R	9.90m	40.09	39.4k	386	2.33k	6.52k	2.79
43k	430R	9.90m	40.09	43.4k	426	2.57k	7.19k	2.79
47k	470R	9.90m	40.09	47.5k	465	2.81k	7.85k	2.79
51k	510R	9.90m	40.09	51.5k	505	3.05k	8.52k	2.79
56k	560R	9.90m	40.09	56.6k	554	3.35k	9.36k	2.79
62k	620R	9.90m	40.09	62.6k	614	3.71k	10.4k	2.79
68k	680R	9.90m	40.09	68.7k	673	4.07k	11.4k	2.79
75k	750R	9.90m	40.09	75.8k	743	4.49k	12.5k	2.79
82k	820R	9.90m	40.09	82.8k	812	4.91k	13.7k	2.79
91k	910R	9.90m	40.09	91.9k	901	5.45k	15.2k	2.79
100k	1k	9.90m	40.09	101.0k	990	5.98k	16.7k	2.79
16k	150R	9.29m	40.64	16.2k	149	957	2.51k	2.62

Table 8.1 Voltage dividers   193

R_1	R_2	A	dB	R_I	R_O	R_{SMAX}	R_{LMIN}	$R_{LS}(1\,dB)$
13k	120R	9.15m	40.78	13.1k	119	777	2.01k	2.58
39k	360R	9.15m	40.78	39.4k	357	2.33k	6.02k	2.58
51k	470R	9.13m	40.79	51.5k	466	3.05k	7.86k	2.58
12k	110R	9.08m	40.84	12.1k	109	718	1.84k	2.56
24k	220R	9.08m	40.84	24.2k	218	1.44k	3.68k	2.56
36k	330R	9.08m	40.84	36.3k	327	2.15k	5.52k	2.56
47k	430R	9.07m	40.85	47.4k	426	2.81k	7.19k	2.56
82k	750R	9.06m	40.85	82.8k	743	4.90k	12.5k	2.56
68k	620R	9.04m	40.88	68.6k	614	4.07k	10.4k	2.55
56k	510R	9.02m	40.89	56.5k	505	3.35k	8.53k	2.55
100k	910R	9.02m	40.90	101.0k	902	5.98k	15.2k	2.55
11k	100R	9.01m	40.91	11.1k	99.1	658	1.67k	2.54
22k	200R	9.01m	40.91	22.2k	198	1.32k	3.34k	2.54
33k	300R	9.01m	40.91	33.3k	297	1.97k	5.02k	2.54
43k	390R	8.99m	40.93	43.4k	386	2.57k	6.52k	2.54
75k	680R	8.99m	40.93	75.7k	674	4.48k	11.4k	2.54
62k	560R	8.95m	40.96	62.6k	555	3.71k	9.37k	2.53
91k	820R	8.93m	40.98	91.8k	813	5.44k	13.7k	2.52
20k	180R	8.92m	40.99	20.2k	178	1.20k	3.01k	2.52
30k	270R	8.92m	40.99	30.3k	268	1.79k	4.52k	2.52
18k	160R	8.81m	41.10	18.2k	159	1.08k	2.68k	2.49
27k	240R	8.81m	41.10	27.2k	238	1.61k	4.01k	2.49
15k	130R	8.59m	41.32	15.1k	129	897	2.18k	2.43
13k	110R	8.39m	41.52	13.1k	109	777	1.84k	2.37
39k	330R	8.39m	41.52	39.3k	327	2.33k	5.52k	2.37
51k	430R	8.36m	41.55	51.4k	426	3.05k	7.20k	2.36
56k	470R	8.32m	41.59	56.5k	466	3.35k	7.87k	2.35
43k	360R	8.30m	41.62	43.4k	357	2.57k	6.03k	2.35
12k	100R	8.26m	41.66	12.1k	99.2	717	1.67k	2.33
18k	150R	8.26m	41.66	18.2k	149	1.08k	2.51k	2.33
24k	200R	8.26m	41.66	24.2k	198	1.43k	3.35k	2.33
36k	300R	8.26m	41.66	36.3k	298	2.15k	5.02k	2.33
47k	390R	8.23m	41.69	47.4k	387	2.81k	6.53k	2.32
82k	680R	8.22m	41.70	82.7k	674	4.90k	11.4k	2.32
75k	620R	8.20m	41.72	75.6k	615	4.48k	10.4k	2.32
91k	750R	8.17m	41.75	91.8k	744	5.44k	12.6k	2.31
68k	560R	8.17m	41.76	68.6k	555	4.06k	9.37k	2.31

R_1	R_2	A	dB	R_I	R_O	R_{SMAX}	R_{LMIN}	R_{LS}(1 dB)
62k	510R	8.16m	41.77	62.5k	506	3.70k	8.54k	2.30
100k	820R	8.13m	41.79	101.0k	813	5.97k	13.7k	2.30
22k	180R	8.12m	41.81	22.2k	179	1.31k	3.01k	2.29
33k	270R	8.12m	41.81	33.3k	268	1.97k	4.52k	2.29
27k	220R	8.08m	41.85	27.2k	218	1.61k	3.68k	2.28
16k	130R	8.06m	41.87	16.1k	129	956	2.18k	2.28
15k	120R	7.94m	42.01	15.1k	119	896	2.01k	2.24
20k	160R	7.94m	42.01	20.2k	159	1.19k	2.68k	2.24
30k	240R	7.94m	42.01	30.2k	238	1.79k	4.02k	2.24
13k	100R	7.63m	42.35	13.1k	99.2	776	1.67k	2.16
39k	300R	7.63m	42.35	39.3k	298	2.33k	5.02k	2.16
56k	430R	7.62m	42.36	56.4k	427	3.34k	7.20k	2.15
43k	330R	7.62m	42.37	43.3k	327	2.57k	5.53k	2.15
47k	360R	7.60m	42.38	47.4k	357	2.81k	6.03k	2.15
51k	390R	7.59m	42.40	51.4k	387	3.05k	6.53k	2.15
62k	470R	7.52m	42.47	62.5k	466	3.70k	7.87k	2.13
82k	620R	7.50m	42.49	82.6k	615	4.90k	10.4k	2.12
16k	120R	7.44m	42.56	16.1k	119	955	2.01k	2.10
20k	150R	7.44m	42.56	20.2k	149	1.19k	2.51k	2.10
24k	180R	7.44m	42.56	24.2k	179	1.43k	3.02k	2.10
36k	270R	7.44m	42.56	36.3k	268	2.15k	4.52k	2.10
68k	510R	7.44m	42.56	68.5k	506	4.06k	8.54k	2.10
100k	750R	7.44m	42.56	101.0k	744	5.97k	12.6k	2.10
91k	680R	7.42m	42.60	91.7k	675	5.43k	11.4k	2.10
75k	560R	7.41m	42.60	75.6k	556	4.48k	9.38k	2.10
27k	200R	7.35m	42.67	27.2k	199	1.61k	3.35k	2.08
15k	110R	7.28m	42.76	15.1k	109	895	1.84k	2.06
30k	220R	7.28m	42.76	30.2k	218	1.79k	3.69k	2.06
22k	160R	7.22m	42.83	22.2k	159	1.31k	2.68k	2.04
33k	240R	7.22m	42.83	33.2k	238	1.97k	4.02k	2.04
18k	130R	7.17m	42.89	18.1k	129	1.07k	2.18k	2.03
51k	360R	7.01m	43.09	51.4k	357	3.04k	6.03k	1.98
47k	330R	6.97m	43.13	47.3k	328	2.80k	5.53k	1.97
43k	300R	6.93m	43.19	43.3k	298	2.57k	5.03k	1.96
56k	390R	6.92m	43.20	56.4k	387	3.34k	6.54k	1.96
62k	430R	6.89m	43.24	62.4k	427	3.70k	7.21k	1.95
39k	270R	6.88m	43.25	39.3k	268	2.33k	4.53k	1.94

Table 8.1 Voltage dividers 195

R_1	R_2	A	dB	R_I	R_O	R_{SMAX}	R_{LMIN}	$R_{LS}(1\,dB)$
68k	470R	6.86m	43.27	68.5k	467	4.06k	7.88k	1.94
16k	110R	6.83m	43.31	16.1k	109	955	1.84k	1.93
82k	560R	6.78m	43.37	82.6k	556	4.89k	9.39k	1.92
22k	150R	6.77m	43.39	22.2k	149	1.31k	2.51k	1.92
91k	620R	6.77m	43.39	91.6k	616	5.43k	10.4k	1.91
75k	510R	6.75m	43.41	75.5k	507	4.47k	8.55k	1.91
100k	680R	6.75m	43.41	101.0k	675	5.97k	11.4k	1.91
15k	100R	6.62m	43.58	15.1k	99.3	895	1.68k	1.87
18k	120R	6.62m	43.58	18.1k	119	1.07k	2.01k	1.87
24k	160R	6.62m	43.58	24.2k	159	1.43k	2.68k	1.87
27k	180R	6.62m	43.58	27.2k	179	1.61k	3.02k	1.87
30k	200R	6.62m	43.58	30.2k	199	1.79k	3.35k	1.87
33k	220R	6.62m	43.58	33.2k	219	1.97k	3.69k	1.87
36k	240R	6.62m	43.58	36.2k	238	2.15k	4.02k	1.87
20k	130R	6.46m	43.80	20.1k	129	1.19k	2.18k	1.83
51k	330R	6.43m	43.84	51.3k	328	3.04k	5.53k	1.82
56k	360R	6.39m	43.89	56.4k	358	3.34k	6.04k	1.81
47k	300R	6.34m	43.95	47.3k	298	2.80k	5.03k	1.80
68k	430R	6.28m	44.04	68.4k	427	4.05k	7.21k	1.78
62k	390R	6.25m	44.08	62.4k	388	3.70k	6.54k	1.77
43k	270R	6.24m	44.10	43.3k	268	2.56k	4.53k	1.77
75k	470R	6.23m	44.11	75.5k	467	4.47k	7.88k	1.76
16k	100R	6.21m	44.14	16.1k	99.4	954	1.68k	1.76
24k	150R	6.21m	44.14	24.2k	149	1.43k	2.52k	1.76
82k	510R	6.18m	44.18	82.5k	507	4.89k	8.55k	1.75
100k	620R	6.16m	44.21	101.0k	616	5.96k	10.4k	1.74
39k	240R	6.12m	44.27	39.2k	239	2.33k	4.03k	1.73
91k	560R	6.12m	44.27	91.6k	557	5.43k	9.39k	1.73
18k	110R	6.07m	44.33	18.1k	109	1.07k	1.85k	1.72
36k	220R	6.07m	44.33	36.2k	219	2.15k	3.69k	1.72
33k	200R	6.02m	44.40	33.2k	199	1.97k	3.35k	1.71
20k	120R	5.96m	44.49	20.1k	119	1.19k	2.01k	1.69
30k	180R	5.96m	44.49	30.2k	179	1.79k	3.02k	1.69
27k	160R	5.89m	44.60	27.2k	159	1.61k	2.68k	1.67
22k	130R	5.87m	44.62	22.1k	129	1.31k	2.18k	1.66
56k	330R	5.86m	44.64	56.3k	328	3.34k	5.54k	1.66
51k	300R	5.85m	44.66	51.3k	298	3.04k	5.03k	1.66

R_1	R_2	A	dB	R_I	R_O	R_{SMAX}	R_{LMIN}	$R_{LS}(1\,dB)$
62k	360R	5.77m	44.77	62.4k	358	3.70k	6.04k	1.63
47k	270R	5.71m	44.86	47.3k	268	2.80k	4.53k	1.62
68k	390R	5.70m	44.88	68.4k	388	4.05k	6.54k	1.61
75k	430R	5.70m	44.88	75.4k	428	4.47k	7.22k	1.61
82k	470R	5.70m	44.88	82.5k	467	4.89k	7.89k	1.61
39k	220R	5.61m	45.02	39.2k	219	2.32k	3.69k	1.59
91k	510R	5.57m	45.08	91.5k	507	5.42k	8.56k	1.58
100k	560R	5.57m	45.08	101.0k	557	5.96k	9.40k	1.58
43k	240R	5.55m	45.11	43.2k	239	2.56k	4.03k	1.57
18k	100R	5.52m	45.15	18.1k	99.4	1.07k	1.68k	1.56
27k	150R	5.52m	45.15	27.2k	149	1.61k	2.52k	1.56
36k	200R	5.52m	45.15	36.2k	199	2.14k	3.36k	1.56
20k	110R	5.47m	45.24	20.1k	109	1.19k	1.85k	1.55
22k	120R	5.42m	45.31	22.1k	119	1.31k	2.01k	1.54
33k	180R	5.42m	45.31	33.2k	179	1.97k	3.02k	1.54
24k	130R	5.39m	45.37	24.1k	129	1.43k	2.18k	1.53
56k	300R	5.33m	45.47	56.3k	298	3.34k	5.04k	1.51
30k	160R	5.31m	45.51	30.2k	159	1.79k	2.69k	1.50
62k	330R	5.29m	45.52	62.3k	328	3.69k	5.54k	1.50
51k	270R	5.27m	45.57	51.3k	269	3.04k	4.53k	1.49
68k	360R	5.27m	45.57	68.4k	358	4.05k	6.04k	1.49
82k	430R	5.22m	45.65	82.4k	428	4.88k	7.22k	1.48
75k	390R	5.17m	45.72	75.4k	388	4.47k	6.55k	1.47
91k	470R	5.14m	45.78	91.5k	468	5.42k	7.89k	1.46
39k	200R	5.10m	45.85	39.2k	199	2.32k	3.36k	1.45
43k	220R	5.09m	45.87	43.2k	219	2.56k	3.69k	1.44
47k	240R	5.08m	45.88	47.2k	239	2.80k	4.03k	1.44
100k	510R	5.07m	45.89	101.0k	507	5.96k	8.56k	1.44
20k	100R	4.98m	46.06	20.1k	99.5	1.19k	1.68k	1.41
22k	110R	4.98m	46.06	22.1k	109	1.31k	1.85k	1.41
24k	120R	4.98m	46.06	24.1k	119	1.43k	2.02k	1.41
30k	150R	4.98m	46.06	30.2k	149	1.79k	2.52k	1.41
36k	180R	4.98m	46.06	36.2k	179	2.14k	3.02k	1.41
68k	330R	4.83m	46.32	68.3k	328	4.05k	5.54k	1.37
33k	160R	4.83m	46.33	33.2k	159	1.96k	2.69k	1.37
62k	300R	4.82m	46.35	62.3k	299	3.69k	5.04k	1.36
56k	270R	4.80m	46.38	56.3k	269	3.33k	4.53k	1.36

Table 8.1 Voltage dividers 197

R_1	R_2	A	dB	R_I	R_O	R_{SMAX}	R_{LMIN}	$R_{LS}(1\,dB)$
27k	130R	4.79m	46.39	27.1k	129	1.61k	2.18k	1.36
75k	360R	4.78m	46.42	75.4k	358	4.47k	6.05k	1.35
82k	390R	4.73m	46.50	82.4k	388	4.88k	6.55k	1.34
91k	430R	4.70m	46.55	91.4k	428	5.42k	7.22k	1.33
51k	240R	4.68m	46.59	51.2k	239	3.04k	4.03k	1.33
100k	470R	4.68m	46.60	100.0k	468	5.95k	7.89k	1.33
47k	220R	4.66m	46.63	47.2k	219	2.80k	3.70k	1.32
43k	200R	4.63m	46.69	43.2k	199	2.56k	3.36k	1.31
39k	180R	4.59m	46.76	39.2k	179	2.32k	3.02k	1.30
24k	110R	4.56m	46.82	24.1k	109	1.43k	1.85k	1.29
22k	100R	4.52m	46.89	22.1k	99.5	1.31k	1.68k	1.28
33k	150R	4.52m	46.89	33.2k	149	1.96k	2.52k	1.28
27k	120R	4.42m	47.08	27.1k	119	1.61k	2.02k	1.25
36k	160R	4.42m	47.08	36.2k	159	2.14k	2.69k	1.25
68k	300R	4.39m	47.15	68.3k	299	4.05k	5.04k	1.25
75k	330R	4.38m	47.17	75.3k	329	4.46k	5.54k	1.24
82k	360R	4.37m	47.19	82.4k	358	4.88k	6.05k	1.24
62k	270R	4.34m	47.26	62.3k	269	3.69k	4.54k	1.23
30k	130R	4.31m	47.30	30.1k	129	1.79k	2.18k	1.22
51k	220R	4.30m	47.34	51.2k	219	3.03k	3.70k	1.22
100k	430R	4.28m	47.37	100.0k	428	5.95k	7.23k	1.21
56k	240R	4.27m	47.40	56.2k	239	3.33k	4.03k	1.21
91k	390R	4.27m	47.40	91.4k	388	5.42k	6.55k	1.21
47k	200R	4.24m	47.46	47.2k	199	2.80k	3.36k	1.20
43k	180R	4.17m	47.60	43.2k	179	2.56k	3.03k	1.18
24k	100R	4.15m	47.64	24.1k	99.6	1.43k	1.68k	1.18
36k	150R	4.15m	47.64	36.2k	149	2.14k	2.52k	1.18
39k	160R	4.09m	47.77	39.2k	159	2.32k	2.69k	1.16
27k	110R	4.06m	47.83	27.1k	110	1.61k	1.85k	1.15
82k	330R	4.01m	47.94	82.3k	329	4.88k	5.55k	1.14
30k	120R	3.98m	47.99	30.1k	120	1.78k	2.02k	1.13
75k	300R	3.98m	47.99	75.3k	299	4.46k	5.04k	1.13
68k	270R	3.95m	48.06	68.3k	269	4.05k	4.54k	1.12
91k	360R	3.94m	48.09	91.4k	359	5.41k	6.05k	1.12
33k	130R	3.92m	48.13	33.1k	129	1.96k	2.19k	1.11
56k	220R	3.91m	48.15	56.2k	219	3.33k	3.70k	1.11
51k	200R	3.91m	48.16	51.2k	199	3.03k	3.36k	1.11

R_1	R_2	A	dB	R_I	R_O	R_{SMAX}	R_{LMIN}	$R_{LS}(1\,dB)$
100k	390R	3.88m	48.21	100.0k	388	5.95k	6.56k	1.10
62k	240R	3.86m	48.28	62.2k	239	3.69k	4.03k	1.09
39k	150R	3.83m	48.33	39.2k	149	2.32k	2.52k	1.09
47k	180R	3.82m	48.37	47.2k	179	2.80k	3.03k	1.08
43k	160R	3.71m	48.62	43.2k	159	2.56k	2.69k	1.05
27k	100R	3.69m	48.66	27.1k	99.6	1.61k	1.68k	1.05
30k	110R	3.65m	48.75	30.1k	110	1.78k	1.85k	1.04
82k	300R	3.65m	48.77	82.3k	299	4.88k	5.04k	1.03
33k	120R	3.62m	48.82	33.1k	120	1.96k	2.02k	1.03
91k	330R	3.61m	48.84	91.3k	329	5.41k	5.55k	1.03
36k	130R	3.60m	48.88	36.1k	130	2.14k	2.19k	1.02
75k	270R	3.59m	48.91	75.3k	269	4.46k	4.54k	1.02
100k	360R	3.59m	48.91	100.0k	359	5.95k	6.05k	1.02
56k	200R	3.56m	48.97	56.2k	199	3.33k	3.36k	1.01
62k	220R	3.54m	49.03	62.2k	219	3.69k	3.70k	1.00
51k	180R	3.52m	49.08	51.2k	179	3.03k	3.03k	998m
68k	240R	3.52m	49.08	68.2k	239	4.04k	4.04k	998m
43k	150R	3.48m	49.18	43.2k	149	2.56k	2.52k	987m
47k	160R	3.39m	49.39	47.2k	159	2.79k	2.69k	963m
30k	100R	3.32m	49.57	30.1k	99.7	1.78k	1.68k	943m
33k	110R	3.32m	49.57	33.1k	110	1.96k	1.85k	943m
36k	120R	3.32m	49.57	36.1k	120	2.14k	2.02k	943m
39k	130R	3.32m	49.57	39.1k	130	2.32k	2.19k	943m
100k	330R	3.29m	49.66	100.0k	329	5.94k	5.55k	934m
91k	300R	3.29m	49.67	91.3k	299	5.41k	5.05k	933m
82k	270R	3.28m	49.68	82.3k	269	4.87k	4.54k	932m
68k	220R	3.22m	49.83	68.2k	219	4.04k	3.70k	916m
62k	200R	3.22m	49.86	62.2k	199	3.69k	3.36k	913m
56k	180R	3.20m	49.89	56.2k	179	3.33k	3.03k	910m
75k	240R	3.19m	49.92	75.2k	239	4.46k	4.04k	906m
47k	150R	3.18m	49.95	47.2k	150	2.79k	2.52k	903m
51k	160R	3.13m	50.10	51.2k	159	3.03k	2.69k	888m
39k	120R	3.07m	50.26	39.1k	120	2.32k	2.02k	871m
36k	110R	3.05m	50.32	36.1k	110	2.14k	1.85k	865m
33k	100R	3.02m	50.40	33.1k	99.7	1.96k	1.68k	858m
43k	130R	3.01m	50.42	43.1k	130	2.56k	2.19k	856m
100k	300R	2.99m	50.48	100.0k	299	5.94k	5.05k	849m

Table 8.1 Voltage dividers 199

R_1	R_2	A	dB	R_I	R_O	R_{SMAX}	R_{LMIN}	$R_{LS}(1\,dB)$
91k	270R	2.96m	50.58	91.3k	269	5.41k	4.54k	840m
51k	150R	2.93m	50.66	51.2k	150	3.03k	2.52k	833m
68k	200R	2.93m	50.66	68.2k	199	4.04k	3.37k	833m
75k	220R	2.92m	50.68	75.2k	219	4.46k	3.70k	831m
82k	240R	2.92m	50.70	82.2k	239	4.87k	4.04k	829m
62k	180R	2.89m	50.77	62.2k	179	3.68k	3.03k	822m
56k	160R	2.85m	50.91	56.2k	160	3.33k	2.69k	809m
39k	110R	2.81m	51.02	39.1k	110	2.32k	1.85k	799m
43k	120R	2.78m	51.11	43.1k	120	2.56k	2.02k	790m
36k	100R	2.77m	51.15	36.1k	99.7	2.14k	1.68k	787m
47k	130R	2.76m	51.19	47.1k	130	2.79k	2.19k	783m
100k	270R	2.69m	51.40	100.0k	269	5.94k	4.54k	765m
82k	220R	2.68m	51.45	82.2k	219	4.87k	3.70k	760m
56k	150R	2.67m	51.47	56.2k	150	3.33k	2.52k	759m
75k	200R	2.66m	51.50	75.2k	199	4.46k	3.37k	755m
68k	180R	2.64m	51.57	68.2k	180	4.04k	3.03k	750m
91k	240R	2.63m	51.60	91.2k	239	5.41k	4.04k	747m
62k	160R	2.57m	51.79	62.2k	160	3.68k	2.69k	731m
39k	100R	2.56m	51.84	39.1k	99.7	2.32k	1.68k	727m
43k	110R	2.55m	51.86	43.1k	110	2.55k	1.85k	725m
47k	120R	2.55m	51.88	47.1k	120	2.79k	2.02k	723m
51k	130R	2.54m	51.89	51.1k	130	3.03k	2.19k	722m
82k	200R	2.43m	52.28	82.2k	200	4.87k	3.37k	691m
62k	150R	2.41m	52.35	62.2k	150	3.68k	2.53k	686m
91k	220R	2.41m	52.35	91.2k	219	5.41k	3.70k	685m
75k	180R	2.39m	52.42	75.2k	180	4.45k	3.03k	680m
100k	240R	2.39m	52.42	100.0k	239	5.94k	4.04k	680m
51k	120R	2.35m	52.59	51.1k	120	3.03k	2.02k	667m
68k	160R	2.35m	52.59	68.2k	160	4.04k	2.69k	667m
47k	110R	2.33m	52.63	47.1k	110	2.79k	1.85k	663m
43k	100R	2.32m	52.69	43.1k	99.8	2.55k	1.68k	659m
56k	130R	2.32m	52.71	56.1k	130	3.33k	2.19k	658m
68k	150R	2.20m	53.15	68.2k	150	4.04k	2.53k	626m
100k	220R	2.20m	53.17	100.0k	220	5.94k	3.70k	624m
91k	200R	2.19m	53.18	91.2k	200	5.40k	3.37k	623m
82k	180R	2.19m	53.19	82.2k	180	4.87k	3.03k	622m
51k	110R	2.15m	53.34	51.1k	110	3.03k	1.85k	612m

R_1	R_2	A	dB	R_I	R_O	R_{SMAX}	R_{LMIN}	$R_{LS}(1\,dB)$
56k	120R	2.14m	53.40	56.1k	120	3.33k	2.02k	608m
75k	160R	2.13m	53.44	75.2k	160	4.45k	2.69k	605m
47k	100R	2.12m	53.46	47.1k	99.8	2.79k	1.68k	603m
62k	130R	2.09m	53.59	62.1k	130	3.68k	2.19k	595m
75k	150R	2.00m	54.00	75.2k	150	4.45k	2.53k	567m
100k	200R	2.00m	54.00	100.0k	200	5.94k	3.37k	567m
91k	180R	1.97m	54.09	91.2k	180	5.40k	3.03k	561m
56k	110R	1.96m	54.15	56.1k	110	3.32k	1.85k	557m
51k	100R	1.96m	54.17	51.1k	99.8	3.03k	1.68k	556m
82k	160R	1.95m	54.21	82.2k	160	4.87k	2.69k	554m
62k	120R	1.93m	54.28	62.1k	120	3.68k	2.02k	549m
68k	130R	1.91m	54.39	68.1k	130	4.04k	2.19k	542m
82k	150R	1.83m	54.77	82.2k	150	4.87k	2.53k	519m
100k	180R	1.80m	54.91	100.0k	180	5.94k	3.03k	511m
56k	100R	1.78m	54.98	56.1k	99.8	3.32k	1.68k	507m
62k	110R	1.77m	55.04	62.1k	110	3.68k	1.85k	504m
68k	120R	1.76m	55.08	68.1k	120	4.04k	2.02k	501m
91k	160R	1.76m	55.11	91.2k	160	5.40k	2.70k	499m
75k	130R	1.73m	55.24	75.1k	130	4.45k	2.19k	492m
91k	150R	1.65m	55.67	91.2k	150	5.40k	2.53k	468m
68k	110R	1.62m	55.84	68.1k	110	4.04k	1.85k	459m
62k	100R	1.61m	55.86	62.1k	99.8	3.68k	1.68k	458m
75k	120R	1.60m	55.93	75.1k	120	4.45k	2.02k	454m
100k	160R	1.60m	55.93	100.0k	160	5.93k	2.70k	454m
82k	130R	1.58m	56.01	82.1k	130	4.87k	2.19k	450m
100k	150R	1.50m	56.49	100.0k	150	5.93k	2.53k	426m
68k	100R	1.47m	56.66	68.1k	99.9	4.04k	1.69k	418m
75k	110R	1.46m	56.69	75.1k	110	4.45k	1.85k	417m
82k	120R	1.46m	56.71	82.1k	120	4.87k	2.02k	416m
91k	130R	1.43m	56.91	91.1k	130	5.40k	2.19k	406m
82k	110R	1.34m	57.46	82.1k	110	4.87k	1.85k	381m
75k	100R	1.33m	57.51	75.1k	99.9	4.45k	1.69k	379m
91k	120R	1.32m	57.61	91.1k	120	5.40k	2.02k	375m
100k	130R	1.30m	57.73	100.0k	130	5.93k	2.19k	369m
82k	100R	1.22m	58.29	82.1k	99.9	4.86k	1.69k	346m
91k	110R	1.21m	58.36	91.1k	110	5.40k	1.85k	343m
100k	120R	1.20m	58.43	100.0k	120	5.93k	2.02k	341m

Table 8.1 Voltage dividers 201

R_1	R_2	A	dB	R_I	R_O	R_{SMAX}	R_{LMIN}	$R_{LS}(1\,dB)$
100k	110R	1.10m	59.18	100.0k	110	5.93k	1.85k	313m
91k	100R	1.10m	59.19	91.1k	99.9	5.40k	1.69k	312m
100k	100R	999u	60.01	100.0k	99.9	5.93k	1.69k	284m

Part Three

Reactive Circuits

9 First-order CR and LR circuits

Introduction

There are two first-order CR and two first-order LR circuits which we will consider here. One circuit of each type forms an integrator, or a low-pass filter, and one a differentiator, or a high-pass filter. All four circuits can be considered to be variations on the voltage dividers which we looked at in Chapter 8. In each case, either the input resistor or the output resistor is replaced with either a capacitor or an inductor.

Where we call them integrators and differentiators we will look at their output waveforms when their input is a pulse. This is called the time response. Then we will consider their action as filters; that is, we will look at the amplitude and phase of their outputs when the input is a sine wave of varying frequency. This is the frequency response.

The two approaches are in fact different ways of looking at the same thing, which is their transfer function – in other words their output for any given input. Both methods can be both understood and quantified without much mathematics, and one, other or both is usually very helpful when we need to use them in a circuit, or understand their use.

Time response

A quick explanation of CR and LR circuit waveforms

The four circuits, with their associated waveforms, are shown in Figure 9.1.

We assume that the resistance of the source is negligible, and the load on the circuit is high enough to be ignored. Both CR circuits have the same current waveform, and so do both LR circuits. (As discussed in Chapter 5, the current in a series circuit is the same through each of its components, and putting them in a different order makes no difference to that current.) Also, the sum of the capacitor's and the resistor's, or the inductor's and the resistor's, voltages always equals the input voltage (by Kirchhoff's voltage law).

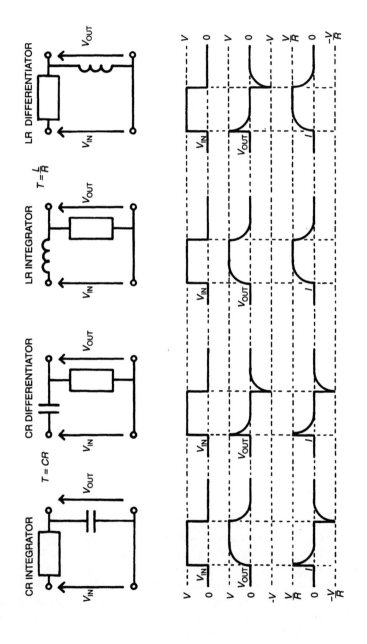

Figure 9.1 First-order circuits – time response

A capacitor will always resist a sudden change in the voltage across its terminals. Hence, when the leading edge of the pulse arrives, its voltage remains, initially, at zero. All of the available voltage is, at that moment, across the resistor. Therefore the current at that moment is $I = \frac{V}{R}$.

The current charges the capacitor, causing its voltage to increase from zero. As this happens the voltage across the resistor decreases. The current (proportional always to the voltage across the resistor) falls accordingly. This in turn slows down the charge rate of the capacitor. Hence the rate of change of the voltage across the capacitor decreases as that voltage approaches the input voltage. The current in the circuit falls from $I = \frac{V}{R}$ towards zero, and the voltage across the resistor falls from V towards zero.

Eventually, the capacitor is fully charged, and the voltage across it is equal to the input voltage. Things will stay like this until the input changes again. When it falls to zero the voltage across the cap cannot change instantly; it must now discharge, back through the resistor, and into the source (the only possible path – remember that the source has a low, and the load a high, resistance). Hence a current of $\frac{-V}{R}$ now flows for an instant (the minus sign indicating that the direction has reversed). This current decreases towards zero again as the voltage which created it, that across the plates of the capacitor, falls. Eventually, there is no charge left in the capacitor, no voltage across its terminals and no current in the circuit. The circuit is once again at rest, as it was in the beginning.

Because an inductor resists a sudden change in current, an LR circuit's current is at zero when the leading edge of the pulse arrives. So then is the voltage across the resistor at that moment. For the sum of the resistor's and the inductor's voltage to equal the input voltage, as it must, the inductor sets up a back EMF across its terminals, opposing the input voltage. The back EMF decays with time, and current begins to flow in the circuit, creating a voltage across the resistor. Eventually there is no back EMF, and the current arrives at a final value of $\frac{V}{R}$. The voltage across the inductor is now zero, and the voltage across the resistor is V.

When the trailing edge of the pulse arrives at the input (i.e. the input voltage falls towards zero again) the current in the circuit also will try to fall to zero. Again the inductor will resist change. To do this it will set up another back EMF, opposing the voltage across the resistor now. This back EMF will also decay with time, allowing the current in the circuit and the voltage across the resistor to fall towards zero.

The only difference between the two CR (or the two LR) circuits is which device, resistor or capacitor (inductor), is on the output side. The voltages across the components, and the currents through them, are identical.

Time constants It can be seen that, essentially, only two waveform shapes exist for both output voltage and for current in all four circuits. These two waveforms are shown in more detail in Figure 9.2.

All of the curves are exponential in shape. How long they take to change from initial to final values is determined by the circuit time constant T (not to be confused with the T used for period of an AC waveform), which is in turn a function of component values in the circuit.

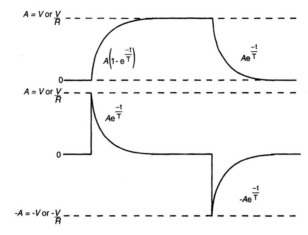

Figure 9.2　Detail of Figure 9.1

The formulae for T are:

for a CR circuit:

$$T = CR \tag{9.1}$$

and for an LR circuit:

$$T = \frac{L}{R} \tag{9.2}$$

T is tabulated for various values of C and R in Table 9.1, and for L and R in Table 9.2. These tables also give values of frequency constants which we will look at later under 'Frequency response'.

The tables are sorted in order of T, increasing for CR and decreasing for LR. T is given in μs. Should we want to obtain a value of T which is outside the range of the table we should do it by scaling one or both components by factors of 10 appropriately. As always, some examples:

Example 9.1:　Find values for a CR circuit with $T = 56$ ns.

From Table 9.1:

f (Hz)	$T(\mu s)$	R	C
2842	56.00	1k	56n
2842	56.00	5k6	10n

We want $T = 56$ ns, so we need to make R and C between them 1000 times smaller. We could for example make $R = 100\,\Omega$ and $C = 560p$, or $R = 56\,\Omega$ and $C = 1n$.

Example 9.2: Find values for an LR circuit with $T = 4\,\mu s$.

From Table 9.2:

f(Hz)	$T(\mu s)$	R	L
4063	39.17	1k2	47m

To make T 10 times smaller we could either make R 10 times larger or L 10 times smaller. So $R = 12k$ and $L = 47m$ or $R = 1k2$ and $L = 4.7m$ would both work.

Values of voltage and current as a function of t and T   Once we have determined a value of T, we can find the value of V_t or I_t, that is output voltage for any moment in time, t, after the rising or falling edge of the input waveform. The formulae are as follows.

For a curve going from zero to a final value:

$$V_t \text{ or } I_t = \left(V \text{ or } \frac{V}{R}\right)\left(1 - e^{\frac{-t}{T}}\right) \tag{9.3}$$

and for a curve going back down to zero:

$$V_t \text{ or } I_t = (-)\left(V \text{ or } \frac{V}{R}\right)e^{\frac{-t}{T}} \tag{9.4}$$

Table 9.3 gives values of Eqns (9.3) and (9.4) for various values of $\frac{t}{T}$.

Example 9.3: Find out to what the voltage out from a 2k2, $1\,\mu F$ differentiator will be after 6 ms, when the input is a rising edge of 6 V:

First we calculate T, from Eqn (9.1).

$$T = CR = 2k2 \times 10u = 22\,ms$$

Now we evaluate $\frac{t}{T}$. If $t = 6$ ms, $\frac{t}{T} = \frac{6}{22} = 0.27$. Referring to Figures 9.1 and 9.2, we see that the waveform is a rising edge, governed by the $1 - e^{\frac{-t}{T}}$ function. So we use Table 9.1 to find a value for that function where $\frac{t}{T} = 0.27$. We find:

$\frac{t}{T}$	$1 - e^{\frac{-t}{T}}$	$e^{\frac{-t}{T}}$
0.3	0.7408	0.2592

which is the nearest to 0.27. The final value of the voltage will be 6 V. So its value after 6 ms will be 0.741 $\times 6 = 4.45$ V.

Example 9.4: Design a circuit whose output voltage will be −7 V 220 ms after the falling edge of a + 10 V pulse input. Give a CR and an LR alternative.

By referring to Figure 9.1, we see that the circuit must be a differentiator, as that circuit has an output voltage which goes from −V to 0 after the falling edge of the input. The equation for that curve is $-Ve^{\frac{-t}{T}}$, where $V = 10$ V in this case. So we can write:

$$-7 = -10\,e^{\frac{-t}{T}}, \text{ or } e^{\frac{-t}{T}} = 0.7$$

and we can use Table 9.1 in reverse to find:

$\frac{t}{T}$	$1 - e^{\frac{-t}{T}}$	$e^{\frac{-t}{T}}$
1.2	0.3012	0.6988

Now we know that $t = 220$ ms, and that $\frac{t}{T} = 1.2$, so we know that $T = \frac{220}{1.2} = 183.3$ ms. Now we can use Table 9.3 to find suitable C and R values, and Table 9.4 to find L and R.

From Table 9.3:

$f(\text{Hz})$	$T(\mu s)$	R	C
8682	18.33	4k7	3n9

T is in μs, so we need to scale up by $\frac{183.3\,\text{m}}{18.33\,\mu} = 10\,000$. We could do this by multiplying both C and R by 100. So we get $R = 470$ k, $C = 390$ n.

From Table 9.4:

$f(\text{Hz})$	$T(\mu s)$	R	L
8681	18.33	1k8	33m

Here we need to scale by making R smaller and L larger. We could say theoretical values of $R = 18$ Ω and $L = 3.3$ H. A 3.3 H inductor would be huge (though not unknown!), which illustrates why CR circuits tend to be more common than LR ones.

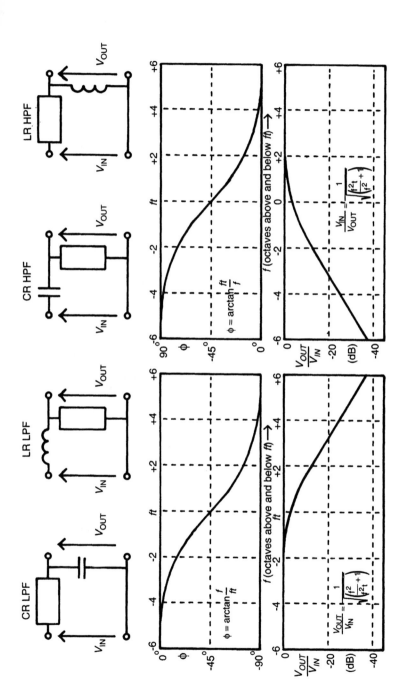

Figure 9.3 First-order circuits – frequency response

Frequency response

Now we look at the behaviour of the same circuits when their input is a continuous sine wave. We are interested in what happens as the frequency of the input is varied.

How first-order filters work

As we saw in Chapter 4, capacitive reactance falls with increasing frequency:

$$X_C = \frac{1}{2\pi f C} \qquad (9.5)$$

Inductive reactance increases with frequency:

$$X_L = 2\pi f L \qquad (9.6)$$

Each obeys Ohm's Law, in that $V_C = IX_C$, or $X_L = IX_L$ – that is the voltage across a capacitor or an inductor is the current through it multiplied by its reactance, like a resistor. However, whereas voltage and current are in phase with a resistor, they are 90° out of phase with inductors and capacitors.

The four basic filter circuits are shown in Figure 9.3. We can view them all as being similar to the voltage divider which we looked at in Chapter 6, except that one component in each has a 'resistance' which varies with frequency. Hence the overall attenuation of each varies with frequency too.

Where the component on the input side of the circuit has higher impedance at higher frequencies, the circuit is a low-pass filter. These are the two circuits on the left hand side of Figure 9.3, a CR with the capacitor across the output or an LR with the inductor on the input. Conversely, a high-pass filter is formed by a CR with the capacitor on the input side or by an LR with the inductor across the output.

Frequency constants   The frequency at which the reactance of the capacitor or inductor is the equal to the resistor is known as the 'turnover frequency' (f_t). The attenuation at this frequency is not 6 dB, as it would be if both components were resistive. Because of the fact that the voltage across the reactive component is 90° out of phase with the voltage across the resistor the attenuation is only 3 dB. The turnover frequency can be calculated from the component values (or component values selected to give a desired f_t). The governing formulae are:

For a CR circuit:

$$R = \frac{1}{2\pi f_t C}$$

which becomes

$$f_t = \frac{1}{2\pi RC} \tag{9.7}$$

For an LR circuit:

$$R = 2\pi f_t L$$

or

$$f_t = \frac{R}{2\pi L} \tag{9.8}$$

Values for f_t are given in Table 9.1 for CR circuits and in Table 9.2 for LR circuits. As before, component values can be scaled to get frequencies which are outside the range of the table.

Example 9.5: Give component values for a CR filter with $f_t = 12\,\text{kHz}$.

From Table 9.1:

$f(\text{Hz})$	$T(\mu s)$	R	C
1205	132.00	1k1	120n

We need to multiply the frequency by 10, so we could divide either C or R by 10 to achieve this. So both $R = 110\,\Omega$, $C = 120\text{n}$ and $R = 1\text{k}1$, $C = 12\text{n}$ would do it.

Example 9.6: Give component values for an LR filter with $f_t = 500\text{kHz}$.

From Table 9.2:

$f(\text{Hz})$	$T(\mu s)$	R	L
5079	31.33	1k5	47m

We need f_t 100 times larger, so we can make L smaller and/or R larger. We could choose, for instance, $R = 15\text{k}$, $L = 4.7\text{m}$.

Determining the level and phase at any frequency The level of the output voltage, relative to the input, is determined by the ratio of the signal frequency and f_t:

For a LPF:

$$\frac{V_{\text{OUT}}}{V_{\text{IN}}} = \frac{1}{\sqrt{\left(\frac{f}{f_t}\right)^2 + 1}} \tag{9.9}$$

We can see that when f is much less than f_t, the value of $\frac{V_{OUT}}{V_{IN}}$ is very close to 1, and hence the signal is not much attenuated by the circuit. When $\frac{f}{f_t}$ is much greater than 1, the value of $\frac{V_{OUT}}{V_{IN}}$ approximates to:

$$\frac{V_{OUT}}{V_{IN}} = \frac{1}{\frac{f}{f_t}} = \frac{f_t}{f} \tag{9.10}$$

From this simplified expression, it can be seen that a doubling in frequency will lead to a halving in output signal, or multiplying the input frequency by 10 will reduce its output level to one-tenth. Hence it is customary to say that the filter slope, above the turnover frequency, is −6 db/octave, or −20 dB/decade.

The equation for the attenuation of a HPF is:

$$\frac{V_{OUT}}{V_{IN}} = \frac{1}{\sqrt{\left(\frac{f_t}{f}\right)^2 + 1}} \tag{9.11}$$

and a similar argument applies; this time the slope is −6 dB/octave or −20 dB/decade below the turnover frequency.

These filters affect the relative phase (ϕ), as well as the level, of the signal. The LPF causes the output signal to be between 0 and 90° lagging the input signal, the HPF 0° to 90° leading. In both cases the phase angle is 45° at f_t. These functions are again determined by the ratio of f and f_t:

For a HPF:

$$\phi \text{ (leading)} = \arctan\left(\frac{f_t}{f}\right) \tag{9.12}$$

For a LPF:

$$\phi \text{ (lagging)} = \arctan\left(\frac{f}{f_t}\right) \tag{9.13}$$

Hence it is possible to predict both attenuation and phase shift through a filter for any given frequency, provided that f_t is known. Table 9.4 gives both at intervals of a third of an octave, for six octaves either side of f_t.

The first column of Table 9.4 gives the frequency in terms of the number of octaves away from f_t. (The minus sign means higher in frequency for an HPF, lower for an LPF.) The second gives the ratio of frequencies, which we call F (f_t/f for an HPF, f/f_t for an LPF). The third column is the attenuation as the ratio V_{OUT}/V_{IN}, the fourth is attenuation in dBs. The fifth column is the phase shift in degrees.

Example 9.6: What is the approximate attenuation and phase shift at a frequency of 5 kHz for a LPF consisting of a 3n3 capacitor and a 15k resistor?

$$f_t = \frac{1}{2\pi \times 15k \times 3n3} = 3215\,Hz$$

$$F = \frac{f}{f_t} = \frac{5000}{3215} = 1.56$$

From Table 9.4:

Octaves	F	Att	dB	$\phi°$
0 2/3	01.59	0.533	05.47	57.8

Nearest value in the table is 1.59 (2/3 octave) where attenuation is 5.5 dB, phase shift is 58°.

Example 9.7: A circuit is needed to low-pass filter a signal such that the level at 100 kHz is 15 dB down on the signal below f_t. What would f_t be (using the table to get the nearest approximation)? What will the attenuation be at 50 kHz?

From Table 9.4:

Octaves	F	Att	dB	$\phi°$
2 1/3	05.04	0.195	14.22	78.8

This tells us that for a 14 dB attenuation at 100 kHz, $F = \frac{f}{f_t} = 5$, so $f_t = \frac{100\,kHz}{5}$, or 20 kHz. At 50 kHz $F = \frac{50}{20}$ or 2.5, and we get:

Octaves	F	Att	dB	$\phi°$
0 2/3	01.59	0.533	05.47	57.8

and the attenuation is 5.5 dB.

Table 9.1 *Time and frequency constants for first-order CR circuits*

f(Hz)	$T(\mu s)$	R	C	f(Hz)	$T(\mu s)$	R	C
1020	156.00	1k3	120n	1326	120.00	1k2	100n
1026	155.10	3k3	47n	1340	118.80	3k6	33n
1026	155.10	4k7	33n	1360	117.00	3k	39n
1046	152.10	3k9	39n	1371	116.10	4k3	27n
1053	151.20	2k7	56n	1411	112.80	2k4	47n
1053	151.20	5k6	27n	1415	112.50	7k5	15n
1061	150.00	1k	150n	1418	112.20	5k1	22n
1061	150.00	1k5	100n	1421	112.00	2k	56n
1064	149.60	2k2	68n	1426	111.60	6k2	18n
1064	149.60	6k8	22n	1447	110.00	1k1	100n
1078	147.60	1k8	82n	1457	109.20	9k1	12n
1078	147.60	8k2	18n	1461	108.90	3k3	33n
1105	144.00	1k2	120n	1463	108.80	1k6	68n
1122	141.90	4k3	33n	1493	106.60	1k3	82n
1129	141.00	3k	47n	1511	105.30	3k9	27n
1134	140.40	3k6	39n	1511	105.30	2k7	39n
1156	137.70	5k1	27n	1539	103.40	2k2	47n
1166	136.50	9k1	15n	1539	103.40	4k7	22n
1167	136.40	6k2	22n	1560	102.00	6k8	15n
1170	136.00	2k	68n	1560	102.00	1k5	68n
1179	135.00	7k5	18n	1579	100.80	1k8	56n
1184	134.40	2k4	56n	1579	100.80	5k6	18n
1206	132.00	1k1	120n	1592	100.00	1k	100n
1213	131.20	1k6	82n	1608	99.00	3k	33n
1224	130.00	1k3	100n	1617	98.40	8k2	12n
1237	128.70	3k3	39n	1617	98.40	1k2	82n
1237	128.70	3k9	33n	1637	97.20	3k6	27n
1254	126.90	2k7	47n	1682	94.60	4k3	22n
1254	126.90	4k7	27n	1693	94.00	2k	47n
1292	123.20	2k2	56n	1700	93.60	2k4	39n
1292	123.20	5k6	22n	1711	93.00	6k2	15n
1294	123.00	1k5	82n	1734	91.80	5k1	18n
1294	123.00	8k2	15n	1749	91.00	9k1	10n
1300	122.40	6k8	18n	1764	90.20	1k1	82n
1300	122.40	1k8	68n	1768	90.00	7k5	12n
1326	120.00	1k	120n	1776	89.60	1k6	56n

Table 9.1 Time and frequency constants for first-order CR circuits 217

$f(\text{Hz})$	$T(\mu s)$	R	C	$f(\text{Hz})$	$T(\mu s)$	R	C
1786	89.10	2k7	33n	2411	66.00	2k	33n
1786	89.10	3k3	27n	2411	66.00	3k	22n
1800	88.40	1k3	68n	2456	64.80	2k4	27n
1855	85.80	2k2	39n	2456	64.80	3k6	18n
1855	85.80	3k9	22n	2468	64.50	4k3	15n
1881	84.60	1k8	47n	2551	62.40	1k6	39n
1881	84.60	4k7	18n	2567	62.00	6k2	10n
1895	84.00	1k5	56n	2572	61.88	9k1	6n8
1895	84.00	5k6	15n	2584	61.60	1k1	56n
1941	82.00	1k	82n	2588	61.50	7k5	8n2
1941	82.00	8k2	10n	2601	61.20	5k1	12n
1950	81.60	6k8	12n	2605	61.10	1k3	47n
1950	81.60	1k2	68n	2679	59.40	2k2	27n
1965	81.00	3k	27n	2679	59.40	3k3	18n
2010	79.20	2k4	33n	2679	59.40	1k8	33n
2010	79.20	3k6	22n	2679	59.40	2k7	22n
2040	78.00	2k	39n	2721	58.50	3k9	15n
2056	77.40	4k3	18n	2721	58.50	1k5	39n
2080	76.50	5k1	15n	2822	56.40	4k7	12n
2116	75.20	1k6	47n	2822	56.40	1k2	47n
2122	75.00	7k5	10n	2842	56.00	1k	56n
2128	74.80	1k1	68n	2842	56.00	5k6	10n
2133	74.62	9k1	8n2	2854	55.76	6k8	8n2
2139	74.40	6k2	12n	2854	55.76	8k2	6n8
2183	72.90	2k7	27n	2947	54.00	2k	27n
2186	72.80	1k3	56n	2947	54.00	3k6	15n
2192	72.60	2k2	33n	2947	54.00	3k	18n
2192	72.60	3k3	22n	3014	52.80	1k6	33n
2258	70.50	1k5	47n	3014	52.80	2k4	22n
2258	70.50	4k7	15n	3078	51.70	1k1	47n
2267	70.20	1k8	39n	3084	51.60	4k3	12n
2267	70.20	3k9	18n	3121	51.00	7k5	6n8
2341	68.00	1k	68n	3121	51.00	5k1	10n
2341	68.00	6k8	10n	3123	50.96	9k1	5n6
2367	67.24	8k2	8n2	3131	50.84	6k2	8n2
2368	67.20	1k2	56n	3139	50.70	1k3	39n
2368	67.20	5k6	12n	3215	49.50	1k5	33n

f(Hz)	$T(\mu s)$	R	C	f(Hz)	$T(\mu s)$	R	C
3215	49.50	3k3	15n	4421	36.00	3k	12n
3275	48.60	1k8	27n	4421	36.00	2k	18n
3275	48.60	2k7	18n	4421	36.00	3k6	10n
3288	48.40	2k2	22n	4485	35.49	9k1	3n9
3386	47.00	1k	47n	4514	35.26	4k3	8n2
3386	47.00	4k7	10n	4515	35.25	7k5	4n7
3401	46.80	1k2	39n	4521	35.20	1k6	22n
3401	46.80	3k9	12n	4534	35.10	1k3	27n
3442	46.24	6k8	6n8	4584	34.72	6k2	5n6
3466	45.92	8k2	5n6	4589	34.68	5k1	6n8
3466	45.92	5k6	8n2	4823	33.00	1k	33n
3537	45.00	3k	15n	4823	33.00	2k2	15n
3617	44.00	2k	22n	4823	33.00	1k5	22n
3684	43.20	1k6	27n	4823	33.00	3k3	10n
3684	43.20	2k4	18n	4912	32.40	1k2	27n
3684	43.20	3k6	12n	4912	32.40	2k7	12n
3701	43.00	4k3	10n	4912	32.40	1k8	18n
3710	42.90	1k1	39n	4977	31.98	3k9	8n2
3710	42.90	1k3	33n	4977	31.98	8k2	3n9
3721	42.77	9k1	4n7	4980	31.96	4k7	6n8
3775	42.16	6k2	6n8	4980	31.96	6k8	4n7
3789	42.00	7k5	5n6	5075	31.36	5k6	5n6
3806	41.82	5k1	8n2	5300	30.03	9k1	3n3
3930	40.50	1k5	27n	5305	30.00	2k	15n
3930	40.50	2k7	15n	5305	30.00	3k	10n
4019	39.60	3k3	12n	5359	29.70	1k1	27n
4019	39.60	1k2	33n	5391	29.52	3k6	8n2
4019	39.60	1k8	22n	5441	29.25	7k5	3n9
4019	39.60	2k2	18n	5443	29.24	4k3	6n8
4081	39.00	1k	39n	5462	29.14	6k2	4n7
4081	39.00	3k9	10n	5526	28.80	2k4	12n
4130	38.54	4k7	8n2	5526	28.80	1k6	18n
4130	38.54	8k2	4n7	5565	28.60	1k3	22n
4179	38.08	6k8	5n6	5573	28.56	5k1	5n6
4179	38.08	5k6	6n8	5882	27.06	3k3	8n2
4384	36.30	1k1	33n	5882	27.06	8k2	3n3
4421	36.00	2k4	15n	5895	27.00	1k	27n

Table 9.1 Time and frequency constants for first-order CR circuits   219

f(Hz)	$T(\mu s)$	R	C	f(Hz)	$T(\mu s)$	R	C
5895	27.00	1k8	15n	7802	20.40	3k	6n8
5895	27.00	1k5	18n	7860	20.25	7k5	2n7
5895	27.00	2k7	10n	7875	20.21	4k3	4n7
6001	26.52	3k9	6n8	7895	20.16	3k6	5n6
6001	26.52	6k8	3n9	7950	20.02	9k1	2n2
6029	26.40	2k2	12n	7958	20.00	2k	10n
6029	26.40	1k2	22n	8002	19.89	5k1	3n9
6047	26.32	4k7	5n6	8038	19.80	1k1	18n
6047	26.32	5k6	4n7	8087	19.68	2k4	8n2
6431	24.75	7k5	3n3	8162	19.50	1k3	15n
6470	24.60	3k	8n2	8289	19.20	1k6	12n
6478	24.57	9k1	2n7	8612	18.48	3k3	5n6
6501	24.48	3k6	6n8	8612	18.48	5k6	3n3
6577	24.20	1k1	22n	8669	18.36	2k7	6n8
6582	24.18	6k2	3n9	8669	18.36	6k8	2n7
6609	24.08	4k3	5n6	8683	18.33	3k9	4n7
6631	24.00	1k6	15n	8683	18.33	4k7	3n9
6631	24.00	2k	12n	8822	18.04	2k2	8n2
6631	24.00	2k4	10n	8822	18.04	8k2	2n2
6640	23.97	5k1	4n7	8842	18.00	1k2	15n
6801	23.40	1k3	18n	8842	18.00	1k5	12n
7074	22.50	1k5	15n	8842	18.00	1k	18n
7092	22.44	3k3	6n8	8842	18.00	1k8	10n
7092	22.44	6k8	3n3	9406	16.92	3k6	4n7
7189	22.14	2k7	8n2	9457	16.83	5k1	3n3
7189	22.14	8k2	2n7	9474	16.80	3k	5n6
7205	22.09	4k7	4n7	9490	16.77	4k3	3n9
7234	22.00	1k	22n	9507	16.74	6k2	2n7
7234	22.00	2k2	10n	9646	16.50	7k5	2n2
7287	21.84	3k9	5n6	9646	16.50	1k1	15n
7287	21.84	5k6	3n9	9705	16.40	2k	8n2
7368	21.60	1k2	18n	9716	16.38	9k1	1n8
7368	21.60	1k8	12n	9752	16.32	2k4	6n8
7779	20.46	6k2	3n3	9947	16.00	1k6	10n

Table 9.2 *Time and frequency constants for first-order LR circuits*

f(Hz)	$T(\mu s)$	R	L	f(Hz)	$T(\mu s)$	R	L
1592	100.00	1k	100m	5418	29.37	1k6	47m
1751	90.91	1k1	100m	5617	28.33	2k4	68m
1910	83.33	1k2	100m	5730	27.78	3k6	100m
2069	76.92	1k3	100m	5787	27.50	1k2	33m
2341	68.00	1k	68m	6095	26.11	1k8	47m
2387	66.67	1k5	100m	6207	25.64	3k9	100m
2546	62.50	1k6	100m	6270	25.38	1k3	33m
2575	61.82	1k1	68m	6319	25.19	2k7	68m
2809	56.67	1k2	68m	6773	23.50	2k	47m
2865	55.56	1k8	100m	6844	23.26	4k3	100m
3043	52.31	1k3	68m	7022	22.67	3k	68m
3183	50.00	2k	100m	7234	22.00	1k	22m
3386	47.00	1k	47m	7234	22.00	1k5	33m
3501	45.45	2k2	100m	7450	21.36	2k2	47m
3511	45.33	1k5	68m	7480	21.28	4k7	100m
3725	42.73	1k1	47m	7717	20.62	1k6	33m
3745	42.50	1k6	68m	7724	20.61	3k3	68m
3820	41.67	2k4	100m	7958	20.00	1k1	22m
4064	39.17	1k2	47m	8117	19.61	5k1	100m
4213	37.78	1k8	68m	8127	19.58	2k4	47m
4297	37.04	2k7	100m	8426	18.89	3k6	68m
4402	36.15	1k3	47m	8681	18.33	1k2	22m
4681	34.00	2k	68m	8681	18.33	1k8	33m
4775	33.33	3k	100m	8913	17.86	5k6	100m
4823	33.00	1k	33m	9128	17.44	3k9	68m
5079	31.33	1k5	47m	9143	17.41	2k7	47m
5149	30.91	2k2	68m	9405	16.92	1k3	22m
5252	30.30	3k3	100m	9646	16.50	2k	33m
5305	30.00	1k1	33m	9868	16.13	6k2	100m

Table 9.3 *Step response of first-order CR and LR circuits*

$\frac{t}{T}$	$e^{\frac{-t}{T}}$	$1 - e^{\frac{-t}{T}}$	$\frac{t}{T}$	$e^{\frac{-t}{T}}$	$1 - e^{\frac{-t}{T}}$
0.1	0.9048	0.0952	2.6	0.0743	0.9257
0.2	0.8187	0.1813	2.7	0.0672	0.9328
0.3	0.7408	0.2592	2.8	0.0608	0.9392
0.4	0.6703	0.3297	2.9	0.0550	0.9450
0.5	0.6065	0.3935	3.0	0.0498	0.9502
0.6	0.5488	0.4512	3.1	0.0450	0.9550
0.7	0.4966	0.5034	3.2	0.0408	0.9592
0.8	0.4493	0.5507	3.3	0.0369	0.9631
0.9	0.4066	0.5934	3.4	0.0334	0.9666
1.0	0.3679	0.6321	3.5	0.0302	0.9698
1.1	0.3329	0.6671	3.6	0.0273	0.9727
1.2	0.3012	0.6988	3.7	0.0247	0.9753
1.3	0.2725	0.7275	3.8	0.0224	0.9776
1.4	0.2466	0.7534	3.9	0.0202	0.9798
1.5	0.2231	0.7769	4.0	0.0183	0.9817
1.6	0.2019	0.7981	4.1	0.0166	0.9834
1.7	0.1827	0.8173	4.2	0.0150	0.9850
1.8	0.1653	0.8347	4.3	0.0136	0.9864
1.9	0.1496	0.8504	4.4	0.0123	0.9877
2.0	0.1353	0.8647	4.5	0.0111	0.9889
2.1	0.1225	0.8775	4.6	0.0101	0.9899
2.2	0.1108	0.8892	4.7	0.0091	0.9909
2.3	0.1003	0.8997	4.8	0.0082	0.9918
2.4	0.0907	0.9093	4.9	0.0074	0.9926
2.5	0.0821	0.9179	5.0	0.0067	0.9933

Table 9.4 *Frequency and phase response of first-order CR and LR circuits*

Octaves	F	Att	dB	$\phi(°)$
−6	00.02	1.000	00.00	00.9
−6 1/3	00.02	1.000	00.00	01.1
−6 2/3	00.02	1.000	00.00	01.4
−5	00.03	1.000	00.00	01.8
−5 1/3	00.04	0.999	00.01	02.3
−5 2/3	00.05	0.999	00.01	02.8
−4	00.06	0.998	00.02	03.6
−4 1/3	00.08	0.997	00.03	04.5
−4 2/3	00.10	0.995	00.04	05.7
−3	00.12	0.992	00.07	07.1
−3 1/3	00.16	0.988	00.11	09.0
−3 2/3	00.20	0.981	00.17	11.2
−2	00.25	0.970	00.26	14.0
−2 1/3	00.31	0.954	00.41	17.5
−2 2/3	00.40	0.929	00.64	21.6
−1	00.50	0.894	00.97	26.6
−1 1/3	00.63	0.846	01.45	32.2
−1 2/3	00.79	0.783	02.12	38.4
0	01.00	0.707	03.01	45.0
0 1/3	01.26	0.622	04.13	51.6
0 2/3	01.59	0.533	05.47	57.8
1	02.00	0.447	06.99	63.4
1 1/3	02.52	0.369	08.66	68.4
1 2/3	03.17	0.300	10.45	72.5
2	04.00	0.243	12.30	76.0
2 1/3	05.04	0.195	14.22	78.8
2 2/3	06.35	0.156	16.16	81.0
3	08.00	0.124	18.13	82.9
3 1/3	10.08	0.099	20.11	84.3
3 2/3	12.70	0.079	22.10	85.5
4	16.00	0.062	24.10	86.4
4 1/3	20.16	0.050	26.10	87.2
4 2/3	25.40	0.039	28.10	87.7
5	32.00	0.031	30.11	88.2
5 1/3	40.32	0.025	32.11	88.6
5 2/3	50.80	0.020	34.12	88.9
6	64.00	0.016	36.12	89.1

10 LC tuned circuits

Introduction

When inductors and capacitors are combined the circuit is called a tuned circuit. Where the circuits in Chapter 9 were termed first-order, we call these second-order, because there are two frequency-dependent elements rather than one. The capacitor and inductor can be combined in either series or parallel, giving rise to two types which we shall consider separately. These circuits both display an effect known as resonance, by which we mean that their impedance is at a minimum or a maximum at one specific frequency.

Series tuned circuit

A series tuned circuit is an inductor and a capacitor in series. A resistive component is also assumed to exist which may be considered to consist solely of the circuit's residual resistances (the inductor's internal resistance being, usually, the only significant one) or perhaps of residual resistances plus a physical resistor. The circuit is shown in Figure 10.1.

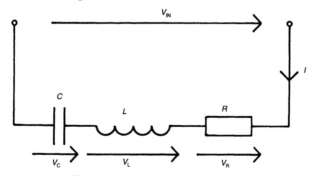

Figure 10.1 Series tuned circuit

What the circuit does

The reactance of the capacitor decreases with frequency, while the reactance of the inductor increases. As we know, we cannot just add the impedance of each component at some frequency together because the phase angles of the voltages will be different. But we can deduce what is happening.

We will take the current in the circuit as being our phase reference, as it is common to all components. The voltage across the capacitor lags it by 90°, the voltage across the inductor leads it by 90° and the voltage across the resistor is in phase with it. The input voltage, which we would divide by input current to find the impedance, will be the phasor sum of these. (If you're a beginner, and this is losing you, look back at 'Calculations on AC quantities' in Chapter 4.)

If X_C decreases with increasing frequency, and X_L increases, then they must be equal at some point. At this frequency the capacitor's and the inductor's voltage are equal, but opposing each other in phase. (Their magnitudes can be calculated by using IX_L and IX_C respectively, and I is the same for each.) This means that they cancel each other – their phasor sum is zero. So the input voltage is equal to the voltage across the resistor – the resistor has the maximum amount of voltage across it that it can have.

This causes the current in the circuit to be at a maximum, and hence impedance is at a minimum, and in phase (resistive). We call this frequency the resonant frequency of the circuit (f_O). Below resonance X_C is greater than X_L, and the impedance is overall capacitive. This means that the current leads the input voltage by some small angle when the impedances are almost equal, and by nearly 90° when the frequency is very low. Above resonance X_L is more significant and the impedance is inductive. Then input voltage leads current by between 0 and 90°.

Calculating resonant frequency

We can deduce the frequency where the two impedances are equal:

$$\frac{1}{2\pi f_O C} = 2\pi f_O L$$

After a little algebra this becomes:

$$f_O = \frac{1}{2\pi\sqrt{LC}} \tag{10.1}$$

The impedance at resonance, which we term Zo is simply:

$$Z_O = R \tag{10.2}$$

Q factor

Although V_C and V_L cancel at resonance, they are still definitely there. In fact they can be larger than the input voltage. This is because the current can be large, given by $\frac{V_{IN}}{R}$, if R is small. The voltage across, say, the capacitor, will be IX_C, or $\frac{V_{IN}}{R} X_C$. We call the ratio of this magnified voltages to the input voltage the Q, or quality factor, of the circuit. Q can be calculated:

$$V_C = \frac{V_{IN}}{R} X_C$$

(because the impedance at resonance is simply R). So:

$$Q = \frac{V_C}{V_{IN}} = \frac{X_C}{R} = \frac{1}{2\pi f_O CR} \qquad (10.3)$$

or

$$V_L = IX_L = \frac{V_{IN}}{R} X_L$$

so

$$Q = \frac{V_L}{V_{IN}} = \frac{X_L}{R} = \frac{2\pi f_O L}{R} \qquad (10.4)$$

The result should be the same whichever equation we use to calculate Q. From these equations it can be seen that Q can be decreased quite easily by simply adding more resistance in series with the circuit. Conversely the maximum obtainable Q is set by the leftover resistance in the circuit when we have no physical resistor present. The inductor will normally possess far more resistance than the capacitor, so it is common to talk about the Q of an inductor at some specified frequency – this is the highest Q obtainable with that inductor in a tuned circuit with that f_O.

Q as an indication of bandwidth

The fact that the impedance of a series tuned circuit drops to a minimum at resonance makes it very useful for applications where we need some mechanism for picking a certain group of frequencies out of a wideband signal – for instance radio tuners, or audio filters. In these situations we need to know how selective the circuit actually is. The Q factor, conveniently, tells us this – the circuit's bandwidth, or BW. The bandwidth is the difference, in Hz, between the frequencies above and below resonance where the current is 3 dB less than at resonance, for the same level of input voltage – i.e. where the impedance drops by 3 dB. The formula we need is this:

$$BW = \frac{f_O}{Q} \qquad (10.5)$$

For circuits of high selectivity (low BW), the requirement is that Q be high, and hence R low. This will influence the selection of components used in the design, particularly that an inductor with low internal resistance be found relative to its impedance at f_O. Often we design the circuit with Q higher than needed, and then add a suitable series resistor, to reduce the bandwidth to the desired value. This is called a 'damping' resistor.

Table 10.1 lists the resonant frequencies from 500 Hz to 50 MHz obtainable with E6 inductors and E12 capacitors. A column is also provided, X, showing the value of X_C or X_L at resonance. This can be used to find the highest possible Q, provided that the resistance of the inductor is known.

Example 10.1: (a) What values of C and L can we use to give a series resonant circuit with $f_O = 8\,kHz$?

From Table 10.1:

f_O	C	L	X
8k	12n	33m	1.66k
8k	18n	22m	1.11k
8k	120n	3m3	166
8k	180n	2m2	111

This gives us a choice of four pairs.

(b) By consulting suppliers catalogues, we source a 3.3 mH inductor with $R = 6.6\,\Omega$, and a 2.2 mH inductor with resistance, R_L, of 4.1 Ω, What is the best possible Q with each of these? What will the bandwidth be?

With 3.3 mH, X is given as 166 Ω, so Q will be $\dfrac{166}{6.6} = 25.2$ at best. The bandwidth will be $\frac{8000}{25.2} = 318$ Hz

With 2.2 mH, X is given as 111 Ω, so Q will be $\frac{111}{4.1} = 27.1$ at best. The bandwidth will be $\frac{8000}{27.1} = 295$ Hz

(c) Now calculate values for a series resistor to reduce Q to around 3 for each.

For 3.3 mH: If $X = 166\,\Omega$, and $Q = 3$, then $R = \frac{X}{Q} = \frac{166}{3} = 55.3\,\Omega$. As the inductor already has 6.6 Ω of resistance, we should add a resistor of $55.3 - 6.6 = 48.7\,\Omega$. (with typical inductor tolerances of 10%, 47 Ω would be fine).

For 2.2 mH: $R = \frac{X}{Q} = \frac{111}{3} = 37\,\Omega$. Resistor should be $37 - 4.1 = 32.9\,\Omega$ – say a standard value of 33 Ω.

Parallel tuned circuits

A parallel tuned circuit consists of a capacitor and an inductor in parallel. The circuit can be understood by using a similar approach to the one which we took for the series tuned circuit. However, the practical circuit is slightly complicated by the fact that the inductive branch possesses more resistance than the capacitive branch, which makes it somewhat asymmetrical. In many cases this effect can be ignored, but not all, so we need to be able to predict whether it will be significant or not. But to make the analysis one step at a time, it makes sense to start by considering an ideal version of the circuit, where the effect does not come into play at all.

Example 10.2: If the 2.2 mH inductor of Example 10.1 has a 10% tolerance, how much might f_0 and Q vary in practice, assuming that both inductance and resistance increase together?

If the inductance and resistance increase by 10%, we would have $L(\text{max}) = 2.2\,\text{mH} \times 1.1 = 2.42\,\text{mH}$, and $R_{L(\text{max})} = 4.1 \times 1.1 = 4.51\,\Omega$. This would give, using Eqn (6.1):

$$f_0 = \frac{1}{2\pi\sqrt{2.42\text{m} \times 180\text{n}}} = 7.63\,\text{kHz}$$

and a corresponding $Q = \frac{2\pi \times 7.63\text{k} \times 2.42\text{m}}{33 + 4.51} = 3.1$
Decreasing by 10% gives $L_{(\text{min})} = 2.2\text{m} \times 0.9 = 1.98\,\text{mH}$ and $R_L(\text{min}) = 4.1 \times 0.9 = 3.69\,\Omega$

$$f_0 = \frac{1}{2\pi\sqrt{1.98\text{m} \times 180\text{n}}} = 8.43\,\text{kHz}$$

with

$$Q = \frac{2\pi \times 8.43\text{k} \times 1.98\text{m}}{33 + 3.69} = 2.9$$

The ideal parallel tuned circuit

This consists of an ideal inductor and capacitor in parallel -- each has no resistance. In this circuit we take the voltage across the circuit, V_{IN}, as our phase reference, as it is common to both components, and look at the current that each branch draws from the source at different frequencies. These currents will always be in antiphase – the capacitor's current, I_{C}, leading V_{IN} by 90°, the inductor's, I_{L}, lagging. The input current, I_{IN}, will be their phasor sum. As they are in antiphase, their phasor sum is simply the difference between them.

We can once again say that X_{C} will fall with increasing frequency, and X_{L} will rise. And once again, X_{C} will equal X_{L} at some frequency, that given by Eqn (10.1). At this frequency, again f_0, the current in each branch will be equal, and hence the input current will be zero. The impedance is $\frac{V_{\text{IN}}}{I_{\text{IN}}}$, and is infinite. The circuit looks like an open circuit to the source at resonance, and its resonant frequency is calculated identically to that of the series tuned circuit.

Below resonance, the capacitor will have the greater impedance of the two devices, and the impedance will appear inductive with current lagging voltage by $90°$. Above, the inductor possesses more impedance and the impedance is capacitive, with V_{IN} in lagging I_{IN}.

At resonance, the input current is zero, while I_C and I_L are $\frac{V_{IN}}{X_C}$ and $\frac{V_{IN}}{X_L}$ respectively. The Q factor, $\frac{I_C}{I_{IN}}$ or $\frac{I_L}{I_{IN}}$ in would therefore be infinite.

The practical parallel tuned circuit

The practical circuit differs from the ideal mostly in that the inductor possesses resistance. Such a circuit is shown in Figure 10.2, with R_L representing the internal resistance of the inductor.

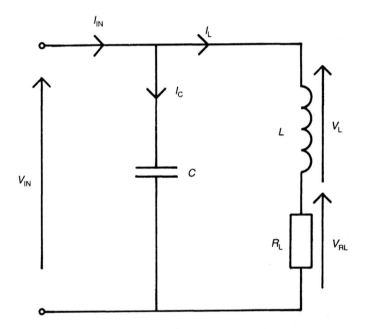

Figure 10.2 Practical parallel tuned circuit

The addition of the resistive element in series with the inductor causes I_L to be less than $90°$ lagging V_{IN}. In turn, this causes I_{IN}, at the frequency at which $X_C = X_L$, to be leading V_{IN} by some angle, rather than in phase, as it was with the ideal circuit—the circuit is overall capacitive, rather than resistive. To compensate, and bring I_{IN} back in phase with V_{IN}, making the circuit truly resonant, the frequency of V_{IN} must be lowered. This has the effect of making I_C smaller and I_L larger. It can be seen that the effect becomes more pronounced the greater the inductor's resistance is, compared to its reactance.

To what degree all this affects Eqn (10.1) depends on the Q of the inductor at the resonant frequency, as defined in Eqn (10.4).

Resonant frequency error

The resonant frequency as found from Eqn (10.1) must be multiplied by the term:

$$k = \sqrt{1 - \frac{CR^2}{L}} = \sqrt{1 - \frac{1}{Q^2}} \qquad (10.6)$$

to find the true resonant frequency. In practice, the Q has to be pretty low to cause much deviation; a Q of 4 gives $k = 0.97$, a Q of 2 gives $k = 0.87$. Eqn (10.1) is often used as it stands.

Dynamic impedance

The practical circuit no longer looks like an open circuit to the source at resonance. Its impedance is purely resistive however, and is often called the dynamic impedance. It is given by:

$$Z_O = \frac{L}{CR_L} \qquad (10.7)$$

From this we can see that dynamic impedance is inversely proportional to the resistance of the inductor.

Q factor

With the practical parallel tuned circuit the Q is, as with the series resonant circuit, the same as the Q of the inductor at the resonant frequency. We might ask ourselves what happens when we wish to deliberately reduce the circuit's Q; for if we placed our damping resistor in series with the inductor, it would also affect our resonant frequency. The solution is that the damping resistor is placed across the whole tuned circuit, as shown in Figure 10.3, where it has the desired effect without affecting f_O. The value of the damping resistor can be calculated as follows:

$$R_D = Z_O \frac{Q_D}{Q_I - Q_D} \qquad (10.8)$$

where R_D is the damping resistor value, Q_D is the desired Q (Q *damped*), $Z_O = \frac{L}{CR}$ as previously, and Q_I is the Q of the inductor at resonance, or the Q of the circuit with no damping resistor. Q_D must obviously be less than Q_I.

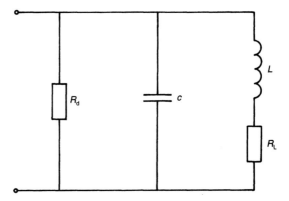

Figure 10.3 Practical parallel tuned circuit with damping resistor

Maximum capacitor value

We can see that as R becomes larger, k Eqn (10.6) goes from 1 towards zero, and the resonant frequency falls, also towards zero. When:

$$\frac{CR_L^2}{L} = 1 \text{ or } C = \frac{L}{R_L^2} \tag{10.9}$$

then $k = 0$, and the circuit ceases to resonate. This tells us that for any practical inductor, there is a maximum value of capacitor which can be placed across it for it to remain resonant. This value is easily calculated from Eqn (10.9).

Example 10.3: design a parallel tuned circuit with $f_O = 120\,\text{kHz} +/-2\,\text{kHz}$, $Q = 20$ and the highest feasible dynamic impedance, with the restriction that C is greater than 100 pF. What is the dynamic impedance, with and without the damping resistor? (Assume that the resistance of the inductor is 5 Ω.)

From Table 10.1, we choose $C = 120\,\text{pF}$, $L = 15\,\text{mH}$, giving a nominal frequency of 119 kHz. We can calculate the Q of the inductor by taking X from the table $-11.2\text{k}-$ and dividing by 5 Ω. This gives $Q = 2240$. Now we calculate Z_O, using Eqn (10.7):

$$Z_O = \frac{15\text{m}}{120\text{p} \times 5} = 25\,\text{M}\Omega$$

and so

$$R_D = 25\text{M} \times \frac{20}{2240 - 20} = 225\,\text{k}\,\Omega$$

As Z_O is so much larger than R_D, we can take the dynamic impedance with R_D fitted to be equal to R_D, 225 k Ω.

Table 10.1 *Frequency constants for LC circuits*

f_O	C	L	X	f_O	C	L	X
503	1u	100m	316	1.40k	390n	33m	291
556	820n	100m	349	1.41k	270n	47m	417
610	680n	100m	383	1.43k	560n	22m	198
610	1u	68m	261	1.44k	820n	15m	135
673	560n	100m	423	1.44k	180n	68m	615
674	820n	68m	288	1.45k	120n	100m	913
734	470n	100m	461	1.53k	330n	33m	316
734	1u	47m	217	1.57k	220n	47m	462
740	680n	68m	316	1.57k	470n	22m	216
806	390n	100m	506	1.58k	680n	15m	149
811	820n	47m	239	1.58k	150n	68m	673
816	560n	68m	348	1.59k	100n	100m	1.00k
876	330n	100m	550	1.59k	1u	10m	100
876	1u	33m	182	1.69k	270n	33m	350
890	680n	47m	263	1.72k	390n	22m	238
890	470n	68m	380	1.73k	180n	47m	511
968	820n	33m	201	1.74k	560n	15m	164
969	270n	100m	609	1.76k	820n	10m	110
977	390n	68m	418	1.76k	82n	100m	1.10k
981	560n	47m	290	1.76k	120n	68m	753
1.06k	330n	68m	454	1.87k	220n	33m	387
1.06k	680n	33m	220	1.87k	330n	22m	258
1.07k	470n	47m	316	1.90k	150n	47m	560
1.07k	220n	100m	674	1.90k	470n	15m	179
1.07k	1u	22m	148	1.93k	68n	100m	1.21k
1.17k	560n	33m	243	1.93k	100n	68m	825
1.17k	270n	68m	502	1.93k	680n	10m	121
1.18k	390n	47m	347	1.93k	1u	6m8	82.5
1.18k	820n	22m	164	2.07k	180n	33m	428
1.19k	180n	100m	745	2.07k	270n	22m	285
1.28k	470n	33m	265	2.08k	390n	15m	196
1.28k	330n	47m	377	2.12k	120n	47m	626
1.30k	150n	100m	816	2.13k	56n	100m	1.34k
1.30k	1u	15m	122	2.13k	560n	10m	134
1.30k	220n	68m	556	2.13k	82n	68m	911
1.30k	680n	22m	180	2.13k	820n	6m8	91.1

f_O	C	L	X	f_O	C	L	X
2.26k	150n	33m	469	3.10k	560n	4m7	91.6
2.26k	330n	15m	213	3.36k	150n	15m	316
2.29k	220n	22m	316	3.36k	33n	68m	1.44k
2.32k	47n	100m	1.46k	3.36k	68n	33m	697
2.32k	100n	47m	686	3.36k	330n	6m8	144
2.32k	470n	10m	146	3.36k	680n	3m3	69.7
2.32k	1u	4m7	68.6	3.39k	47n	47m	1.00k
2.34k	68n	68m	1.00k	3.39k	470n	4m7	100
2.34k	680n	6m8	100	3.39k	22n	100m	2.13k
2.50k	270n	15m	236	3.39k	100n	22m	469
2.53k	120n	33m	524	3.39k	220n	10m	213
2.53k	180n	22m	350	3.39k	1u	2m2	46.9
2.55k	39n	100m	1.60k	3.70k	56n	33m	768
2.55k	390n	10m	160	3.70k	560n	3m3	76.8
2.56k	82n	47m	757	3.71k	27n	68m	1.59k
2.56k	820n	4m7	75.7	3.71k	270n	6m8	159
2.58k	56n	68m	1.10k	3.72k	39n	47m	1.10k
2.58k	560n	6m8	110	3.72k	390n	4m7	110
2.77k	33n	100m	1.74k	3.75k	82n	22m	518
2.77k	100n	33m	574	3.75k	820n	2m2	51.8
2.77k	150n	22m	383	3.75k	18n	100m	2.36k
2.77k	220n	15m	261	3.75k	120n	15m	354
2.77k	330n	10m	174	3.75k	180n	10m	236
2.77k	1u	3m3	57.4	4.04k	33n	47m	1.19k
2.82k	680n	4m7	83.1	4.04k	47n	33m	838
2.82k	47n	68m	1.20k	4.04k	330n	4m7	119
2.82k	68n	47m	831	4.04k	470n	3m3	83.8
2.82k	470n	6m8	120	4.11k	15n	100m	2.58k
3.06k	82n	33m	634	4.11k	100n	15m	387
3.06k	820n	3m3	63.4	4.11k	150n	10m	258
3.06k	27n	100m	1.92k	4.11k	1u	1m5	38.7
3.06k	180n	15m	289	4.11k	22n	68m	1.76k
3.06k	270n	10m	192	4.11k	68n	22m	569
3.09k	390n	6m8	132	4.11k	220n	6m8	176
3.09k	39n	68m	1.32k	4.11k	680n	2m2	56.9
3.10k	120n	22m	428	4.44k	390n	3m3	92.0
3.10k	56n	47m	916	4.44k	39n	33m	920

Table 10.1 Frequency constants for LC circuits 233

f_O	C	L	X	f_O	C	L	X
4.47k	27n	47m	1.32k	5.91k	33n	22m	816
4.47k	270n	4m7	132	5.91k	220n	3m3	122
4.53k	56n	22m	627	5.91k	330n	2m2	81.6
4.53k	560n	2m2	62.7	5.99k	15n	47m	1.77k
4.54k	820n	1m5	42.8	5.99k	47n	15m	565
4.54k	82n	15m	428	5.99k	150n	4m7	177
4.55k	18n	68m	1.94k	5.99k	470n	1m5	56.5
4.55k	180n	6m8	194	6.10k	6n8	100m	3.83k
4.59k	12n	100m	2.89k	6.10k	100n	6m8	261
4.59k	120n	10m	289	6.10k	680n	1m	38.3
4.82k	33n	33m	1.00k	6.10k	10n	68m	2.61k
4.82k	330n	3m3	100	6.10k	68n	10m	383
4.95k	22n	47m	1.46k	6.10k	1u	680u	26.1
4.95k	47n	22m	684	6.53k	18n	33m	1.35k
4.95k	220n	4m7	146	6.53k	180n	3m3	135
4.95k	470n	2m2	68.4	6.53k	27n	22m	903
4.98k	680n	1m5	47.0	6.53k	270n	2m2	90.3
4.98k	15n	68m	2.13k	6.58k	390n	1m5	62.0
4.98k	68n	15m	470	6.58k	39n	15m	620
4.98k	150n	6m8	213	6.70k	12n	47m	1.98k
5.03k	10n	100m	3.16k	6.70k	120n	4m7	198
5.03k	100n	10m	316	6.73k	5n6	100m	4.23k
5.03k	1u	1m	31.6	6.73k	56n	10m	423
5.33k	27n	33m	1.11k	6.73k	560n	1m	42.3
5.33k	270n	3m3	111	6.74k	8n2	68m	2.88k
5.43k	39n	22m	751	6.74k	82n	6m8	288
5.43k	390n	2m2	75.1	6.74k	820n	680u	28.8
5.47k	180n	4m7	162	7.15k	15n	33m	1.48k
5.47k	18n	47m	1.62k	7.15k	33n	15m	674
5.49k	56n	15m	518	7.15k	150n	3m3	148
5.49k	560n	1m5	51.8	7.15k	330n	1m5	67.4
5.56k	8n2	100m	3.49k	7.23k	22n	22m	1.00k
5.56k	82n	10m	349	7.23k	220n	2m2	100
5.56k	820n	1m	34.9	7.34k	4n7	100m	4.61k
5.57k	12n	68m	2.38k	7.34k	10n	47m	2.17k
5.57k	120n	6m8	238	7.34k	47n	10m	461
5.91k	22n	33m	1.22k	7.34k	100n	4m7	217

f_O	C	L	X	f_O	C	L	X
7.34k	470n	1m	46.1	9.68k	82n	3m3	201
7.34k	1u	470u	21.7	9.68k	820n	330u	20.1
7.40k	6n8	68m	3.16k	9.69k	2n7	100m	6.09k
7.40k	68n	6m8	316	9.69k	18n	15m	913
7.40k	680n	680u	31.6	9.69k	180n	1m5	91.3
7.91k	27n	15m	745	9.69k	27n	10m	609
7.91k	270n	1m5	74.5	9.69k	270n	1m	60.9
8.00k	12n	33m	1.66k	9.77k	3n9	68m	4.18k
8.00k	18n	22m	1.11k	9.77k	39n	6m8	418
8.00k	120n	3m3	166	9.77k	390n	680u	41.8
8.00k	180n	2m2	111	9.80k	12n	22m	1.35k
8.06k	3n9	100m	5.06k	9.80k	120n	2m2	135
8.06k	39n	10m	506	9.81k	5n6	47m	2.90k
8.06k	390n	1m	50.6	9.81k	56n	4m7	290
8.11k	8n2	47m	2.39k	9.81k	560n	470u	29.0
8.11k	82n	4m7	239	10.6k	15n	15m	1.00k
8.11k	820n	470u	23.9	10.6k	150n	1m5	100
8.16k	5n6	68m	3.48k	10.6k	3n3	68m	4.54k
8.16k	56n	6m8	348	10.6k	6n8	33m	2.20k
8.16k	560n	680u	34.8	10.6k	33n	6m8	454
8.76k	3n3	100m	5.50k	10.6k	68n	3m3	220
8.76k	10n	33m	1.82k	10.6k	330n	680u	45.4
8.76k	15n	22m	1.21k	10.6k	680n	330u	22.0
8.76k	22n	15m	826	10.7k	4n7	47m	3.16k
8.76k	33n	10m	550	10.7k	47n	4m7	316
8.76k	100n	3m3	182	10.7k	470n	470u	31.6
8.76k	150n	2m2	121	10.7k	2n2	100m	6.74k
8.76k	220n	1m5	82.6	10.7k	10n	22m	1.48k
8.76k	330n	1m	55.0	10.7k	22n	10m	674
8.76k	1u	330u	18.2	10.7k	100n	2m2	148
8.90k	6n8	47m	2.63k	10.7k	220n	1m	67.4
8.90k	68n	4m7	263	10.7k	1u	220u	14.8
8.90k	680n	470u	26.3	11.7k	5n6	33m	2.43k
8.90k	4n7	68m	3.80k	11.7k	56n	3m3	243
8.90k	47n	6m8	380	11.7k	560n	330u	24.3
8.90k	470n	680u	38.0	11.7k	2n7	68m	5.02k
9.68k	8n2	33m	2.01k	11.7k	27n	6m8	502

Table 10.1 Frequency constants for LC circuits 235

f_O	C	L	X	f_O	C	L	X
11.7k	270n	680u	50.2	14.3k	56n	2m2	198
11.8k	3n9	47m	3.47k	14.3k	560n	220u	19.8
11.8k	39n	4m7	347	14.4k	8n2	15m	1.35k
11.8k	390n	470u	34.7	14.4k	82n	1m5	135
11.8k	8n2	22m	1.64k	14.4k	820n	150u	13.5
11.8k	82n	2m2	164	14.4k	1n8	68m	6.15k
11.8k	820n	220u	16.4	14.4k	18n	6m8	615
11.9k	1n8	100m	7.45k	14.4k	180n	680u	61.5
11.9k	12n	15m	1.12k	14.5k	12n	10m	913
11.9k	18n	10m	745	14.5k	1n2	100m	9.13k
11.9k	120n	1m5	112	14.5k	120n	1m	91.3
11.9k	180n	1m	74.5	15.3k	3n3	33m	3.16k
12.8k	3n3	47m	3.77k	15.3k	33n	3m3	316
12.8k	4n7	33m	2.65k	15.3k	330n	330u	31.6
12.8k	33n	4m7	377	15.7k	2n2	47m	4.62k
12.8k	47n	3m3	265	15.7k	4n7	22m	2.16k
12.8k	330n	470u	37.7	15.7k	22n	4m7	462
12.8k	470n	330u	26.5	15.7k	47n	2m2	216
13.0k	1n5	100m	8.16k	15.7k	220n	470u	46.2
13.0k	10n	15m	1.22k	15.7k	470n	220u	21.6
13.0k	15n	10m	816	15.8k	1n5	68m	6.73k
13.0k	100n	1m5	122	15.8k	6n8	15m	1.49k
13.0k	150n	1m	81.6	15.8k	15n	6m8	673
13.0k	1u	150u	12.2	15.8k	68n	1m5	149
13.0k	2n2	68m	5.56k	15.8k	150n	680u	67.3
13.0k	6n8	22m	1.80k	15.8k	680n	150u	14.9
13.0k	22n	6m8	556	15.9k	1n	100m	10.0k
13.0k	220n	680u	55.6	15.9k	10n	10m	1.00k
13.0k	680n	220u	18.0	15.9k	100n	1m	100
13.0k	68n	2m2	180	15.9k	1u	100u	10.0
14.0k	3n9	33m	2.91k	16.9k	2n7	33m	3.50k
14.0k	390n	330u	29.1	16.9k	27n	3m3	350
14.0k	39n	3m3	291	16.9k	270n	330u	35.0
14.1k	2n7	47m	4.17k	17.2k	3n9	22m	2.38k
14.1k	27n	4m7	417	17.2k	39n	2m2	238
14.1k	270n	470u	41.7	17.2k	390n	220u	23.8
14.3k	5n6	22m	1.98k	17.3k	1n8	47m	5.11k

f_O	C	L	X	f_O	C	L	X
17.3k	18n	4m7	511	20.7k	270n	220u	28.5
17.3k	180n	470u	51.1	20.8k	3n9	15m	1.96k
17.4k	5n6	15m	1.64k	20.8k	39n	1m5	196
17.4k	56n	1m5	164	20.8k	390n	150u	19.6
17.4k	560n	150u	16.4	21.2k	1n2	47m	6.26k
17.6k	820p	100m	11.0k	21.2k	12n	4m7	626
17.6k	8n2	10m	1.10k	21.2k	120n	470u	62.6
17.6k	82n	1m	110	21.3k	560p	100m	13.4k
17.6k	820n	100u	11.0	21.3k	5n6	10m	1.34k
17.6k	1n2	68m	7.53k	21.3k	56n	1m	134
17.6k	12n	6m8	753	21.3k	560n	100u	13.4
17.6k	120n	680u	75.3	21.3k	8n2	6m8	911
18.7k	2n2	33m	3.87k	21.3k	820n	68u	9.11
18.7k	3n3	22m	2.58k	21.3k	820p	68m	9.11k
18.7k	22n	3m3	387	21.3k	82n	680u	91.1
18.7k	33n	2m2	258	22.6k	1n5	33m	4.69k
18.7k	220n	330u	38.7	22.6k	3n3	15m	2.13k
18.7k	330n	220u	25.8	22.6k	15n	3m3	469
19.0k	1n5	47m	5.60k	22.6k	33n	1m5	213
19.0k	4n7	15m	1.79k	22.6k	150n	330u	46.9
19.0k	15n	4m7	560	22.6k	330n	150u	21.3
19.0k	47n	1m5	179	22.9k	2n2	22m	3.16k
19.0k	150n	470u	56.0	22.9k	22n	2m2	316
19.0k	470n	150u	17.9	22.9k	220n	220u	31.6
19.3k	680p	100m	12.1k	23.2k	470p	100m	14.6k
19.3k	1n	68m	8.25k	23.2k	1n	47m	6.86k
19.3k	6n8	10m	1.21k	23.2k	4n7	10m	1.46k
19.3k	10n	6m8	825	23.2k	10n	4m7	686
19.3k	100n	680u	82.5	23.2k	47n	1m	146
19.3k	680n	100u	12.1	23.2k	100n	470u	68.6
19.3k	1u	68u	8.25	23.2k	470n	100u	14.6
19.3k	68n	1m	121	23.2k	1u	47u	6.86
20.7k	1n8	33m	4.28k	23.4k	6n8	6m8	1.00k
20.7k	2n7	22m	2.85k	23.4k	680n	68u	10.0
20.7k	18n	3m3	428	23.4k	680p	68m	10.0k
20.7k	27n	2m2	285	23.4k	68n	680u	100
20.7k	180n	330u	42.8	25.0k	2n7	15m	2.36k

Table 10.1 Frequency constants for LC circuits 237

f_0	C	L	X	f_0	C	L	X
25.0k	27n	1m5	236	28.2k	6n8	4m7	831
25.0k	270n	150u	23.6	28.2k	47n	680u	120
25.3k	1n2	33m	5.24k	28.2k	68n	470u	83.1
25.3k	1n8	22m	3.50k	28.2k	470n	68u	12.0
25.3k	12n	3m3	524	28.2k	680n	47u	8.31
25.3k	18n	2m2	350	30.6k	820p	33m	6.34k
25.3k	120n	330u	52.4	30.6k	8n2	3m3	634
25.3k	180n	220u	35.0	30.6k	82n	330u	63.4
25.5k	3n9	10m	1.60k	30.6k	820n	33u	6.34
25.5k	390n	100u	16.0	30.6k	270p	100m	19.2k
25.5k	390p	100m	16.0k	30.6k	1n8	15m	2.89k
25.5k	39n	1m	160	30.6k	2n7	10m	1.92k
25.6k	820p	47m	7.57k	30.6k	18n	1m5	289
25.6k	8n2	4m7	757	30.6k	27n	1m	192
25.6k	82n	470u	75.7	30.6k	180n	150u	28.9
25.6k	820n	47u	7.57	30.6k	270n	100u	19.2
25.8k	560p	68m	11.0k	30.9k	390p	68m	13.2k
25.8k	5n6	6m8	1.10k	30.9k	3n9	6m8	1.32k
25.8k	56n	680u	110	30.9k	39n	680u	132
25.8k	560n	68u	11.0	30.9k	390n	68u	13.2
27.7k	330p	100m	17.4k	31.0k	1n2	22m	4.28k
27.7k	1n	33m	5.74k	31.0k	12n	2m2	428
27.7k	1n5	22m	3.83k	31.0k	120n	220u	42.8
27.7k	2n2	15m	2.61k	31.0k	560p	47m	9.16k
27.7k	3n3	10m	1.74k	31.0k	5n6	4m7	916
27.7k	10n	3m3	574	31.0k	56n	470u	91.6
27.7k	15n	2m2	383	31.0k	560n	47u	9.16
27.7k	22n	1m5	261	33.6k	1n5	15m	3.16k
27.7k	33n	1m	174	33.6k	15n	1m5	316
27.7k	100n	330u	57.4	33.6k	150n	150u	31.6
27.7k	150n	220u	38.3	33.6k	330p	68m	14.4k
27.7k	220n	150u	26.1	33.6k	680p	33m	6.97k
27.7k	330n	100u	17.4	33.6k	3n3	6m8	1.44k
27.7k	1u	33u	5.74	33.6k	6n8	3m3	697
28.2k	470p	68m	12.0k	33.6k	33n	680u	144
28.2k	680p	47m	8.31k	33.6k	330n	68u	14.4
28.2k	4n7	6m8	1.20k	33.6k	680n	33u	6.97

f_O	C	L	X	f_O	C	L	X
33.6k	68n	330u	69.7	40.4k	3n3	4m7	1.19k
33.9k	470p	47m	10.0k	40.4k	330p	47m	11.9k
33.9k	4n7	4m7	1.00k	40.4k	4n7	3m3	838
33.9k	47n	470u	100	40.4k	33n	470u	119
33.9k	470n	47u	10.0	40.4k	47n	330u	83.8
33.9k	220p	100m	21.3k	40.4k	330n	47u	11.9
33.9k	1n	22m	4.69k	40.4k	470n	33u	8.38
33.9k	2n2	10m	2.13k	41.1k	150p	100m	25.8k
33.9k	10n	2m2	469	41.1k	1n	15m	3.87k
33.9k	22n	1m	213	41.1k	1n5	10m	2.58k
33.9k	100n	220u	46.9	41.1k	10n	1m5	387
33.9k	220n	100u	21.3	41.1k	15n	1m	258
33.9k	1u	22u	4.69	41.1k	100n	150u	38.7
37.0k	5n6	3m3	768	41.1k	150n	100u	25.8
37.0k	560p	33m	7.68k	41.1k	1u	15u	3.87
37.0k	56n	330u	76.8	41.1k	220p	68m	17.6k
37.0k	560n	33u	7.68	41.1k	680p	22m	5.69k
37.1k	2n7	6m8	1.59k	41.1k	2n2	6m8	1.76k
37.1k	270n	68u	15.9	41.1k	6n8	2m2	569
37.1k	270p	68m	15.9k	41.1k	22n	680u	176
37.1k	27n	680u	159	41.1k	68n	220u	56.9
37.2k	3n9	4m7	1.10k	41.1k	220n	68u	17.6
37.2k	390n	47u	11.0	41.1k	680n	22u	5.69
37.2k	390p	47m	11.0k	44.4k	390p	33m	9.20k
37.2k	39n	470u	110	44.4k	3n9	3m3	920
37.5k	820p	22m	5.18k	44.4k	39n	330u	92.0
37.5k	8n2	2m2	518	44.4k	390n	33u	9.20
37.5k	82n	220u	51.8	44.7k	270p	47m	13.2k
37.5k	820n	22u	5.18	44.7k	2n7	4m7	1.32k
37.5k	180p	100m	23.6k	44.7k	27n	470u	132
37.5k	1n2	15m	3.54k	44.7k	270n	47u	13.2
37.5k	1n8	10m	2.36k	45.3k	560p	22m	6.27k
37.5k	12n	1m5	354	45.3k	5n6	2m2	627
37.5k	18n	1m	236	45.3k	56n	220u	62.7
37.5k	120n	150u	35.4	45.3k	560n	22u	6.27
37.5k	180n	100u	23.6	45.4k	820p	15m	4.28k
40.4k	470p	33m	8.38k	45.4k	8n2	1m5	428

Table 10.1 Frequency constants for LC circuits 239

f_O	C	L	X	f_O	C	L	X
45.4k	82n	150u	42.8	53.3k	27n	330u	111
45.4k	820n	15u	4.28	53.3k	270n	33u	11.1
45.5k	1n8	6m8	1.94k	54.3k	390n	22u	7.51
45.5k	180n	68u	19.4	54.3k	390p	22m	7.51k
45.5k	180p	68m	19.4k	54.3k	3n9	2m2	751
45.5k	18n	680u	194	54.3k	39n	220u	75.1
45.9k	120p	100m	28.9k	54.7k	180p	47m	16.2k
45.9k	1n2	10m	2.89k	54.7k	1n8	4m7	1.62k
45.9k	12n	1m	289	54.7k	18n	470u	162
45.9k	120n	100u	28.9	54.7k	180n	47u	16.2
48.2k	330p	33m	10.0k	54.9k	560p	15m	5.18k
48.2k	3n3	3m3	1.00k	54.9k	5n6	1m5	518
48.2k	33n	330u	100	54.9k	56n	150u	51.8
48.2k	330n	33u	10.0	54.9k	560n	15u	5.18
49.5k	2n2	4m7	1.46k	55.6k	82p	100m	34.9k
49.5k	220n	47u	14.6	55.6k	820p	10m	3.49k
49.5k	220p	47m	14.6k	55.6k	8n2	1m	349
49.5k	470p	22m	6.84k	55.6k	82n	100u	34.9
49.5k	22n	470u	146	55.6k	820n	10u	3.49
49.5k	47n	220u	68.4	55.7k	120p	68m	23.8k
49.5k	470n	22u	6.84	55.7k	1n2	6m8	2.38k
49.5k	4n7	2m2	684	55.7k	12n	680u	238
49.8k	680p	15m	4.70k	55.7k	120n	68u	23.8
49.8k	1n5	6m8	2.13k	59.1k	220p	33m	12.2k
49.8k	6n8	1m5	470	59.1k	2n2	3m3	1.22k
49.8k	680n	15u	4.70	59.1k	3n3	2m2	816
49.8k	150p	68m	21.3k	59.1k	330p	22m	8.16k
49.8k	15n	680u	213	59.1k	22n	330u	122
49.8k	68n	150u	47.0	59.1k	33n	220u	81.6
49.8k	150n	68u	21.3	59.1k	220n	33u	12.2
50.3k	100n	100u	31.6	59.1k	330n	22u	8.16
50.3k	100p	100m	31.6k	59.9k	150p	47m	17.7k
50.3k	1n	10m	3.16k	59.9k	470p	15m	5.65k
50.3k	10n	1m	316	59.9k	1n5	4m7	1.77k
50.3k	1u	10u	3.16	59.9k	4n7	1m5	565
53.3k	270p	33m	11.1k	59.9k	15n	470u	177
53.3k	2n7	3m3	1.11k	59.9k	47n	150u	56.5

f_O	C	L	X	f_O	C	L	X
59.9k	470n	15u	5.65	67.4k	82n	68u	28.8
59.9k	150n	47u	17.7	71.5k	150p	33m	14.8k
61.0k	100p	68m	26.1k	71.5k	330p	15m	6.74k
61.0k	680p	10m	3.83k	71.5k	1n5	3m3	1.48k
61.0k	1n	6m8	2.61k	71.5k	3n3	1m5	674
61.0k	6n8	1m	383	71.5k	15n	330u	148
61.0k	68n	100u	38.3	71.5k	33n	150u	67.4
61.0k	100n	68u	26.1	71.5k	330n	15u	6.74
61.0k	680n	10u	3.83	71.5k	150n	33u	14.8
61.0k	1u	6u8	2.61	72.3k	220p	22m	10.0k
61.0k	68p	100m	38.3k	72.3k	2n2	2m2	1.00k
61.0k	10n	680u	261	72.3k	22n	220u	100
65.3k	180p	33m	13.5k	72.3k	220n	22u	10.0
65.3k	270p	22m	9.03k	73.4k	100p	47m	21.7k
65.3k	1n8	3m3	1.35k	73.4k	470p	10m	4.61k
65.3k	2n7	2m2	903	73.4k	1n	4m7	2.17k
65.3k	18n	330u	135	73.4k	10n	470u	217
65.3k	27n	220u	90.3	73.4k	47n	100u	46.1
65.3k	180n	33u	13.5	73.4k	100n	47u	21.7
65.3k	270n	22u	9.03	73.4k	470n	10u	4.61
65.8k	3n9	1m5	620	73.4k	1u	4u7	2.17
65.8k	390p	15m	6.20k	73.4k	47p	100m	46.1k
65.8k	39n	150u	62.0	73.4k	4n7	1m	461
65.8k	390n	15u	6.20	74.0k	6n8	680u	316
67.0k	1n2	4m7	1.98k	74.0k	680n	6u8	3.16
67.0k	12n	470u	198	74.0k	68p	68m	31.6k
67.0k	120p	47m	19.8k	74.0k	680p	6m8	3.16k
67.0k	120n	47u	19.8	74.0k	68n	68u	31.6
67.3k	56p	100m	42.3k	79.1k	2n7	1m5	745
67.3k	560p	10m	4.23k	79.1k	270p	15m	7.45k
67.3k	5n6	1m	423	79.1k	27n	150u	74.5
67.3k	56n	100u	42.3	79.1k	270n	15u	7.45
67.3k	560n	10u	4.23	80.0k	120p	33m	16.6k
67.4k	820n	6u8	2.88	80.0k	180p	22m	11.1k
67.4k	82p	68m	28.8k	80.0k	1n2	3m3	1.66k
67.4k	820p	6m8	2.88k	80.0k	1n8	2m2	1.11k
67.4k	8n2	680u	288	80.0k	12n	330u	166

Table 10.1 Frequency constants for LC circuits 241

f_O	C	L	X	f_O	C	L	X
80.0k	18n	220u	111	89.0k	6n8	470u	263
80.0k	120n	33u	16.6	89.0k	47n	68u	38.0
80.0k	180n	22u	11.1	89.0k	470n	6u8	3.80
80.6k	39p	100m	50.6k	89.0k	680n	4u7	2.63
80.6k	390p	10m	5.06k	89.0k	47p	68m	38.0k
80.6k	3n9	1m	506	89.0k	68p	47m	26.3k
80.6k	39n	100u	50.6	89.0k	470p	6m8	3.80k
80.6k	390n	10u	5.06	89.0k	4n7	680u	380
81.1k	820p	4m7	2.39k	89.0k	68n	47u	26.3
81.1k	8n2	470u	239	96.8k	82p	33m	20.1k
81.1k	820n	4u7	2.39	96.8k	820p	3m3	2.01k
81.1k	82p	47m	23.9k	96.8k	8n2	330u	201
81.1k	82n	47u	23.9	96.8k	820n	3u3	2.01
81.6k	5n6	680u	348	96.8k	82n	33u	20.1
81.6k	56n	68u	34.8	96.9k	1n8	1m5	913
81.6k	56p	68m	34.8k	96.9k	180p	15m	9.13k
81.6k	560p	6m8	3.48k	96.9k	270p	10m	6.09k
81.6k	560n	6u8	3.48	96.9k	18n	150u	91.3
87.6k	2n2	1m5	826	96.9k	180n	15u	9.13
87.6k	100p	33m	18.2k	96.9k	27p	100m	60.9k
87.6k	150p	22m	12.1k	96.9k	2n7	1m	609
87.6k	220p	15m	8.26k	96.9k	27n	100u	60.9
87.6k	1n	3m3	1.82k	96.9k	270n	10u	6.09
87.6k	1n5	2m2	1.21k	97.7k	3n9	680u	418
87.6k	15n	220u	121	97.7k	39n	68u	41.8
87.6k	22n	150u	82.6	97.7k	390n	6u8	4.18
87.6k	220n	15u	8.26	97.7k	39p	68m	41.8k
87.6k	330n	10u	5.50	97.7k	390p	6m8	4.18k
87.6k	330p	10m	5.50k	98.0k	120p	22m	13.5k
87.6k	3n3	1m	550	98.0k	1n2	2m2	1.35k
87.6k	33n	100u	55.0	98.0k	12n	220u	135
87.6k	150n	22u	12.1	98.0k	120n	22u	13.5
87.6k	1u	3u3	1.82	98.1k	56p	47m	29.0k
87.6k	33p	100m	55.0k	98.1k	560p	4m7	2.90k
87.6k	10n	330u	182	98.1k	5n6	470u	290
87.6k	100n	33u	18.2	98.1k	56n	47u	29.0
89.0k	680p	4m7	2.63k	98.1k	560n	4u7	2.90

f_O	C	L	X	f_O	C	L	X
106k	1n5	1m5	1.00k	117k	2n7	680u	502
106k	150p	15m	10.0k	117k	270n	6u8	5.02
106k	15n	150u	100	118k	39p	47m	34.7k
106k	150n	15u	10.0	118k	390p	4m7	3.47k
106k	33p	68m	45.4k	118k	3n9	470u	347
106k	68p	33m	22.0k	118k	39n	47u	34.7
106k	330p	6m8	4.54k	118k	390n	4u7	3.47
106k	680p	3m3	2.20k	118k	82p	22m	16.4k
106k	3n3	680u	454	118k	820p	2m2	1.64k
106k	6n8	330u	220	118k	8n2	220u	164
106k	33n	68u	45.4	118k	82n	22u	16.4
106k	330n	6u8	4.54	118k	820n	2u2	1.64
106k	680n	3u3	2.20	119k	1n2	1m5	1.12k
106k	68n	33u	22.0	119k	18p	100m	74.5k
107k	4n7	470u	316	119k	120p	15m	11.2k
107k	47p	47m	31.6k	119k	180p	10m	7.45k
107k	470p	4m7	3.16k	119k	1n8	1m	745
107k	47n	47u	31.6	119k	12n	150u	112
107k	470n	4u7	3.16	119k	18n	100u	74.5
107k	22p	100m	67.4k	119k	120n	15u	11.2
107k	100p	22m	14.8k	119k	180n	10u	7.45
107k	220p	10m	6.74k	128k	33p	47m	37.7k
107k	1n	2m2	1.48k	128k	47p	33m	26.5k
107k	2n2	1m	674	128k	330p	4m7	3.77k
107k	10n	220u	148	128k	470p	3m3	2.65k
107k	22n	100u	67.4	128k	3n3	470u	377
107k	100n	22u	14.8	128k	330n	4u7	3.77
107k	220n	10u	6.74	128k	470n	3u3	2.65
107k	1u	2u2	1.48	128k	4n7	330u	265
117k	56p	33m	24.3k	128k	33n	47u	37.7
117k	560p	3m3	2.43k	128k	47n	33u	26.5
117k	5n6	330u	243	130k	100p	15m	12.2k
117k	56n	33u	24.3	130k	15p	100m	81.6k
117k	560n	3u3	2.43	130k	150p	10m	8.16k
117k	27n	68u	50.2	130k	1n	1m5	1.22k
117k	27p	68m	50.2k	130k	1n5	1m	816
117k	270p	6m8	5.02k	130k	10n	150u	122

Table 10.1 Frequency constants for LC circuits 243

f_O	C	L	X	f_O	C	L	X
130k	15n	100u	81.6	144k	18n	68u	61.5
130k	100n	15u	12.2	144k	180n	6u8	6.15
130k	150n	10u	8.16	145k	12p	100m	91.3k
130k	1u	1u5	1.22	145k	120p	10m	9.13k
130k	2n2	680u	556	145k	1n2	1m	913
130k	6n8	220u	180	145k	12n	100u	91.3
130k	22n	68u	55.6	145k	120n	10u	9.13
130k	220n	6u8	5.56	153k	33p	33m	31.6k
130k	680n	2u2	1.80	153k	330p	3m3	3.16k
130k	22p	68m	55.6k	153k	3n3	330u	316
130k	68p	22m	18.0k	153k	33n	33u	31.6
130k	220p	6m8	5.56k	153k	330n	3u3	3.16
130k	680p	2m2	1.80k	157k	22p	47m	46.2k
130k	68n	22u	18.0	157k	47p	22m	21.6k
140k	39p	33m	29.1k	157k	220p	4m7	4.62k
140k	390p	3m3	2.91k	157k	470p	2m2	2.16k
140k	3n9	330u	291	157k	2n2	470u	462
140k	390n	3u3	2.91	157k	4n7	220u	216
140k	39n	33u	29.1	157k	22n	47u	46.2
141k	27p	47m	41.7k	157k	47n	22u	21.6
141k	270p	4m7	4.17k	157k	220n	4u7	4.62
141k	2n7	470u	417	157k	470n	2u2	2.16
141k	27n	47u	41.7	158k	15p	68m	67.3k
141k	270n	4u7	4.17	158k	68p	15m	14.9k
143k	5n6	220u	198	158k	150p	6m8	6.73k
143k	56n	22u	19.8	158k	680p	1m5	1.49k
143k	56p	22m	19.8k	158k	1n5	680u	673
143k	560p	2m2	1.98k	158k	6n8	150u	149
143k	560n	2u2	1.98	158k	15n	68u	67.3
144k	82p	15m	13.5k	158k	150n	6u8	6.73
144k	820p	1m5	1.35k	158k	680n	1u5	1.49
144k	8n2	150u	135	158k	68n	15u	14.9
144k	82n	15u	13.5	159k	10p	100m	100k
144k	820n	1u5	1.35	159k	100p	10m	10.0k
144k	18p	68m	61.5k	159k	1n	1m	1.00k
144k	180p	6m8	6.15k	159k	10n	100u	100
144k	1n8	680u	615	159k	100n	10u	10.0

f_O	C	L	X	f_O	C	L	X
159k	1u	1u	1.00	187k	330n	2u2	2.58
169k	27p	33m	35.0k	187k	33p	22m	25.8k
169k	270p	3m3	3.50k	187k	22n	33u	38.7
169k	2n7	330u	350	187k	33n	22u	25.8
169k	270n	3u3	3.50	190k	15p	47m	56.0k
169k	27n	33u	35.0	190k	47p	15m	17.9k
172k	3n9	220u	238	190k	150p	4m7	5.60k
172k	390n	2u2	2.38	190k	470p	1m5	1.79k
172k	39p	22m	23.8k	190k	1n5	470u	560
172k	390p	2m2	2.38k	190k	4n7	150u	179
172k	39n	22u	23.8	190k	15n	47u	56.0
173k	18p	47m	51.1k	190k	47n	15u	17.9
173k	180p	4m7	5.11k	190k	150n	4u7	5.60
173k	1n8	470u	511	190k	470n	1u5	1.79
173k	180n	4u7	5.11	193k	10p	68m	82.5k
173k	18n	47u	51.1	193k	100p	6m8	8.25k
174k	5n6	150u	164	193k	6n8	100u	121
174k	56p	15m	16.4k	193k	10n	68u	82.5
174k	560p	1m5	1.64k	193k	100n	6u8	8.25
174k	56n	15u	16.4	193k	680n	1u	1.21
174k	560n	1u5	1.64	193k	68p	10m	12.1k
176k	8n2	100u	110	193k	680p	1m	1.21k
176k	82p	10m	11.0k	193k	1n	680u	825
176k	820p	1m	1.10k	193k	68n	10u	12.1
176k	82n	10u	11.0	207k	18p	33m	42.8k
176k	820n	1u	1.10	207k	27p	22m	28.5k
176k	120p	6m8	7.53k	207k	180p	3m3	4.28k
176k	1n2	680u	753	207k	1n8	330u	428
176k	12n	68u	75.3	207k	2n7	220u	285
176k	120n	6u8	7.53	207k	27n	22u	28.5
176k	12p	68m	75.3k	207k	180n	3u3	4.28
187k	22p	33m	38.7k	207k	270p	2m2	2.85k
187k	220p	3m3	3.87k	207k	18n	33u	42.8
187k	330p	2m2	2.58k	207k	270n	2u2	2.85
187k	2n2	330u	387	208k	390n	1u5	1.96
187k	3n3	220u	258	208k	39p	15m	19.6k
187k	220n	3u3	3.87	208k	390p	1m5	1.96k

Table 10.1 Frequency constants for LC circuits **245**

f_O	C	L	X	f_O	C	L	X
208k	3n9	150u	196	232k	1n	470u	686
208k	39n	15u	19.6	232k	10n	47u	68.6
212k	12p	47m	62.6k	232k	47n	10u	14.6
212k	120p	4m7	6.26k	232k	470n	1u	1.46
212k	1n2	470u	626	234k	6n8	68u	100
212k	12n	47u	62.6	234k	68p	6m8	10.0k
212k	120n	4u7	6.26	234k	680p	680u	1.00k
213k	56p	10m	13.4k	234k	68n	6u8	10.0
213k	5n6	100u	134	250k	27p	15m	23.6k
213k	56n	10u	13.4	250k	270p	1m5	2.36k
213k	560p	1m	1.34k	250k	2n7	150u	236
213k	560n	1u	1.34	250k	27n	15u	23.6
213k	8n2	68u	91.1	250k	270n	1u5	2.36
213k	82p	6m8	9.11k	253k	12p	33m	52.4k
213k	820p	680u	911	253k	18p	22m	35.0k
213k	82n	6u8	9.11	253k	120p	3m3	5.24k
226k	15p	33m	46.9k	253k	180p	2m2	3.50k
226k	33p	15m	21.3k	253k	1n2	330u	524
226k	150p	3m3	4.69k	253k	1n8	220u	350
226k	330p	1m5	2.13k	253k	12n	33u	52.4
226k	1n5	330u	469	253k	18n	22u	35.0
226k	3n3	150u	213	253k	120n	3u3	5.24
226k	33n	15u	21.3	253k	180n	2u2	3.50
226k	150n	3u3	4.69	255k	39p	10m	16.0k
226k	330n	1u5	2.13	255k	390p	1m	1.60k
226k	15n	33u	46.9	255k	3n9	100u	160
229k	22p	22m	31.6k	255k	39n	10u	16.0
229k	220p	2m2	3.16k	255k	390n	1u	1.60
229k	2n2	220u	316	256k	82p	4m7	7.57k
229k	22n	22u	31.6	256k	820p	470u	757
229k	220n	2u2	3.16	256k	8n2	47u	75.7
232k	10p	47m	68.6k	256k	82n	4u7	7.57
232k	100p	4m7	6.86k	258k	5n6	68u	110
232k	4n7	100u	146	258k	56n	6u8	11.0
232k	100n	4u7	6.86	258k	56p	6m8	11.0k
232k	47p	10m	14.6k	258k	560p	680u	1.10k
232k	470p	1m	1.46k	277k	10p	33m	57.4k

f_O	C	L	X	f_O	C	L	X
277k	22p	15m	26.1k	306k	2n7	100u	192
277k	150p	2m2	3.83k	306k	18n	15u	28.9
277k	220p	1m5	2.61k	306k	27n	10u	19.2
277k	1n5	220u	383	306k	270n	1u	1.92
277k	2n2	150u	261	309k	3n9	68u	132
277k	3n3	100u	174	309k	39n	6u8	13.2
277k	100n	3u3	5.74	309k	39p	6m8	13.2k
277k	220n	1u5	2.61	309k	390p	680u	1.32k
277k	15p	22m	38.3k	310k	12p	22m	42.8k
277k	33p	10m	17.4k	310k	120p	2m2	4.28k
277k	100p	3m3	5.74k	310k	1n2	220u	428
277k	330p	1m	1.74k	310k	12n	22u	42.8
277k	1n	330u	574	310k	120n	2u2	4.28
277k	10n	33u	57.4	310k	56p	4m7	9.16k
277k	15n	22u	38.3	310k	560p	470u	916
277k	22n	15u	26.1	310k	5n6	47u	91.6
277k	33n	10u	17.4	310k	56n	4u7	9.16
277k	150n	2u2	3.83	336k	15p	15m	31.6k
277k	330n	1u	1.74	336k	150p	1m5	3.16k
282k	47p	6m8	12.0k	336k	1n5	150u	316
282k	68p	4m7	8.31k	336k	15n	15u	31.6
282k	470p	680u	1.20k	336k	150n	1u5	3.16
282k	680p	470u	831	336k	3n3	68u	144
282k	4n7	68u	120	336k	33p	6m8	14.4k
282k	6n8	47u	83.1	336k	68p	3m3	6.97k
282k	47n	6u8	12.0	336k	330p	680u	1.44k
282k	68n	4u7	8.31	336k	680p	330u	697
306k	82p	3m3	6.34k	336k	6n8	33u	69.7
306k	820p	330u	634	336k	33n	6u8	14.4
306k	8n2	33u	63.4	336k	68n	3u3	6.97
306k	82n	3u3	6.34	339k	47p	4m7	10.0k
306k	18p	15m	28.9k	339k	470p	470u	1.00k
306k	180p	1m5	2.89k	339k	4n7	47u	100
306k	1n8	150u	289	339k	47n	4u7	10.0
306k	180n	1u5	2.89	339k	10p	22m	46.9k
306k	27p	10m	19.2k	339k	22p	10m	21.3k
306k	270p	1m	1.92k	339k	100p	2m2	4.69k

Table 10.1 Frequency constants for LC circuits 247

f_O	C	L	X	f_O	C	L	X
339k	2n2	100u	213	404k	33n	4u7	11.9
339k	100n	2u2	4.69	404k	47n	3u3	8.38
339k	220n	1u	2.13	404k	470p	330u	838
339k	220p	1m	2.13k	404k	4n7	33u	83.8
339k	1n	220u	469	411k	10p	15m	38.7k
339k	10n	22u	46.9	411k	15p	10m	25.8k
339k	22n	10u	21.3	411k	100p	1m5	3.87k
370k	56p	3m3	7.68k	411k	150p	1m	2.58k
370k	5n6	33u	76.8	411k	1n	150u	387
370k	56n	3u3	7.68	411k	1n5	100u	258
370k	560p	330u	768	411k	10n	15u	38.7
371k	27p	6m8	15.9k	411k	15n	10u	25.8
371k	270p	680u	1.59k	411k	100n	1u5	3.87
371k	2n7	68u	159	411k	150n	1u	2.58
371k	27n	6u8	15.9	411k	2n2	68u	176
372k	39p	4m7	11.0k	411k	22p	6m8	17.6k
372k	390p	470u	1.10k	411k	68p	2m2	5.69k
372k	3n9	47u	110	411k	680p	220u	569
372k	39n	4u7	11.0	411k	6n8	22u	56.9
375k	82p	2m2	5.18k	411k	22n	6u8	17.6
375k	820p	220u	518	411k	220p	680u	1.76k
375k	8n2	22u	51.8	411k	68n	2u2	5.69
375k	82n	2u2	5.18	444k	39p	3m3	9.20k
375k	12p	15m	35.4k	444k	390p	330u	920
375k	18p	10m	23.6k	444k	39n	3u3	9.20
375k	120p	1m5	3.54k	444k	3n9	33u	92.0
375k	1n2	150u	354	447k	27p	4m7	13.2k
375k	1n8	100u	236	447k	270p	470u	1.32k
375k	12n	15u	35.4	447k	2n7	47u	132
375k	18n	10u	23.6	447k	27n	4u7	13.2
375k	120n	1u5	3.54	453k	5n6	22u	62.7
375k	180n	1u	2.36	453k	56n	2u2	6.27
375k	180p	1m	2.36k	453k	56p	2m2	6.27k
404k	33p	4m7	11.9k	453k	560p	220u	627
404k	47p	3m3	8.38k	454k	82p	1m5	4.28k
404k	330p	470u	1.19k	454k	820p	150u	428
404k	3n3	47u	119	454k	8n2	15u	42.8

f_O	C	L	X	f_O	C	L	X
454k	82n	1u5	4.28	533k	270p	330u	1.11k
455k	18p	6m8	19.4k	533k	2n7	33u	111
455k	180p	680u	1.94k	543k	39p	2m2	7.51k
455k	1n8	68u	194	543k	390p	220u	751
455k	18n	6u8	19.4	543k	3n9	22u	75.1
459k	1n2	100u	289	543k	39n	2u2	7.51
459k	12n	10u	28.9	547k	18p	4m7	16.2k
459k	12p	10m	28.9k	547k	180p	470u	1.62k
459k	120n	1u	2.89	547k	1n8	47u	162
459k	120p	1m	2.89k	547k	18n	4u7	16.2
482k	33p	3m3	10.0k	549k	56p	1m5	5.18k
482k	330p	330u	1.00k	549k	560p	150u	518
482k	3n3	33u	100	549k	5n6	15u	51.8
482k	33n	3u3	10.0	549k	56n	1u5	5.18
495k	22p	4m7	14.6k	556k	82p	1m	3.49k
495k	220p	470u	1.46k	556k	820p	100u	349
495k	2n2	47u	146	556k	8n2	10u	34.9
495k	47p	2m2	6.84k	556k	82n	1u	3.49
495k	470p	220u	684	557k	120p	680u	2.38k
495k	4n7	22u	68.4	557k	1n2	68u	238
495k	22n	4u7	14.6	557k	12n	6u8	23.8
495k	47n	2u2	6.84	557k	12p	6m8	23.8k
498k	1n5	68u	213	591k	3n3	22u	81.6
498k	15p	6m8	21.3k	591k	22p	3m3	12.2k
498k	68p	1m5	4.70k	591k	33p	2m2	8.16k
498k	150p	680u	2.13k	591k	220p	330u	1.22k
498k	680p	150u	470	591k	330p	220u	816
498k	6n8	15u	47.0	591k	2n2	33u	122
498k	15n	6u8	21.3	591k	22n	3u3	12.2
498k	68n	1u5	4.70	591k	33n	2u2	8.16
503k	10p	10m	31.6k	599k	15p	4m7	17.7k
503k	100p	1m	3.16k	599k	47p	1m5	5.65k
503k	1n	100u	316	599k	150p	470u	1.77k
503k	10n	10u	31.6	599k	470p	150u	565
503k	100n	1u	3.16	599k	1n5	47u	177
533k	27p	3m3	11.1k	599k	4n7	15u	56.5
533k	27n	3u3	11.1	599k	15n	4u7	17.7

Table 10.1 Frequency constants for LC circuits 249

f_O	C	L	X	f_O	C	L	X
599k	47n	1u5	5.65	715k	33n	1u5	6.74
610k	680p	100u	383	715k	15p	3m3	14.8k
610k	1n	68u	261	715k	1n5	33u	148
610k	6n8	10u	38.3	723k	22p	2m2	10.0k
610k	10n	6u8	26.1	723k	220p	220u	1.00k
610k	10p	6m8	26.1k	723k	2n2	22u	100
610k	68p	1m	3.83k	723k	22n	2u2	10.0
610k	100p	680u	2.61k	734k	10p	4m7	21.7k
610k	68n	1u	3.83	734k	47p	1m	4.61k
653k	18p	3m3	13.5k	734k	100p	470u	2.17k
653k	27p	2m2	9.03k	734k	470p	100u	461
653k	180p	330u	1.35k	734k	1n	47u	217
653k	270p	220u	903	734k	4n7	10u	46.1
653k	1n8	33u	135	734k	10n	4u7	21.7
653k	2n7	22u	90.3	734k	47n	1u	4.61
653k	18n	3u3	13.5	740k	680p	68u	316
653k	27n	2u2	9.03	740k	6n8	6u8	31.6
658k	39p	1m5	6.20k	740k	68p	680u	3.16k
658k	390p	150u	620	791k	27p	1m5	7.45k
658k	3n9	15u	62.0	791k	270p	150u	745
658k	39n	1u5	6.20	791k	2n7	15u	74.5
670k	12p	4m7	19.8k	791k	27n	1u5	7.45
670k	120p	470u	1.98k	800k	12p	3m3	16.6k
670k	1n2	47u	198	800k	18p	2m2	11.1k
670k	12n	4u7	19.8	800k	180p	220u	1.11k
673k	56p	1m	4.23k	800k	1n8	22u	111
673k	560p	100u	423	800k	12n	3u3	16.6
673k	5n6	10u	42.3	800k	18n	2u2	11.1
673k	56n	1u	4.23	800k	120p	330u	1.66k
674k	820p	68u	288	800k	1n2	33u	166
674k	8n2	6u8	28.8	806k	3n9	10u	50.6
674k	82p	680u	2.88k	806k	39p	1m	5.06k
715k	33p	1m5	6.74k	806k	390p	100u	506
715k	150p	330u	1.48k	806k	39n	1u	5.06
715k	330p	150u	674	811k	82p	470u	2.39k
715k	3n3	15u	67.4	811k	820p	47u	239
715k	15n	3u3	14.8	811k	8n2	4u7	23.9

f_O	C	L	X	f_O	C	L	X
816k	560p	68u	348	977k	39p	680u	4.18k
816k	5n6	6u8	34.8	977k	390p	68u	418
816k	56p	680u	3.48k	980k	12p	2m2	13.5k
876k	10p	3m3	18.2k	980k	120p	220u	1.35k
876k	15p	2m2	12.1k	980k	1n2	22u	135
876k	22p	1m5	8.26k	980k	12n	2u2	13.5
876k	33p	1m	5.50k	981k	56p	470u	2.90k
876k	100p	330u	1.82k	981k	5n6	4u7	29.0
876k	150p	220u	1.21k	981k	560p	47u	290
876k	220p	150u	826	1.06M	15p	1m5	10.0k
876k	330p	100u	550	1.06M	150p	150u	1.00k
876k	1n	33u	182	1.06M	1n5	15u	100
876k	1n5	22u	121	1.06M	15n	1u5	10.0
876k	2n2	15u	82.6	1.06M	33p	680u	4.54k
876k	3n3	10u	55.0	1.06M	68p	330u	2.20k
876k	10n	3u3	18.2	1.06M	330p	68u	454
876k	15n	2u2	12.1	1.06M	680p	33u	220
876k	22n	1u5	8.26	1.06M	3n3	6u8	45.4
876k	33n	1u	5.50	1.06M	6n8	3u3	22.0
890k	47p	680u	3.80k	1.07M	47p	470u	3.16k
890k	68p	470u	2.63k	1.07M	4n7	4u7	31.6
890k	470p	68u	380	1.07M	470p	47u	316
890k	680p	47u	263	1.07M	220p	100u	674
890k	4n7	6u8	38.0	1.07M	2n2	10u	67.4
890k	6n8	4u7	26.3	1.07M	10p	2m2	14.8k
968k	82p	330u	2.01k	1.07M	22p	1m	6.74k
968k	820p	33u	201	1.07M	100p	220u	1.48k
968k	8n2	3u3	20.1	1.07M	1n	22u	148
969k	18p	1m5	9.13k	1.07M	10n	2u2	14.8
969k	180p	150u	913	1.07M	22n	1u	6.74
969k	270p	100u	609	1.17M	5n6	3u3	24.3
969k	1n8	15u	91.3	1.17M	56p	330u	2.43k
969k	2n7	10u	60.9	1.17M	560p	33u	243
969k	18n	1u5	9.13	1.17M	270p	68u	502
969k	27p	1m	6.09k	1.17M	2n7	6u8	50.2
969k	27n	1u	6.09	1.17M	27p	680u	5.02k
977k	3n9	6u8	41.8	1.18M	39p	470u	3.47k

Table 10.1 Frequency constants for LC circuits 251

f_O	C	L	X	f_O	C	L	X
1.18M	390p	47u	347	1.41M	27p	470u	4.17k
1.18M	3n9	4u7	34.7	1.41M	270p	47u	417
1.18M	82p	220u	1.64k	1.43M	56p	220u	1.98k
1.18M	820p	22u	164	1.43M	560p	22u	198
1.18M	8n2	2u2	16.4	1.43M	5n6	2u2	19.8
1.19M	12p	1m5	11.2k	1.44M	820p	15u	135
1.19M	18p	1m	7.45k	1.44M	8n2	1u5	13.5
1.19M	120p	150u	1.12k	1.44M	82p	150u	1.35k
1.19M	180p	100u	745	1.44M	18p	680u	6.15k
1.19M	1n2	15u	112	1.44M	180p	68u	615
1.19M	1n8	10u	74.5	1.44M	1n8	6u8	61.5
1.19M	12n	1u5	11.2	1.45M	120p	100u	913
1.19M	18n	1u	7.45	1.45M	1n2	10u	91.3
1.28M	3n3	4u7	37.7	1.45M	12p	1m	9.13k
1.28M	33p	470u	3.77k	1.45M	12n	1u	9.13
1.28M	47p	330u	2.65k	1.53M	3n3	3u3	31.6
1.28M	330p	47u	377	1.53M	33p	330u	3.16k
1.28M	470p	33u	265	1.53M	330p	33u	316
1.28M	4n7	3u3	26.5	1.57M	2n2	4u7	46.2
1.30M	10p	1m5	12.2k	1.57M	22p	470u	4.62k
1.30M	100p	150u	1.22k	1.57M	47p	220u	2.16k
1.30M	150p	100u	816	1.57M	220p	47u	462
1.30M	1n	15u	122	1.57M	470p	22u	216
1.30M	1n5	10u	81.6	1.57M	4n7	2u2	21.6
1.30M	10n	1u5	12.2	1.58M	1n5	6u8	67.3
1.30M	15n	1u	8.16	1.58M	6n8	1u5	14.9
1.30M	15p	1m	8.16k	1.58M	15p	680u	6.73k
1.30M	2n2	6u8	55.6	1.58M	68p	150u	1.49k
1.30M	22p	680u	5.56k	1.58M	150p	68u	673
1.30M	68p	220u	1.80k	1.58M	680p	15u	149
1.30M	220p	68u	556	1.59M	10p	1m	10.0k
1.30M	680p	22u	180	1.59M	100p	100u	1.00k
1.30M	6n8	2u2	18.0	1.59M	1n	10u	100
1.40M	39p	330u	2.91k	1.59M	10n	1u	10.0
1.40M	3n9	3u3	29.1	1.69M	27p	330u	3.50k
1.40M	390p	33u	291	1.69M	270p	33u	350
1.41M	2n7	4u7	41.7	1.69M	2n7	3u3	35.0

f_O	C	L	X	f_O	C	L	X
1.72M	3n9	2u2	23.8	2.07M	270p	22u	285
1.72M	39p	220u	2.38k	2.07M	180p	33u	428
1.72M	390p	22u	238	2.08M	3n9	1u5	19.6
1.73M	18p	470u	5.11k	2.08M	39p	150u	1.96k
1.73M	1n8	4u7	51.1	2.08M	390p	15u	196
1.73M	180p	47u	511	2.12M	12p	470u	6.26k
1.74M	56p	150u	1.64k	2.12M	120p	47u	626
1.74M	560p	15u	164	2.12M	1n2	4u7	62.6
1.74M	5n6	1u5	16.4	2.13M	56p	100u	1.34k
1.76M	82p	100u	1.10k	2.13M	560p	10u	134
1.76M	820p	10u	110	2.13M	5n6	1u	13.4
1.76M	8n2	1u	11.0	2.13M	82p	68u	911
1.76M	12p	680u	7.53k	2.13M	820p	6u8	91.1
1.76M	120p	68u	753	2.26M	33p	150u	2.13k
1.76M	1n2	6u8	75.3	2.26M	330p	15u	213
1.87M	330p	22u	258	2.26M	1n5	3u3	46.9
1.87M	2n2	3u3	38.7	2.26M	3n3	1u5	21.3
1.87M	3n3	2u2	25.8	2.26M	15p	330u	4.69k
1.87M	22p	330u	3.87k	2.26M	150p	33u	469
1.87M	33p	220u	2.58k	2.29M	220p	22u	316
1.87M	220p	33u	387	2.29M	2n2	2u2	31.6
1.90M	15p	470u	5.60k	2.29M	22p	220u	3.16k
1.90M	150p	47u	560	2.32M	10p	470u	6.86k
1.90M	470p	15u	179	2.32M	47p	100u	1.46k
1.90M	1n5	4u7	56.0	2.32M	100p	47u	686
1.90M	4n7	1u5	17.9	2.32M	470p	10u	146
1.90M	47p	150u	1.79k	2.32M	1n	4u7	68.6
1.93M	10p	680u	8.25k	2.32M	4n7	1u	14.6
1.93M	100p	68u	825	2.34M	68p	68u	1.00k
1.93M	680p	10u	121	2.34M	680p	6u8	100
1.93M	1n	6u8	82.5	2.50M	27p	150u	2.36k
1.93M	6n8	1u	12.1	2.50M	270p	15u	236
1.93M	68p	100u	1.21k	2.50M	2n7	1u5	23.6
2.07M	1n8	3u3	42.8	2.53M	12p	330u	5.24k
2.07M	2n7	2u2	28.5	2.53M	18p	220u	3.50k
2.07M	18p	330u	4.28k	2.53M	180p	22u	350
2.07M	27p	220u	2.85k	2.53M	1n2	3u3	52.4

Table 10.1 Frequency constants for LC circuits 253

f_O	C	L	X	f_O	C	L	X
2.53M	1n8	2u2	35.0	3.10M	120p	22u	428
2.53M	120p	33u	524	3.10M	56p	47u	916
2.55M	39p	100u	1.60k	3.10M	560p	4u7	91.6
2.55M	390p	10u	160	3.36M	150p	15u	316
2.55M	3n9	1u	16.0	3.36M	1n5	1u5	31.6
2.56M	82p	47u	757	3.36M	15p	150u	3.16k
2.56M	820p	4u7	75.7	3.36M	33p	68u	1.44k
2.58M	56p	68u	1.10k	3.36M	330p	6u8	144
2.58M	560p	6u8	110	3.36M	680p	3u3	69.7
2.77M	2n2	1u5	26.1	3.36M	68p	33u	697
2.77M	22p	150u	2.61k	3.39M	47p	47u	1.00k
2.77M	150p	22u	383	3.39M	470p	4u7	100
2.77M	220p	15u	261	3.39M	10p	220u	4.69k
2.77M	1n5	2u2	38.3	3.39M	22p	100u	2.13k
2.77M	3n3	1u	17.4	3.39M	100p	22u	469
2.77M	10p	330u	5.74k	3.39M	220p	10u	213
2.77M	15p	220u	3.83k	3.39M	1n	2u2	46.9
2.77M	33p	100u	1.74k	3.39M	2n2	1u	21.3
2.77M	100p	33u	574	3.70M	56p	33u	768
2.77M	330p	10u	174	3.70M	560p	3u3	76.8
2.77M	1n	3u3	57.4	3.71M	27p	68u	1.59k
2.82M	47p	68u	1.20k	3.71M	270p	6u8	159
2.82M	68p	47u	831	3.72M	39p	47u	1.10k
2.82M	470p	6u8	120	3.72M	390p	4u7	110
2.82M	680p	4u7	83.1	3.75M	820p	2u2	51.8
3.06M	820p	3u3	63.4	3.75M	82p	22u	518
3.06M	82p	33u	634	3.75M	1n2	1u5	35.4
3.06M	18p	150u	2.89k	3.75M	12p	150u	3.54k
3.06M	27p	100u	1.92k	3.75M	18p	100u	2.36k
3.06M	180p	15u	289	3.75M	120p	15u	354
3.06M	270p	10u	192	3.75M	180p	10u	236
3.06M	1n8	1u5	28.9	3.75M	1n8	1u	23.6
3.06M	2n7	1u	19.2	4.04M	33p	47u	1.19k
3.09M	39p	68u	1.32k	4.04M	330p	4u7	119
3.09M	390p	6u8	132	4.04M	470p	3u3	83.8
3.10M	12p	220u	4.28k	4.04M	47p	33u	838
3.10M	1n2	2u2	42.8	4.11M	10p	150u	3.87k

f_O	C	L	X	f_O	C	L	X
4.11M	15p	100u	2.58k	5.43M	39p	22u	751
4.11M	100p	15u	387	5.43M	390p	2u2	75.1
4.11M	150p	10u	258	5.47M	18p	47u	1.62k
4.11M	1n	1u5	38.7	5.47M	180p	4u7	162
4.11M	1n5	1u	25.8	5.49M	56p	15u	518
4.11M	22p	68u	1.76k	5.49M	560p	1u5	51.8
4.11M	68p	22u	569	5.56M	82p	10u	349
4.11M	220p	6u8	176	5.56M	820p	1u	34.9
4.11M	680p	2u2	56.9	5.57M	12p	68u	2.38k
4.44M	39p	33u	920	5.57M	120p	6u8	238
4.44M	390p	3u3	92.0	5.91M	22p	33u	1.22k
4.47M	27p	47u	1.32k	5.91M	33p	22u	816
4.47M	270p	4u7	132	5.91M	220p	3u3	122
4.53M	56p	22u	627	5.91M	330p	2u2	81.6
4.53M	560p	2u2	62.7	5.99M	15p	47u	1.77k
4.54M	82p	15u	428	5.99M	47p	15u	565
4.54M	820p	1u5	42.8	5.99M	150p	4u7	177
4.55M	18p	68u	1.94k	5.99M	470p	1u5	56.5
4.55M	180p	6u8	194	6.10M	10p	68u	2.61k
4.59M	12p	100u	2.89k	6.10M	100p	6u8	261
4.59M	120p	10u	289	6.10M	68p	10u	383
4.59M	1n2	1u	28.9	6.10M	680p	1u	38.3
4.82M	33p	33u	1.00k	6.53M	27p	22u	903
4.82M	330p	3u3	100	6.53M	180p	3u3	135
4.95M	22p	47u	1.46k	6.53M	270p	2u2	90.3
4.95M	47p	22u	684	6.53M	18p	33u	1.35k
4.95M	220p	4u7	146	6.58M	39p	15u	620
4.95M	470p	2u2	68.4	6.58M	390p	1u5	62.0
4.98M	15p	68u	2.13k	6.70M	120p	4u7	198
4.98M	68p	15u	470	6.70M	12p	47u	1.98k
4.98M	150p	6u8	213	6.73M	56p	10u	423
4.98M	680p	1u5	47.0	6.73M	560p	1u	42.3
5.03M	10p	100u	3.16k	6.74M	82p	6u8	288
5.03M	100p	10u	316	7.15M	150p	3u3	148
5.03M	1n	1u	31.6	7.15M	330p	1u5	67.4
5.33M	270p	3u3	111	7.15M	33p	15u	674
5.33M	27p	33u	1.11k	7.15M	15p	33u	1.48k

Table 10.1 Frequency constants for LC circuits 255

f$_O$	C	L	X	f$_O$	C	L	X
7.23M	22p	22u	1.00k	10.6M	15p	15u	1.00k
7.23M	220p	2u2	100	10.6M	33p	6u8	454
7.34M	10p	47u	2.17k	10.6M	68p	3u3	220
7.34M	47p	10u	461	10.7M	47p	4u7	316
7.34M	100p	4u7	217	10.7M	10p	22u	1.48k
7.34M	470p	1u	46.1	10.7M	22p	10u	674
7.40M	68p	6u8	316	10.7M	100p	2u2	148
7.91M	27p	15u	745	10.7M	220p	1u	67.4
7.91M	270p	1u5	74.5	11.7M	56p	3u3	243
8.00M	18p	22u	1.11k	11.7M	27p	6u8	502
8.00M	120p	3u3	166	11.8M	39p	4u7	347
8.00M	180p	2u2	111	11.8M	82p	2u2	164
8.00M	12p	33u	1.66k	11.9M	12p	15u	1.12k
8.06M	39p	10u	506	11.9M	18p	10u	745
8.06M	390p	1u	50.6	11.9M	120p	1u5	112
8.11M	82p	4u7	239	11.9M	180p	1u	74.5
8.16M	56p	6u8	348	12.8M	33p	4u7	377
8.76M	22p	15u	826	12.8M	47p	3u3	265
8.76M	100p	3u3	182	13.0M	10p	15u	1.22k
8.76M	150p	2u2	121	13.0M	15p	10u	816
8.76M	220p	1u5	82.6	13.0M	100p	1u5	122
8.76M	10p	33u	1.82k	13.0M	150p	1u	81.6
8.76M	15p	22u	1.21k	13.0M	22p	6u8	556
8.76M	33p	10u	550	13.0M	68p	2u2	180
8.76M	330p	1u	55.0	14.0M	39p	3u3	291
8.90M	47p	6u8	380	14.1M	27p	4u7	417
8.90M	68p	4u7	263	14.3M	56p	2u2	198
9.68M	82p	3u3	201	14.4M	82p	1u5	135
9.69M	18p	15u	913	14.4M	18p	6u8	615
9.69M	180p	1u5	91.3	14.5M	12p	10u	913
9.69M	27p	10u	609	14.5M	120p	1u	91.3
9.69M	270p	1u	60.9	15.3M	33p	3u3	316
9.77M	39p	6u8	418	15.7M	22p	4u7	462
9.80M	12p	22u	1.35k	15.7M	47p	2u2	216
9.80M	120p	2u2	135	15.8M	15p	6u8	673
9.81M	56p	4u7	290	15.8M	68p	1u5	149
10.6M	150p	1u5	100	15.9M	10p	10u	1.00k

f_O	C	L	X	f_O	C	L	X
15.9M	100p	1u	100	23.2M	47p	1u	146
16.9M	27p	3u3	350	25.0M	27p	1u5	236
17.2M	39p	2u2	238	25.3M	12p	3u3	524
17.3M	18p	4u7	511	25.3M	18p	2u2	350
17.4M	56p	1u5	164	25.5M	39p	1u	160
17.6M	82p	1u	110	27.7M	10p	3u3	574
17.6M	12p	6u8	753	27.7M	15p	2u2	383
18.7M	22p	3u3	387	27.7M	22p	1u5	261
18.7M	33p	2u2	258	27.7M	33p	1u	174
19.0M	15p	4u7	560	30.6M	18p	1u5	289
19.0M	47p	1u5	179	30.6M	27p	1u	192
19.3M	10p	6u8	825	31.0M	12p	2u2	428
19.3M	68p	1u	121	33.6M	15p	1u5	316
20.7M	18p	3u3	428	33.9M	10p	2u2	469
20.7M	27p	2u2	285	33.9M	22p	1u	213
20.8M	39p	1u5	196	37.5M	12p	1u5	354
21.2M	12p	4u7	626	37.5M	18p	1u	236
21.3M	56p	1u	134	41.1M	10p	1u5	387
22.6M	15p	3u3	469	41.1M	15p	1u	258
22.6M	33p	1u5	213	45.9M	12p	1u	289
22.9M	22p	2u2	316	50.3M	10p	1u	316
23.2M	10p	4u7	686				

Part Four

Operational Amplifier Circuits

11 Amplifier gains

Introduction

Operational amplifiers, or op-amps, are perhaps the most common building blocks found in analogue circuits. They can be used to create amplification stages very easily, and many other types of circuit besides. There are very many 'op-amp cookbooks' around that give lots of these circuits, so we won't add to them here. The main purpose of this chapter is to present you with a pair of tables to enable easy selection of resistor values for any desired gain.

We'll take a moment, however, to state the two rules that govern an op-amp's behaviour for the purpose of achieving a first-off analysis of their circuits:

1. The input terminals of op-amps draw no current.
2. The output of the op-amp does whatever it can to keep the voltage between the input terminals at zero.

These two statements are something of a generalization, but they are fine if you want to work out what an op-amp circuit is doing.

There are two common ways to use op-amps as amplifier stages. We look at them next.

Inverting op-amp

Figure 11.1 shows a basic inverting op-amp circuit. This circuit can be used to give either attenuation or gain with two resistors. The output is in antiphase with the input. The gain is given by:

$$A_V = \frac{R_2}{R_1} \qquad (11.1)$$

Values of R_1 and R_2 can be found from Table 11.1, where gain is given both as a ratio and in dBs. R_1 and R_2 may, of course, be scaled by factors of 10 without affecting the gain. The Table only shows values above unity gain (0 dB). If you need to select a pair of resistors to give attenuation, then the Table can still be used, but the positions of R_1 and R_2 should be switched. The dB column then reads attenuation directly, or the A_V column the reciprocal of gain expressed as a ratio.

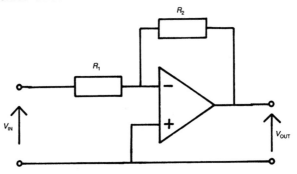

Figure 11.1 Inverting op-amp

The input resistance of the circuit is equal to the value of the input resistor, R_1.

The output resistance is, for practical purposes, negligible. (Limitations on its ability to drive a load will be dependent on maximum values of output current, voltage swing and slew rate, for which the manufacturer's data on the chip used should be consulted.)

It is sometimes desirable to limit the frequency response of the circuit. The standard method is to add suitable value capacitors, as in Figure 11.2. C_1, in series with the input resistor, limits low frequency response, while C_2, in parallel with R_2, limits high frequency response. Each gives a 6 dB/octave slope, as for the first order LPF and HPF described in Chapter 9. The equations for the lower −3 dB point, f_{t-}, and the upper −3 dB point, f_{t+}, are also similar:

$$f_{t-} = \frac{1}{2\pi R_1 C_1} \quad \text{and} \quad f_{t+} = \frac{1}{2\pi R_2 C_2} \tag{11.2a}$$

Example 11.1: give suitable values for an inverting op-amp circuit with a gain of 15 dB, and an input resistance of between 20k and 30k. Determine suitable values for C_1 and C_2 to make the upper and lower −3 dB points as near as possible to 100 Hz and 15 kHz respectively, using standard E12 capacitors.

Table 11.1 gives many combinations, close to 15 dB, but to meet the requirement for input resistance R_1 should be between 20k and 30k. We take the pair 2k7/15k, and multiply by 10, so that $R_1 = 27$k and $R_2 = 150$k.

To find values for C_1 and C_2, we transpose Eqn (11.2a):

$$C = \frac{1}{2\pi f_t R} \tag{11.2b}$$

and get:

1. $C_1 = 59$ nF (ideal) − nearest is 56 nF, substitute back into Eqn (11.2) to get $f_{t-} = 105$ Hz
2. $C_2 = 71$ pF (ideal) − nearest is 68 pF, giving $f_{t+} = 15.6$ kHz

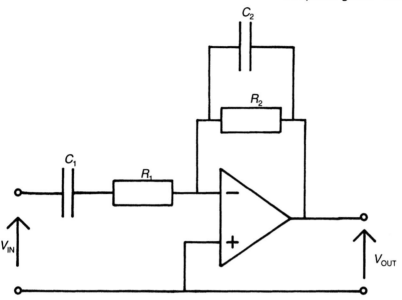

Figure 11.2 Inverting op-amp with restricted bandwidth

Non-inverting op-amp

Figure 11.3 shows a non-inverting op-amp circuit, in its most basic configuration. This circuit has a gain as follows:

$$A_V = \frac{R_1 + R_2}{R_1} \tag{11.3}$$

This means that the circuit can only provide gain of greater than 1. Its input impedance is very large, and is not usually taken into account. As with the inverting op-amp, its output impedance is very low, with the same restrictions on output drive applying (Figure 11.4).

Where it is desired to restrict the bandwidth of the stage, capacitors C_1 and C_2 are added. The calculations are identical to those for the inverting stage. Note that with C_1 in this position the input signal must have some defined DC level. If the input is connected using a DC block capacitor then C_1 must be omitted so that the op-amp has a DC reference. If it is desired to limit the low frequency response of the stage it is sometimes better to connect the input via a suitable high-pass filter, as described in Chapter 9.

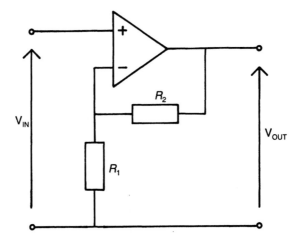

Figure 11.3 Non-inverting op-amp

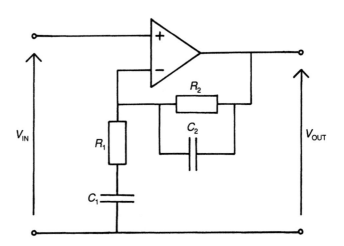

Figure 11.4 Non-inverting op-amp with restricted bandwidth

The main advantage of this stage (other than its being non-inverting) is that of its input impedance. It makes an ideal buffer stage, after circuits which are sensitive to load impedance (such as the reactive circuits described in Part Two).

Where it is desired to use this circuit without gain, purely as a buffer stage, it is customary to omit R_1 and make R_2 short circuit – i.e. to link the output terminal to the inverting input terminal, as in Figure 11.5. This is a 'voltage follower'. The output is identical to the input, but comes from a very low source impedance.

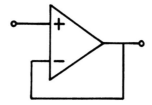

Figure 11.5   A voltage follower

Values for R_1 and R_2 for the non-inverting amplifier can be selected using Table 8.1, in the 'Voltage dividers' chapter. The dB column gives values of gain directly, and the A column values of the reciprocal of gain as a ratio. R_1 from the table should be fitted as position R_2 from Figure 11.3 and R_2 as R_1. As usual, R_1 and R_2 can be scaled by factors of 10 if desired.

Table 11.1 *Inverting op-amp gains*

A_V	dB	R1	R2	A_V	dB	R1	R2
1.07	00.56	1k5	1k6	1.21	01.65	6k2	7k5
1.08	00.70	1k2	1k3	1.21	01.68	7k5	9k1
1.08	00.70	3k6	3k9	1.21	01.69	5k6	6k8
1.09	00.71	4k7	5k1	1.22	01.70	5k1	6k2
1.09	00.76	1k1	1k2	1.22	01.72	8k2	10k
1.09	00.76	2k2	2k4	1.22	01.74	1k8	2k2
1.09	00.76	3k3	3k6	1.22	01.74	2k7	3k3
1.09	00.77	4k3	4k7	1.23	01.78	2k2	2k7
1.09	00.78	7k5	8k2	1.23	01.80	1k3	1k6
1.10	00.80	6k2	6k8	1.25	01.94	1k2	1k5
1.10	00.81	5k1	5k6	1.25	01.94	1k6	2k
1.10	00.82	9k1	10k	1.25	01.94	2k4	3k
1.10	00.83	1k	1k1	1.30	02.28	1k	1k3
1.10	00.83	2k	2k2	1.30	02.28	3k	3k9
1.10	00.83	3k	3k3	1.30	02.29	4k3	5k6
1.10	00.85	3k9	4k3	1.30	02.30	3k3	4k3
1.10	00.85	6k8	7k5	1.31	02.32	3k6	4k7
1.11	00.88	5k6	6k2	1.31	02.33	3k9	5k1
1.11	00.90	8k2	9k1	1.32	02.40	9k1	12k
1.11	00.92	1k8	2k	1.32	02.41	4k7	6k2
1.11	00.92	2k7	3k	1.32	02.43	6k2	8k2
1.13	01.02	1k6	1k8	1.33	02.50	1k8	2k4
1.13	01.02	2k4	2k7	1.33	02.50	7k5	10k
1.15	01.24	1k3	1k5	1.33	02.50	1k2	1k6
1.18	01.45	1k1	1k3	1.33	02.50	1k5	2k
1.18	01.45	3k3	3k9	1.33	02.50	2k7	3k6
1.19	01.48	4k3	5k1	1.33	02.50	5k1	6k8
1.19	01.52	4k7	5k6	1.34	02.53	6k8	9k1
1.19	01.54	3k6	4k3	1.34	02.54	5k6	7k5
1.20	01.58	1k	1k2	1.34	02.55	8k2	11k
1.20	01.58	2k	2k4	1.35	02.61	2k	2k7
1.20	01.58	1k5	1k8	1.36	02.69	1k1	1k5
1.20	01.58	3k	3k6	1.36	02.69	2k2	3k
1.21	01.62	3k9	4k7	1.38	02.77	1k6	2k2
1.21	01.63	6k8	8k2	1.38	02.77	2k4	3k3
1.21	01.65	9k1	11k	1.38	02.83	1k3	1k8

Table 11.1 Inverting op-amp gains 265

A_V	dB	R1	R2	A_V	dB	R1	R2
1.42	03.03	3k6	5k1	1.62	04.18	6k8	11k
1.42	03.07	3k3	4k7	1.63	04.22	2k4	3k9
1.43	03.10	9k1	13k	1.63	04.22	5k6	9k1
1.43	03.13	3k	4k3	1.64	04.28	1k1	1k8
1.44	03.14	3k9	5k6	1.64	04.28	2k2	3k6
1.44	03.18	4k3	6k2	1.65	04.34	9k1	15k
1.44	03.19	2k7	3k9	1.65	04.35	2k	3k3
1.45	03.21	4k7	6k8	1.67	04.44	1k2	2k
1.45	03.25	1k1	1k6	1.67	04.44	1k8	3k
1.46	03.31	8k2	12k	1.69	04.54	1k6	2k7
1.46	03.31	5k6	8k2	1.69	04.57	1k3	2k2
1.47	03.33	1k5	2k2	1.70	04.59	3k3	5k6
1.47	03.33	7k5	11k	1.70	04.61	3k	5k1
1.47	03.33	6k2	9k1	1.72	04.72	3k6	6k2
1.47	03.35	5k1	7k5	1.73	04.78	7k5	13k
1.47	03.35	6k8	10k	1.74	04.81	2k7	4k7
1.50	03.52	1k	1k5	1.74	04.83	3k9	6k8
1.50	03.52	1k2	1k8	1.74	04.83	4k3	7k5
1.50	03.52	1k6	2k4	1.74	04.83	4k7	8k2
1.50	03.52	1k8	2k7	1.76	04.90	9k1	16k
1.50	03.52	2k	3k	1.76	04.93	6k8	12k
1.50	03.52	2k2	3k3	1.77	04.97	2k2	3k9
1.50	03.52	2k4	3k6	1.77	04.98	6k2	11k
1.54	03.74	1k3	2k	1.78	05.03	5k1	9k1
1.55	03.78	3k3	5k1	1.79	05.04	5k6	10k
1.56	03.84	3k6	5k6	1.79	05.07	2k4	4k3
1.57	03.90	3k	4k7	1.80	05.11	1k	1k8
1.58	03.98	4k3	6k8	1.80	05.11	1k5	2k7
1.59	04.00	8k2	13k	1.80	05.11	2k	3k6
1.59	04.03	3k9	6k2	1.82	05.19	1k1	2k
1.59	04.04	2k7	4k3	1.83	05.25	8k2	15k
1.60	04.06	4k7	7k5	1.83	05.26	1k2	2k2
1.60	04.08	1k	1k6	1.83	05.26	1k8	3k3
1.60	04.08	1k5	2k4	1.85	05.33	1k3	2k4
1.60	04.08	7k5	12k	1.87	05.42	3k	5k6
1.61	04.12	5k1	8k2	1.88	05.46	1k6	3k
1.61	04.15	6k2	10k	1.88	05.48	3k3	6k2

A_V	dB	R1	R2	A_V	dB	R1	R2
1.89	05.52	2k7	5k1	2.18	06.78	1k1	2k4
1.89	05.52	3k6	6k8	2.20	06.83	8k2	18k
1.91	05.61	4k3	8k2	2.20	06.84	9k1	20k
1.91	05.63	6k8	13k	2.20	06.85	1k	2k2
1.92	05.68	3k9	7k5	2.20	06.85	1k5	3k3
1.94	05.74	6k2	12k	2.21	06.87	6k8	15k
1.94	05.74	4k7	9k1	2.25	07.04	1k2	2k7
1.95	05.80	2k	3k9	2.25	07.04	1k6	3k6
1.95	05.81	8k2	16k	2.27	07.11	3k	6k8
1.95	05.82	2k2	4k3	2.27	07.13	3k3	7k5
1.96	05.84	2k4	4k7	2.28	07.15	3k6	8k2
1.96	05.85	5k1	10k	2.30	07.22	2k7	6k2
1.96	05.86	5k6	11k	2.31	07.26	1k3	3k
1.98	05.92	9k1	18k	2.32	07.30	2k2	5k1
2.00	06.02	1k	2k	2.32	07.32	5k6	13k
2.00	06.02	1k1	2k2	2.33	07.33	4k3	10k
2.00	06.02	1k2	2k4	2.33	07.36	2k4	5k6
2.00	06.02	1k5	3k	2.33	07.36	3k9	9k1
2.00	06.02	1k8	3k6	2.34	07.39	4k7	11k
2.00	06.02	7k5	15k	2.35	07.42	2k	4k7
2.06	06.28	3k3	6k8	2.35	07.43	5k1	12k
2.06	06.29	1k6	3k3	2.35	07.43	6k8	16k
2.07	06.31	3k	6k2	2.39	07.56	1k8	4k3
2.07	06.34	2k7	5k6	2.40	07.60	1k	2k4
2.08	06.35	1k3	2k7	2.40	07.60	7k5	18k
2.08	06.38	3k6	7k5	2.40	07.60	1k5	3k6
2.10	06.43	6k2	13k	2.42	07.67	9k1	22k
2.10	06.45	3k9	8k2	2.42	07.67	6k2	15k
2.12	06.51	4k3	9k1	2.44	07.74	1k6	3k9
2.13	06.55	2k4	5k1	2.44	07.74	8k2	20k
2.13	06.56	4k7	10k	2.45	07.80	1k1	2k7
2.13	06.58	7k5	16k	2.48	07.91	3k3	8k2
2.14	06.59	2k2	4k7	2.50	07.96	1k2	3k
2.14	06.62	5k6	12k	2.50	07.96	3k	7k5
2.15	06.65	2k	4k3	2.52	08.02	2k7	6k8
2.16	06.68	5k1	11k	2.53	08.05	3k6	9k1
2.17	06.72	1k8	3k9	2.54	08.09	1k3	3k3

Table 11.1 Inverting op-amp gains 267

A_V	dB	R1	R2	A_V	dB	R1	R2
2.55	08.12	2k2	5k6	2.94	09.37	5k1	15k
2.55	08.13	5k1	13k	2.94	09.37	6k8	20k
2.55	08.13	2k	5k1	2.97	09.45	9k1	27k
2.55	08.14	4k7	12k	3.00	09.54	1k	3k
2.56	08.16	4k3	11k	3.00	09.54	1k1	3k3
2.56	08.18	3k9	10k	3.00	09.54	1k2	3k6
2.58	08.23	6k2	16k	3.00	09.54	1k3	3k9
2.58	08.24	2k4	6k2	3.02	09.61	4k3	13k
2.60	08.30	1k5	3k9	3.03	09.63	3k3	10k
2.61	08.34	1k8	4k7	3.03	09.64	3k	9k1
2.64	08.42	9k1	24k	3.04	09.65	2k7	8k2
2.65	08.46	6k8	18k	3.06	09.70	3k6	11k
2.67	08.52	7k5	20k	3.08	09.76	3k9	12k
2.68	08.56	5k6	15k	3.09	09.80	2k2	6k8
2.68	08.57	8k2	22k	3.10	09.83	2k	6k2
2.69	08.59	1k6	4k3	3.11	09.86	1k8	5k6
2.70	08.63	1k	2k7	3.13	09.90	2k4	7k5
2.73	08.71	1k1	3k	3.13	09.92	1k5	4k7
2.73	08.73	3k	8k2	3.14	09.93	5k1	16k
2.75	08.79	1k2	3k3	3.19	10.07	1k6	5k1
2.76	08.81	3k3	9k1	3.19	10.08	4k7	15k
2.77	08.84	4k7	13k	3.20	10.10	7k5	24k
2.77	08.85	1k3	3k6	3.21	10.14	5k6	18k
2.78	08.87	2k7	7k5	3.23	10.17	6k2	20k
2.78	08.87	3k6	10k	3.24	10.20	6k8	22k
2.79	08.91	4k3	12k	3.25	10.24	1k2	3k9
2.80	08.94	2k	5k6	3.27	10.30	1k1	3k6
2.82	09.00	2k2	6k2	3.29	10.35	8k2	27k
2.82	09.01	3k9	11k	3.30	10.36	9k1	30k
2.83	09.05	1k8	5k1	3.30	10.37	1k	3k3
2.83	09.05	2k4	6k8	3.31	10.39	1k3	4k3
2.86	09.12	5k6	16k	3.33	10.46	3k	10k
2.87	09.15	1k5	4k3	3.33	10.46	3k3	11k
2.90	09.26	6k2	18k	3.33	10.46	3k6	12k
2.93	09.33	8k2	24k	3.33	10.46	3k9	13k
2.93	09.35	7k5	22k	3.37	10.55	2k7	9k1
2.94	09.36	1k6	4k7	3.40	10.63	1k5	5k1

A_V	dB	R1	R2	A_V	dB	R1	R2
3.40	10.63	2k	6k8	3.93	11.88	5k6	22k
3.40	10.64	4k7	16k	3.94	11.91	3k3	13k
3.41	10.65	2k2	7k5	3.96	11.95	9k1	36k
3.42	10.67	2k4	8k2	3.97	11.98	6k8	27k
3.44	10.74	1k8	6k2	4.00	12.04	3k	12k
3.49	10.85	4k3	15k	4.00	12.04	7k5	30k
3.50	10.88	1k6	5k6	4.02	12.09	8k2	33k
3.53	10.95	5k1	18k	4.07	12.20	2k7	11k
3.53	10.95	6k8	24k	4.10	12.26	2k	8k2
3.55	10.99	1k1	3k9	4.10	12.26	3k9	16k
3.55	11.00	6k2	22k	4.13	12.33	1k5	6k2
3.57	11.06	5k6	20k	4.14	12.33	2k2	9k1
3.58	11.09	1k2	4k3	4.17	12.40	1k8	7k5
3.60	11.13	1k	3k6	4.17	12.40	2k4	10k
3.60	11.13	7k5	27k	4.17	12.40	3k6	15k
3.61	11.15	3k6	13k	4.19	12.44	4k3	18k
3.62	11.16	1k3	4k7	4.25	12.57	1k2	5k1
3.63	11.19	9k1	33k	4.25	12.57	1k6	6k8
3.64	11.21	3k3	12k	4.26	12.58	4k7	20k
3.66	11.27	8k2	30k	4.27	12.61	1k1	4k7
3.67	11.29	3k	11k	4.29	12.64	5k6	24k
3.70	11.37	2k7	10k	4.29	12.64	9k1	39k
3.72	11.41	4k3	16k	4.30	12.67	1k	4k3
3.73	11.43	2k2	8k2	4.31	12.68	1k3	5k6
3.73	11.44	1k5	5k6	4.31	12.70	5k1	22k
3.75	11.48	2k	7k5	4.33	12.74	3k	13k
3.78	11.54	1k8	6k8	4.35	12.78	6k2	27k
3.79	11.58	2k4	9k1	4.39	12.85	8k2	36k
3.83	11.66	4k7	18k	4.40	12.87	7k5	33k
3.85	11.70	3k9	15k	4.41	12.89	6k8	30k
3.87	11.76	6k2	24k	4.44	12.96	2k7	12k
3.88	11.77	1k6	6k2	4.44	12.96	3k6	16k
3.90	11.82	1k	3k9	4.53	13.13	1k5	6k8
3.91	11.84	1k1	4k3	4.55	13.15	2k2	10k
3.92	11.86	1k2	4k7	4.55	13.15	3k3	15k
3.92	11.87	5k1	20k	4.55	13.16	2k	9k1
3.92	11.87	1k3	5k1	4.56	13.17	1k8	8k2

Table 11.1 Inverting op-amp gains 269

A_V	dB	R1	R2	A_V	dB	R1	R2
4.58	13.22	2k4	11k	5.29	14.48	6k8	36k
4.62	13.28	3k9	18k	5.32	14.52	6k2	33k
4.64	13.32	1k1	5k1	5.33	14.54	3k	16k
4.65	13.35	4k3	20k	5.36	14.58	5k6	30k
4.67	13.38	1k2	5k6	5.42	14.67	2k4	13k
4.68	13.41	4k7	22k	5.45	14.74	3k3	18k
4.69	13.42	1k6	7k5	5.45	14.74	2k2	12k
4.70	13.44	1k	4k7	5.47	14.75	1k5	8k2
4.71	13.45	5k1	24k	5.50	14.81	2k	11k
4.73	13.49	9k1	43k	5.56	14.89	1k8	10k
4.76	13.55	8k2	39k	5.56	14.89	2k7	15k
4.77	13.57	1k3	6k2	5.56	14.89	3k6	20k
4.80	13.62	7k5	36k	5.58	14.93	4k3	24k
4.81	13.65	2k7	13k	5.60	14.96	1k	5k6
4.82	13.66	5k6	27k	5.60	14.97	9k1	51k
4.84	13.69	6k2	30k	5.64	15.02	1k1	6k2
4.85	13.71	3k3	16k	5.64	15.03	3k9	22k
4.85	13.72	6k8	33k	5.67	15.07	1k2	6k8
5.00	13.98	1k5	7k5	5.69	15.10	1k6	9k1
5.00	13.98	2k	10k	5.73	15.17	8k2	47k
5.00	13.98	2k2	11k	5.73	15.17	7k5	43k
5.00	13.98	2k4	12k	5.74	15.17	6k8	39k
5.00	13.98	3k	15k	5.74	15.19	4k7	27k
5.00	13.98	3k6	18k	5.77	15.22	1k3	7k5
5.06	14.08	1k8	9k1	5.81	15.28	6k2	36k
5.09	14.14	1k1	5k6	5.88	15.39	5k1	30k
5.10	14.15	1k	5k1	5.89	15.41	5k6	33k
5.11	14.16	4k7	24k	5.91	15.43	2k2	13k
5.12	14.18	4k3	22k	5.93	15.46	2k7	16k
5.13	14.19	1k6	8k2	6.00	15.56	2k	12k
5.13	14.20	3k9	20k	6.00	15.56	3k	18k
5.16	14.26	9k1	47k	6.06	15.65	3k3	20k
5.17	14.26	1k2	6k2	6.07	15.66	1k5	9k1
5.20	14.32	7k5	39k	6.11	15.72	1k8	11k
5.23	14.37	1k3	6k8	6.11	15.72	3k6	22k
5.24	14.39	8k2	43k	6.15	15.78	3k9	24k
5.29	14.48	5k1	27k	6.15	15.78	9k1	56k

A_V	dB	R1	R2	A_V	dB	R1	R2
6.18	15.82	1k1	6k8	7.06	16.97	5k1	36k
6.20	15.85	1k	6k2	7.22	17.17	1k8	13k
6.22	15.88	8k2	51k	7.27	17.23	3k3	24k
6.25	15.92	1k2	7k5	7.27	17.23	2k2	16k
6.25	15.92	1k6	10k	7.33	17.31	1k5	11k
6.25	15.92	2k4	15k	7.33	17.31	3k	22k
6.27	15.94	7k5	47k	7.41	17.39	2k7	20k
6.28	15.96	4k3	27k	7.45	17.45	1k1	8k2
6.29	15.97	6k2	39k	7.47	17.46	7k5	56k
6.31	16.00	1k3	8k2	7.47	17.47	9k1	68k
6.32	16.02	6k8	43k	7.50	17.50	1k	7k5
6.38	16.10	4k7	30k	7.50	17.50	1k6	12k
6.43	16.16	5k6	36k	7.50	17.50	2k	15k
6.47	16.22	5k1	33k	7.50	17.50	2k4	18k
6.50	16.26	2k	13k	7.50	17.50	3k6	27k
6.67	16.48	1k5	10k	7.50	17.50	6k8	51k
6.67	16.48	1k8	12k	7.56	17.57	8k2	62k
6.67	16.48	2k4	16k	7.58	17.59	6k2	47k
6.67	16.48	2k7	18k	7.58	17.60	1k2	9k1
6.67	16.48	3k	20k	7.65	17.67	5k1	39k
6.67	16.48	3k3	22k	7.66	17.68	4k7	36k
6.67	16.48	3k6	24k	7.67	17.70	4k3	33k
6.80	16.65	1k	6k8	7.68	17.71	5k6	43k
6.80	16.65	7k5	51k	7.69	17.72	1k3	10k
6.81	16.67	9k1	62k	7.69	17.72	3k9	30k
6.82	16.67	1k1	7k5	8.00	18.06	1k5	12k
6.82	16.67	2k2	15k	8.00	18.06	2k	16k
6.83	16.69	8k2	56k	8.00	18.06	3k	24k
6.83	16.69	1k2	8k2	8.13	18.20	1k6	13k
6.88	16.75	1k6	11k	8.15	18.22	2k7	22k
6.91	16.79	6k8	47k	8.18	18.26	2k2	18k
6.92	16.81	3k9	27k	8.18	18.26	3k3	27k
6.94	16.82	6k2	43k	8.20	18.28	1k	8k2
6.96	16.86	5k6	39k	8.23	18.30	6k2	51k
6.98	16.87	4k3	30k	8.24	18.31	6k8	56k
7.00	16.90	1k3	9k1	8.24	18.32	9k1	75k
7.02	16.93	4k7	33k	8.27	18.35	7k5	62k

Table 11.1 Inverting op-amp gains 271

Av	dB	R1	R2	Av	dB	R1	R2
8.27	18.35	1k1	9k1	10.0	20.00	1k1	11k
8.29	18.37	8k2	68k	10.0	20.00	1k2	12k
8.30	18.38	4k7	39k	10.0	20.00	1k3	13k
8.33	18.42	1k2	10k	10.0	20.00	1k5	15k
8.33	18.42	1k8	15k	10.0	20.00	1k6	16k
8.33	18.42	2k4	20k	10.0	20.00	1k8	18k
8.33	18.42	3k6	30k	10.0	20.00	2k	20k
8.37	18.46	4k3	36k	10.0	20.00	2k2	22k
8.39	18.48	5k6	47k	10.0	20.00	2k4	24k
8.43	18.52	5k1	43k	10.0	20.00	2k7	27k
8.46	18.55	1k3	11k	10.0	20.00	3k	30k
8.46	18.55	3k9	33k	10.0	20.00	3k3	33k
8.67	18.76	1k5	13k	10.0	20.00	3k6	36k
8.89	18.98	1k8	16k	10.0	20.00	3k9	39k
8.89	18.98	2k7	24k	10.0	20.00	4k3	43k
9.00	19.08	2k	18k	10.0	20.00	4k7	47k
9.00	19.08	3k	27k	10.0	20.00	5k1	51k
9.01	19.10	9k1	82k	10.0	20.00	5k6	56k
9.03	19.12	6k2	56k	10.0	20.00	6k2	62k
9.07	19.15	7k5	68k	10.0	20.00	6k8	68k
9.07	19.15	4k3	39k	10.0	20.00	7k5	75k
9.09	19.17	1k1	10k	10.0	20.00	8k2	82k
9.09	19.17	2k2	20k	10.0	20.00	9k1	91k
9.09	19.17	3k3	30k	10.7	20.56	1k5	16k
9.10	19.18	1k	9k1	10.8	20.70	1k2	13k
9.11	19.19	5k6	51k	10.8	20.70	3k6	39k
9.12	19.20	6k8	62k	10.9	20.71	4k7	51k
9.15	19.22	8k2	75k	10.9	20.76	3k3	36k
9.15	19.23	4k7	43k	10.9	20.76	1k1	12k
9.17	19.24	3k6	33k	10.9	20.76	2k2	24k
9.17	19.24	1k2	11k	10.9	20.77	4k3	47k
9.17	19.24	2k4	22k	10.9	20.78	7k5	82k
9.22	19.29	5k1	47k	11.0	20.80	6k2	68k
9.23	19.30	1k3	12k	11.0	20.81	5k1	56k
9.23	19.30	3k9	36k	11.0	20.82	9k1	100k
9.38	19.44	1k6	15k	11.0	20.82	910R	10k
10.0	20.00	1k	10k	11.0	20.83	1k	11k

A_V	dB	R1	R2	A_V	dB	R1	R2
11.0	20.83	2k	22k	13.0	22.28	3k	39k
11.0	20.83	3k	33k	13.0	22.29	4k3	56k
11.0	20.85	3k9	43k	13.0	22.30	3k3	43k
11.0	20.85	6k8	75k	13.1	22.32	3k6	47k
11.1	20.88	5k6	62k	13.1	22.33	3k9	51k
11.1	20.90	8k2	91k	13.2	22.40	910R	12k
11.1	20.92	1k8	20k	13.2	22.41	4k7	62k
11.1	20.92	2k7	30k	13.2	22.43	6k2	82k
11.3	21.02	1k6	18k	13.3	22.50	1k2	16k
11.3	21.02	2k4	27k	13.3	22.50	1k5	20k
11.5	21.24	1k3	15k	13.3	22.50	1k8	24k
11.8	21.45	1k1	13k	13.3	22.50	2k7	36k
11.8	21.45	3k3	39k	13.3	22.50	5k1	68k
11.9	21.48	4k3	51k	13.3	22.50	7k5	100k
11.9	21.52	4k7	56k	13.3	22.50	750R	10k
11.9	21.54	3k6	43k	13.4	22.53	6k8	91k
12.0	21.58	1k	12k	13.4	22.54	5k6	75k
12.0	21.58	1k5	18k	13.4	22.55	820R	11k
12.0	21.58	2k	24k	13.5	22.61	2k	27k
12.0	21.58	3k	36k	13.6	22.69	1k1	15k
12.1	21.62	3k9	47k	13.6	22.69	2k2	30k
12.1	21.63	6k8	82k	13.8	22.77	1k6	22k
12.1	21.65	910R	11k	13.8	22.77	2k4	33k
12.1	21.65	6k2	75k	13.8	22.83	1k3	18k
12.1	21.68	7k5	91k	14.2	23.03	3k6	51k
12.1	21.69	5k6	68k	14.2	23.07	3k3	47k
12.2	21.70	5k1	62k	14.3	23.10	910R	13k
12.2	21.72	8k2	100k	14.3	23.13	3k	43k
12.2	21.72	820R	10k	14.4	23.14	3k9	56k
12.2	21.74	1k8	22k	14.4	23.18	4k3	62k
12.2	21.74	2k7	33k	14.4	23.19	2k7	39k
12.3	21.78	2k2	27k	14.5	23.21	4k7	68k
12.3	21.80	1k3	16k	14.5	23.25	1k1	16k
12.5	21.94	1k2	15k	14.6	23.31	820R	12k
12.5	21.94	1k6	20k	14.6	23.31	5k6	82k
12.5	21.94	2k4	30k	14.7	23.33	1k5	22k
13.0	22.28	1k	13k	14.7	23.33	750R	11k

Table 11.1 Inverting op-amp gains 273

A_V	dB	R1	R2	A_V	dB	R1	R2
14.7	23.33	6k2	91k	17.0	24.59	3k3	56k
14.7	23.35	5k1	75k	17.0	24.61	3k	51k
14.7	23.35	6k8	100k	17.2	24.72	3k6	62k
14.7	23.35	680R	10k	17.3	24.78	750R	13k
15.0	23.52	1k	15k	17.4	24.81	2k7	47k
15.0	23.52	1k2	18k	17.4	24.83	3k9	68k
15.0	23.52	1k6	24k	17.4	24.83	4k3	75k
15.0	23.52	1k8	27k	17.4	24.83	4k7	82k
15.0	23.52	2k	30k	17.6	24.90	910R	16k
15.0	23.52	2k2	33k	17.6	24.93	680R	12k
15.0	23.52	2k4	36k	17.7	24.97	2k2	39k
15.4	23.74	1k3	20k	17.7	24.98	620R	11k
15.5	23.78	3k3	51k	17.8	25.03	5k1	91k
15.6	23.84	3k6	56k	17.9	25.04	5k6	100k
15.7	23.90	3k	47k	17.9	25.04	560R	10k
15.8	23.98	4k3	68k	17.9	25.07	2k4	43k
15.9	24.00	820R	13k	18.0	25.11	1k	18k
15.9	24.03	3k9	62k	18.0	25.11	1k5	27k
15.9	24.04	2k7	43k	18.0	25.11	2k	36k
16.0	24.06	4k7	75k	18.2	25.19	1k1	20k
16.0	24.08	1k	16k	18.3	25.25	820R	15k
16.0	24.08	1k5	24k	18.3	25.26	1k8	33k
16.0	24.08	750R	12k	18.3	25.26	1k2	22k
16.1	24.12	5k1	82k	18.5	25.33	1k3	24k
16.1	24.15	6k2	100k	18.7	25.42	3k	56k
16.1	24.15	620R	10k	18.8	25.46	1k6	30k
16.2	24.18	680R	11k	18.8	25.48	3k3	62k
16.3	24.22	2k4	39k	18.9	25.52	3k6	68k
16.3	24.22	5k6	91k	18.9	25.52	2k7	51k
16.4	24.28	1k1	18k	19.1	25.61	4k3	82k
16.4	24.28	2k2	36k	19.1	25.63	680R	13k
16.5	24.34	910R	15k	19.2	25.68	3k9	75k
16.5	24.35	2k	33k	19.4	25.74	620R	12k
16.7	24.44	1k2	20k	19.4	25.74	4k7	91k
16.7	24.44	1k8	30k	19.5	25.80	2k	39k
16.9	24.54	1k6	27k	19.5	25.81	820R	16k
16.9	24.57	1k3	22k	19.5	25.82	2k2	43k

Av	dB	R1	R2	Av	dB	R1	R2
19.6	25.84	2k4	47k	22.7	27.11	3k	68k
19.6	25.85	510R	10k	22.7	27.13	3k3	75k
19.6	25.85	5k1	100k	22.8	27.15	3k6	82k
19.6	25.86	560R	11k	23.0	27.22	2k7	62k
19.8	25.92	910R	18k	23.1	27.26	1k3	30k
20.0	26.02	1k	20k	23.2	27.30	2k2	51k
20.0	26.02	1k1	22k	23.2	27.32	560R	13k
20.0	26.02	1k2	24k	23.3	27.33	430R	10k
20.0	26.02	1k5	30k	23.3	27.33	4k3	100k
20.0	26.02	1k8	36k	23.3	27.36	3k9	91k
20.0	26.02	750R	15k	23.3	27.36	2k4	56k
20.6	26.28	3k3	68k	23.4	27.39	470R	11k
20.6	26.29	1k6	33k	23.5	27.42	2k	47k
20.7	26.31	3k	62k	23.5	27.43	510R	12k
20.7	26.34	2k7	56k	23.5	27.43	680R	16k
20.8	26.35	1k3	27k	23.9	27.56	1k8	43k
20.8	26.38	3k6	75k	24.0	27.60	1k	24k
21.0	26.43	620R	13k	24.0	27.60	1k5	36k
21.0	26.45	3k9	82k	24.0	27.60	750R	18k
21.2	26.51	4k3	91k	24.2	27.67	910R	22k
21.3	26.55	2k4	51k	24.2	27.67	620R	15k
21.3	26.56	470R	10k	24.4	27.74	1k6	39k
21.3	26.56	4k7	100k	24.4	27.74	820R	20k
21.3	26.58	750R	16k	24.5	27.80	1k1	27k
21.4	26.59	2k2	47k	24.8	27.91	3k3	82k
21.4	26.62	560R	12k	25.0	27.96	1k2	30k
21.5	26.65	2k	43k	25.0	27.96	3k	75k
21.6	26.68	510R	11k	25.2	28.02	2k7	68k
21.7	26.72	1k8	39k	25.3	28.05	3k6	91k
21.8	26.78	1k1	24k	25.4	28.09	1k3	33k
22.0	26.83	820R	18k	25.5	28.12	2k2	56k
22.0	26.84	910R	20k	25.5	28.13	510R	13k
22.0	26.85	1k	22k	25.5	28.13	2k	51k
22.0	26.85	1k5	33k	25.5	28.14	470R	12k
22.1	26.87	680R	15k	25.6	28.16	430R	11k
22.5	27.04	1k2	27k	25.6	28.18	3k9	100k
22.5	27.04	1k6	36k	25.6	28.18	390R	10k

Table 11.1 Inverting op-amp gains 275

A$_V$	dB	R1	R2	A$_V$	dB	R1	R2
25.8	28.23	620R	16k	30.0	29.54	1k2	36k
25.8	28.24	2k4	62k	30.0	29.54	1k3	39k
26.0	28.30	1k5	39k	30.2	29.61	430R	13k
26.1	28.34	1k8	47k	30.3	29.63	3k3	100k
26.4	28.42	910R	24k	30.3	29.63	330R	10k
26.5	28.46	680R	18k	30.3	29.64	3k	91k
26.7	28.52	750R	20k	30.4	29.65	2k7	82k
26.8	28.56	560R	15k	30.6	29.70	360R	11k
26.8	28.57	820R	22k	30.8	29.76	390R	12k
26.9	28.59	1k6	43k	30.9	29.80	2k2	68k
27.0	28.63	1k	27k	31.0	29.83	2k	62k
27.3	28.71	1k1	30k	31.1	29.86	1k8	56k
27.3	28.73	3k	82k	31.3	29.90	2k4	75k
27.5	28.79	1k2	33k	31.3	29.92	1k5	47k
27.6	28.81	3k3	91k	31.4	29.93	510R	16k
27.7	28.84	470R	13k	31.9	30.07	1k6	51k
27.7	28.85	1k3	36k	31.9	30.08	470R	15k
27.8	28.87	2k7	75k	32.0	30.10	750R	24k
27.8	28.87	3k6	100k	32.1	30.14	560R	18k
27.8	28.87	360R	10k	32.3	30.17	620R	20k
27.9	28.91	430R	12k	32.4	30.20	680R	22k
28.0	28.94	2k	56k	32.5	30.24	1k2	39k
28.2	29.00	2k2	62k	32.7	30.30	1k1	36k
28.2	29.01	390R	11k	32.9	30.35	820R	27k
28.3	29.05	1k8	51k	33.0	30.36	910R	30k
28.3	29.05	2k4	68k	33.0	30.37	1k	33k
28.6	29.12	560R	16k	33.1	30.39	1k3	43k
28.7	29.15	1k5	43k	33.3	30.46	3k	100k
29.0	29.26	620R	18k	33.3	30.46	300R	10k
29.3	29.33	820R	24k	33.3	30.46	330R	11k
29.3	29.35	750R	22k	33.3	30.46	360R	12k
29.4	29.36	1k6	47k	33.3	30.46	390R	13k
29.4	29.37	510R	15k	33.7	30.55	2k7	91k
29.4	29.37	680R	20k	34.0	30.63	1k5	51k
29.7	29.45	910R	27k	34.0	30.63	2k	68k
30.0	29.54	1k	30k	34.0	30.64	470R	16k
30.0	29.54	1k1	33k	34.1	30.65	2k2	75k

A_V	dB	R1	R2	A_V	dB	R1	R2
34.2	30.67	2k4	82k	39.6	31.95	910R	36k
34.4	30.74	1k8	62k	39.7	31.98	680R	27k
34.9	30.85	430R	15k	40.0	32.04	300R	12k
35.0	30.88	1k6	56k	40.0	32.04	750R	30k
35.3	30.95	510R	18k	40.2	32.09	820R	33k
35.3	30.95	680R	24k	40.7	32.20	270R	11k
35.5	30.99	1k1	39k	41.0	32.26	2k	82k
35.5	31.00	620R	22k	41.0	32.26	390R	16k
35.7	31.06	560R	20k	41.3	32.33	1k5	62k
35.8	31.09	1k2	43k	41.4	32.33	2k2	91k
36.0	31.13	1k	36k	41.7	32.40	1k8	75k
36.0	31.13	750R	27k	41.7	32.40	240R	10k
36.1	31.15	360R	13k	41.7	32.40	2k4	100k
36.2	31.16	1k3	47k	41.7	32.40	360R	15k
36.3	31.19	910R	33k	41.9	32.44	430R	18k
36.4	31.21	330R	12k	42.5	32.57	1k2	51k
36.6	31.27	820R	30k	42.5	32.57	1k6	68k
36.7	31.29	300R	11k	42.6	32.58	470R	20k
37.0	31.37	2k7	100k	42.7	32.61	1k1	47k
37.0	31.37	270R	10k	42.9	32.64	910R	39k
37.2	31.41	430R	16k	42.9	32.64	560R	24k
37.3	31.43	2k2	82k	43.0	32.67	1k	43k
37.3	31.44	1k5	56k	43.1	32.68	1k3	56k
37.5	31.48	2k	75k	43.1	32.70	510R	22k
37.8	31.54	1k8	68k	43.3	32.74	300R	13k
37.9	31.58	2k4	91k	43.5	32.78	620R	27k
38.3	31.66	470R	18k	43.9	32.85	820R	36k
38.5	31.70	390R	15k	44.0	32.87	750R	33k
38.7	31.76	620R	24k	44.1	32.89	680R	30k
38.8	31.77	1k6	62k	44.4	32.96	270R	12k
39.0	31.82	1k	39k	44.4	32.96	360R	16k
39.1	31.84	1k1	43k	45.3	33.13	1k5	68k
39.2	31.86	1k2	47k	45.5	33.15	220R	10k
39.2	31.87	510R	20k	45.5	33.15	2k2	100k
39.2	31.87	1k3	51k	45.5	33.15	330R	15k
39.3	31.88	560R	22k	45.5	33.16	2k	91k
39.4	31.91	330R	13k	45.6	33.17	1k8	82k

Table 11.1 Inverting op-amp gains 277

A_V	dB	R1	R2	A_V	dB	R1	R2
45.8	33.22	240R	11k	52.9	34.48	510R	27k
46.2	33.28	390R	18k	52.9	34.48	680R	36k
46.4	33.32	1k1	51k	53.2	34.52	620R	33k
46.5	33.35	430R	20k	53.3	34.54	300R	16k
46.7	33.38	1k2	56k	53.6	34.58	560R	30k
46.8	33.41	470R	22k	54.2	34.67	240R	13k
46.9	33.42	1k6	75k	54.5	34.74	220R	12k
47.0	33.44	1k	47k	54.5	34.74	330R	18k
47.1	33.45	510R	24k	54.7	34.75	1k5	82k
47.3	33.49	910R	43k	55.0	34.81	200R	11k
47.6	33.55	820R	39k	55.6	34.89	1k8	100k
47.7	33.57	1k3	62k	55.6	34.89	180R	10k
48.0	33.62	750R	36k	55.6	34.89	360R	20k
48.1	33.65	270R	13k	55.6	34.89	270R	15k
48.2	33.66	560R	27k	55.8	34.93	430R	24k
48.4	33.69	620R	30k	56.0	34.96	1k	56k
48.5	33.71	330R	16k	56.0	34.97	910R	51k
48.5	33.72	680R	33k	56.4	35.02	1k1	62k
50.0	33.98	1k5	75k	56.4	35.03	390R	22k
50.0	33.98	2k	100k	56.7	35.07	1k2	,68k
50.0	33.98	200R	10k	56.9	35.10	1k6	91k
50.0	33.98	220R	11k	57.3	35.17	820R	47k
50.0	33.98	240R	12k	57.3	35.17	750R	43k
50.0	33.98	300R	15k	57.4	35.17	680R	39k
50.0	33.98	360R	18k	57.4	35.19	470R	27k
50.6	34.08	1k8	91k	57.7	35.22	1k3	75k
50.9	34.14	1k1	56k	58.1	35.28	620R	36k
51.0	34.15	1k	51k	58.8	35.39	510R	30k
51.1	34.16	470R	24k	58.9	35.41	560R	33k
51.2	34.18	430R	22k	59.1	35.43	220R	13k
51.3	34.19	1k6	82k	59.3	35.46	270R	16k
51.3	34.20	390R	20k	60.0	35.56	200R	12k
51.6	34.26	910R	47k	60.0	35.56	300R	18k
51.7	34.26	1k2	62k	60.6	35.65	330R	20k
52.0	34.32	750R	39k	60.7	35.66	1k5	91k
52.3	34.37	1k3	68k	61.1	35.72	180R	11k
52.4	34.39	820R	43k	61.1	35.72	360R	22k

A_V	dB	R1	R2	A_V	dB	R1	R2
61.5	35.78	390R	24k	69.6	36.86	560R	39k
61.5	35.78	910R	56k	69.8	36.87	430R	30k
61.8	35.82	1k1	68k	70.0	36.90	1k3	91k
62.0	35.85	1k	62k	70.2	36.93	470R	33k
62.2	35.88	820R	51k	70.6	36.97	510R	36k
62.5	35.92	1k2	75k	72.2	37.17	180R	13k
62.5	35.92	1k6	100k	72.7	37.23	220R	16k
62.5	35.92	160R	10k	72.7	37.23	330R	24k
62.5	35.92	240R	15k	73.3	37.31	150R	11k
62.7	35.94	750R	47k	73.3	37.31	300R	22k
62.8	35.96	430R	27k	74.1	37.39	270R	20k
62.9	35.97	620R	39k	74.5	37.45	1k1	82k
63.1	36.00	1k3	82k	74.7	37.46	750R	56k
63.2	36.02	680R	43k	74.7	37.47	910R	68k
63.8	36.10	470R	30k	75.0	37.50	1k	75k
64.3	36.16	560R	36k	75.0	37.50	160R	12k
64.7	36.22	510R	33k	75.0	37.50	200R	15k
65.0	36.26	200R	13k	75.0	37.50	240R	18k
66.7	36.48	1k5	100k	75.0	37.50	360R	27k
66.7	36.48	150R	10k	75.0	37.50	680R	51k
66.7	36.48	180R	12k	75.6	37.57	820R	62k
66.7	36.48	240R	16k	75.8	37.59	620R	47k
66.7	36.48	270R	18k	75.8	37.60	1k2	91k
66.7	36.48	300R	20k	76.5	37.67	510R	39k
66.7	36.48	330R	22k	76.6	37.68	470R	36k
66.7	36.48	360R	24k	76.7	37.70	430R	33k
68.0	36.65	1k	68k	76.8	37.71	560R	43k
68.0	36.65	750R	51k	76.9	37.72	1k3	100k
68.1	36.67	910R	62k	76.9	37.72	130R	10k
68.2	36.67	1k1	75k	76.9	37.72	390R	30k
68.2	36.67	220R	15k	80.0	38.06	150R	12k
68.3	36.69	820R	56k	80.0	38.06	200R	16k
68.3	36.69	1k2	82k	80.0	38.06	300R	24k
68.8	36.75	160R	11k	81.3	38.20	160R	13k
69.1	36.79	680R	47k	81.5	38.22	270R	22k
69.2	36.81	390R	27k	81.8	38.26	220R	18k
69.4	36.82	620R	43k	81.8	38.26	330R	27k

Table 11.1 *Inverting op-amp gains* 279

A_V	dB	R1	R2	A_V	dB	R1	R2
82.0	38.28	1k	82k	91.7	39.24	240R	22k
82.3	38.30	620R	51k	91.7	39.24	360R	33k
82.4	38.31	680R	56k	92.2	39.29	510R	47k
82.4	38.32	910R	75k	92.3	39.30	390R	36k
82.7	38.35	750R	62k	92.3	39.30	130R	12k
82.7	38.35	1k1	91k	93.8	39.44	160R	15k
82.9	38.37	820R	68k	100	40.00	1k	100k
83.0	38.38	470R	39k	100	40.00	100R	10k
83.3	38.42	120R	10k	100	40.00	110R	11k
83.3	38.42	240R	20k	100	40.00	120R	12k
83.3	38.42	1k2	100k	100	40.00	130R	13k
83.3	38.42	180R	15k	100	40.00	150R	15k
83.3	38.42	360R	30k	100	40.00	160R	16k
83.7	38.46	430R	36k	100	40.00	180R	18k
83.9	38.48	560R	47k	100	40.00	200R	20k
84.3	38.52	510R	43k	100	40.00	220R	22k
84.6	38.55	390R	33k	100	40.00	240R	24k
84.6	38.55	130R	11k	100	40.00	270R	27k
86.7	38.76	150R	13k	100	40.00	300R	30k
88.9	38.98	180R	16k	100	40.00	330R	33k
88.9	38.98	270R	24k	100	40.00	360R	36k
90.0	39.08	200R	18k	100	40.00	390R	39k
90.0	39.08	300R	27k	100	40.00	430R	43k
90.1	39.10	910R	82k	100	40.00	470R	47k
90.3	39.12	620R	56k	100	40.00	510R	51k
90.7	39.15	750R	68k	100	40.00	560R	56k
90.7	39.15	430R	39k	100	40.00	620R	62k
90.9	39.17	110R	10k	100	40.00	680R	68k
90.9	39.17	220R	20k	100	40.00	750R	75k
90.9	39.17	1k1	100k	100	40.00	820R	82k
90.9	39.17	330R	30k	100	40.00	910R	91k
91.0	39.18	1k	91k	107	40.56	150R	16k
91.1	39.19	560R	51k	108	40.70	120R	13k
91.2	39.20	680R	62k	108	40.70	360R	39k
91.5	39.22	820R	75k	109	40.71	470R	51k
91.5	39.23	470R	43k	109	40.76	110R	12k
91.7	39.24	120R	11k	109	40.76	220R	24k

Av	dB	R1	R2	Av	dB	R1	R2
109	40.76	330R	36k	123	41.80	130R	16k
109	40.77	430R	47k	125	41.94	120R	15k
109	40.78	750R	82k	125	41.94	160R	20k
110	40.80	620R	68k	125	41.94	240R	30k
110	40.81	510R	56k	130	42.28	100R	13k
110	40.82	910R	100k	130	42.28	300R	39k
110	40.83	100R	11k	130	42.29	430R	56k
110	40.83	200R	22k	130	42.30	330R	43k
110	40.83	300R	33k	131	42.32	360R	47k
110	40.85	390R	43k	131	42.33	390R	51k
110	40.85	680R	75k	132	42.41	470R	62k
111	40.88	560R	62k	132	42.43	620R	82k
111	40.90	820R	91k	133	42.50	120R	16k
111	40.92	180R	20k	133	42.50	150R	20k
111	40.92	270R	30k	133	42.50	180R	24k
113	41.02	160R	18k	133	42.50	270R	36k
113	41.02	240R	27k	133	42.50	510R	68k
115	41.24	130R	15k	133	42.50	750R	100k
118	41.45	110R	13k	134	42.53	680R	91k
118	41.45	330R	39k	134	42.54	560R	75k
119	41.48	430R	51k	135	42.61	200R	27k
119	41.52	470R	56k	136	42.69	110R	15k
119	41.54	360R	43k	136	42.69	220R	30k
120	41.58	100R	12k	138	42.77	160R	22k
120	41.58	150R	18k	138	42.77	240R	33k
120	41.58	200R	24k	138	42.83	130R	18k
120	41.58	300R	36k	142	43.03	360R	51k
121	41.62	390R	47k	142	43.07	330R	47k
121	41.63	680R	82k	143	43.13	300R	43k
121	41.65	620R	75k	144	43.14	390R	56k
121	41.68	750R	91k	144	43.18	430R	62k
121	41.69	560R	68k	144	43.19	270R	39k
122	41.70	510R	62k	145	43.21	470R	68k
122	41.72	820R	100k	145	43.25	110R	16k
122	41.74	180R	22k	146	43.31	560R	82k
122	41.74	270R	33k	147	43.33	150R	22k
123	41.78	220R	27k	147	43.33	620R	91k

Table 11.1 Inverting op-amp gains 281

A_V	dB	R1	R2	A_V	dB	R1	R2
147	43.35	510R	75k	177	44.97	220R	39k
147	43.35	680R	100k	178	45.03	510R	91k
150	43.52	100R	15k	179	45.04	560R	100k
150	43.52	120R	18k	179	45.07	240R	43k
150	43.52	160R	24k	180	45.11	100R	18k
150	43.52	180R	27k	180	45.11	150R	27k
150	43.52	200R	30k	180	45.11	200R	36k
150	43.52	220R	33k	182	45.19	110R	20k
150	43.52	240R	36k	183	45.26	120R	22k
154	43.74	130R	20k	183	45.26	180R	33k
155	43.78	330R	51k	185	45.33	130R	24k
156	43.84	360R	56k	187	45.42	300R	56k
157	43.90	300R	47k	188	45.46	160R	30k
158	43.98	430R	68k	188	45.48	330R	62k
159	44.03	390R	62k	189	45.52	270R	51k
159	44.04	270R	43k	189	45.52	360R	68k
160	44.06	470R	75k	191	45.61	430R	82k
160	44.08	100R	16k	192	45.68	390R	75k
160	44.08	150R	24k	194	45.74	470R	91k
161	44.12	510R	82k	195	45.80	200R	39k
161	44.15	620R	100k	195	45.82	220R	43k
163	44.22	240R	39k	196	45.84	240R	47k
163	44.22	560R	91k	196	45.85	510R	100k
164	44.28	110R	18k	200	46.02	100R	20k
164	44.28	220R	36k	200	46.02	110R	22k
165	44.35	200R	33k	200	46.02	120R	24k
167	44.44	120R	20k	200	46.02	150R	30k
167	44.44	180R	30k	200	46.02	180R	36k
169	44.54	160R	27k	206	46.28	330R	68k
169	44.57	130R	22k	206	46.29	160R	33k
170	44.59	330R	56k	207	46.31	300R	62k
170	44.61	300R	51k	207	46.34	270R	56k
172	44.72	360R	62k	208	46.35	130R	27k
174	44.81	270R	47k	208	46.38	360R	75k
174	44.83	390R	68k	210	46.45	390R	82k
174	44.83	430R	75k	212	46.51	430R	91k
174	44.83	470R	82k	213	46.55	240R	51k

Av	dB	R1	R2	Av	dB	R1	R2
213	46.56	470R	100k	270	48.63	100R	27k
214	46.59	220R	47k	273	48.71	110R	30k
215	46.65	200R	43k	273	48.73	300R	82k
217	46.72	180R	39k	275	48.79	120R	33k
218	46.78	110R	24k	276	48.81	330R	91k
220	46.85	100R	22k	277	48.85	130R	36k
220	46.85	150R	33k	278	48.87	270R	75k
225	47.04	120R	27k	278	48.87	360R	100k
225	47.04	160R	36k	280	48.94	200R	56k
227	47.11	300R	68k	282	49.00	220R	62k
227	47.13	330R	75k	283	49.05	240R	68k
228	47.15	360R	82k	283	49.05	180R	51k
230	47.22	270R	62k	287	49.15	150R	43k
231	47.26	130R	30k	294	49.36	160R	47k
232	47.30	220R	51k	300	49.54	100R	30k
233	47.33	430R	100k	300	49.54	110R	33k
233	47.36	240R	56k	300	49.54	120R	36k
233	47.36	390R	91k	300	49.54	130R	39k
235	47.42	200R	47k	303	49.63	330R	100k
239	47.56	180R	43k	303	49.64	300R	91k
240	47.60	100R	24k	304	49.65	270R	82k
240	47.60	150R	36k	309	49.80	220R	68k
244	47.74	160R	39k	310	49.83	200R	62k
245	47.80	110R	27k	311	49.86	180R	56k
248	47.91	330R	82k	313	49.90	240R	75k
250	47.96	120R	30k	313	49.92	150R	47k
250	47.96	300R	75k	319	50.07	160R	51k
252	48.02	270R	68k	325	50.24	120R	39k
253	48.05	360R	91k	327	50.30	110R	36k
254	48.09	130R	33k	330	50.37	100R	33k
255	48.12	220R	56k	331	50.39	130R	43k
255	48.13	200R	51k	333	50.46	300R	100k
256	48.18	390R	100k	337	50.55	270R	91k
258	48.24	240R	62k	340	50.63	150R	51k
260	48.30	150R	39k	340	50.63	200R	68k
261	48.34	180R	47k	341	50.65	220R	75k
269	48.59	160R	43k	342	50.67	240R	82k

Table 11.1 Inverting op-amp gains 283

A_V	dB	R1	R2	A_V	dB	R1	R2
344	50.74	180R	62k	500	53.98	150R	75k
350	50.88	160R	56k	500	53.98	200R	100k
355	50.99	110R	39k	506	54.08	180R	91k
358	51.09	120R	43k	509	54.14	110R	56k
360	51.13	100R	36k	510	54.15	100R	51k
362	51.16	130R	47k	513	54.19	160R	82k
370	51.37	270R	100k	517	54.26	120R	62k
373	51.43	220R	82k	523	54.37	130R	68k
373	51.44	150R	56k	547	54.75	150R	82k
375	51.48	200R	75k	556	54.89	180R	100k
378	51.54	180R	68k	560	54.96	100R	56k
379	51.58	240R	91k	564	55.02	110R	62k
388	51.77	160R	62k	567	55.07	120R	68k
390	51.82	100R	39k	569	55.10	160R	91k
391	51.84	110R	43k	577	55.22	130R	75k
392	51.86	120R	47k	607	55.66	150R	91k
392	51.87	130R	51k	618	55.82	110R	68k
410	52.26	200R	82k	620	55.85	100R	62k
413	52.33	150R	62k	625	55.92	120R	75k
414	52.33	220R	91k	625	55.92	160R	100k
417	52.40	180R	75k	631	56.00	130R	82k
417	52.40	240R	100k	667	56.48	150R	100k
425	52.57	120R	51k	680	56.65	100R	68k
425	52.57	160R	68k	682	56.67	110R	75k
427	52.61	110R	47k	683	56.69	120R	82k
430	52.67	100R	43k	700	56.90	130R	91k
431	52.68	130R	56k	745	57.45	110R	82k
453	53.13	150R	68k	750	57.50	100R	75k
455	53.15	220R	100k	758	57.60	120R	91k
455	53.16	200R	91k	769	57.72	130R	100k
456	53.17	180R	82k	820	58.28	100R	82k
464	53.32	110R	51k	827	58.35	110R	91k
467	53.38	120R	56k	833	58.42	120R	100k
469	53.42	160R	75k	909	59.17	110R	100k
470	53.44	100R	47k	910	59.18	100R	91k
477	53.57	130R	62k	1.00k	60.00	100R	100k

Appendix 1

Units, symbols and suffixes

I have used standard units, symbols and suffixes throughout this work. Here they are:

Units
Ω or R: ohms (resistance)
F: farads (capacitance)
H: henries (inductance)
s: seconds (time)
Hz: hertz (frequency)
°: degrees (phase angle)
V: volts (electrical potential)
A: amperes (electrical current)
W: watts (power)
J: joules (energy or work)

Symbols
R: resistance
C: capacitance
L: inductance
V: voltage
I: current
X_C: capacitive reactance
X_L: inductive reactance
Z: impedance
f: frequency
ϕ: phase angle
f_O: resonant frequency
Z_O: dynamic (resonant) impedance

Electronics engineers tend to deal in a very wide range of numbers, from million-millionths of farads to millions of ohms. Hence it is usual to use a range of letter suffixes to indicate the order of magnitude of a number.

These are the standard suffixes used in this book:

p: $\times 10^{-12}$

n: $\times 10^{-9}$

μ : $\times 10^{-6}$

m: $\times 10^{-3}$

k: $\times 10^{3}$

M: $\times 10^{6}$

So, for instance, $\frac{1V}{1k\Omega} = 1mA$; $1M \times 1\,nF = 1ms$; and so on.

Appendix 2

Colour codes

The following colour codes are used on resistors, and sometimes on capacitors and inductors, to indicate value:

Number bands
Black 0
Brown 1
Red 2
Orange 3
Yellow 4
Green 5
Blue 6
Violet 7
Grey 8
White 9

Tolerance bands
Brown 1%
Red 2%
Gold 5%
Silver 10%

Figure A2 shows the significance of the bands for a resistor.

For example; a four band resistor with colours yellow–violet–red–red (the last one thick) is $47 \times 10^2 = 4k7, 2\%$. A five band with brown–grey–black–brown–brown (again the last thick) is $180 \times 10 = 1k8, 1\%$.

4 band

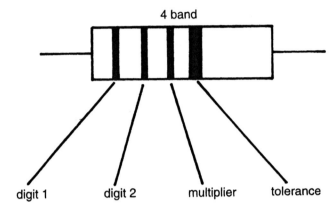

digit 1   digit 2   multiplier   tolerance

5 band

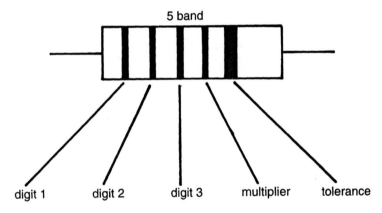

digit 1   digit 2   digit 3   multiplier   tolerance

Figure A2   Resistor colour codes

Appendix 3

The decibel

The decibel, or dB, is a logarithmic unit for expressing ratios, much beloved of engineers. Its popularity is due to the fact that it enables a wide range of numbers to be expressed conveniently.

Where we are talking about a ratio of two voltages then:

$$dB = 20\, log_{10}\left(\frac{V1}{V2}\right) \qquad (A3.1)$$

which means that 'A is twice B' is often expressed as 'A is 6dB greater than B', and 'A is ten times B' becomes 'A is 20dB up on B'. Gain or attenuation of circuit blocks are very well expressed like this (a negative gain is an attenuation, or a ratio of less than one). The ratio is 'output/input'.

dB's are also used to express signal levels a lot. Here some reference level should be stated to make the situation clear. The most common are:

$$dBV = 20\, log_{10}(V_{rms}) - \text{reference to } 1V_{rms}$$

$$dBu = 20\, log_{10}\left(\frac{V_{rms}}{0.775}\right) - \text{reference to } 1mW \text{ into } 600\ \Omega$$

though you may very well come across others. Where there is doubt it is often wise to try to find out what reference was used, as it can be a source of confusion.

The beauty of this system is that the level of a signal at various point in a chain can be found just by adding the dB values together. See Figure A. 3. It is apparent that the calculation in dB's is trivial compared to that using V_{RMS} and gain/attenuation as ratios.

dB's are also used sometimes to express relative power. The equation then used is:

$$dB = 10 \log_{10}\left(\frac{P1}{P2}\right) \tag{A3.2}$$

This is because power varies proportionally to the square of voltage, as implied by Eqs 2.4d and 2.4e.

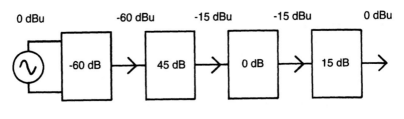

$$0.775\ \text{V} \times \frac{1}{1000} = 0.775\ \text{mV} \times 177.8 = 138\ \text{mV} \times 1 = 138\ \text{mV} \times 5.62 = 0.775\ \text{V}_{\text{RMS}}$$

Figure A3   Illustration of signal level calculations using dB's

Appendix 4

Complex number analysis

The standard way to analyse a circuit containing AC quantities at a fixed frequency is to use complex numbers. A complex number has two parts, real and imaginary. The imaginary part is some function of the quantity j (i to pure mathematicians, but that would be confused with current by us), where j represents $\sqrt{(-1)}$. This is clearly a 'nonsense number', or at least unquantifiable, yet it has some useful properties:

$$j^2 = -1 \qquad \text{(A4.1a)}$$
$$j^3 = -j \qquad \text{(A4.1b)}$$
$$j^4 = 1 \qquad \text{(A4.1c)}$$

We use multiplication by j to signify advancing its phase by $90°$. Complex numbers owe their usefulness to the fact that they can be seen as representations of phasors. For example:

$$V = a + jb$$

is a voltage which can be represented in the j-plane as in Figure A. 4.

This number could also be represented as 'magnitude and phase' (as in Chapter 4) by using some trigonometry:

$$M = \sqrt{a^2 + b^2} \qquad \text{(A4.2)}$$
$$\phi = \arctan\left(\frac{b}{a}\right) \qquad \text{(A4.3)}$$

Then we are able to write the phasor in polar form:

$$V = M < \phi$$

Should we wish to convert a quantity from polar to complex form we need a little more trigonometry:

$$a = M\cos\phi \qquad (A4.4)$$

$$b = M\sin\phi \qquad (A4.5)$$

Mathematical operations with complex numbers

We treat complex numbers like any other algebraic quantity, keeping the j part separate as we would with x or y, but also applying Eqs A4.1 as necessary. So:

$$\text{Addition}: (a + jb) + (c + jd) = (a + b) + j(c + d) \qquad (A4.6)$$

$$\text{Subtraction}: (a + jb) - (c + jd) = (a - c) + j(b - d) \qquad (A4.7)$$

$$\text{Multiplication}: (a + jb)(c + jd) = (ac - bd) + j(bc + ad) \qquad (A4.8)$$

Division: a useful tool for separating fractions is multiplication by the 'complex conjugate', which makes the denominator wholly real:
as:

$$(c + jd)(c - jd) = c^2 + jcd - jcd - j^2d^2 = c^2 + d^2$$

then:

$$\frac{a + jb}{c + jd} = \frac{(a + jb)(c - jd)}{c^2 + d^2} \qquad (A4.9)$$

However, multiplication and division are sometimes more easily performed by converting to polar form and then:
   Multiplication in polar form:

$$M_1 < \phi_1.M_2 < \phi_2 = M_1M_2 < (\phi_1 + \phi_2) \qquad (A4.10)$$

Division in polar form:

$$\frac{M_1 < \phi_1}{M_2 < \phi_2} = \frac{M_1}{M_2} < (\phi_1 - \phi_2) \qquad (A4.11)$$

Representation of capacitors and inductors

Complex numbers come into their own when we use them to represent reactive components in equations. The impedance of a capacitor becomes:

$$X_C = \frac{-j}{2\pi fC} \qquad (A4.12)$$

and an inductor:

$$X_L = j2\pi f L \tag{A4.13}$$

and all of the analysis techniques of Chapter 5 can be used, following the algebraic rules given above.

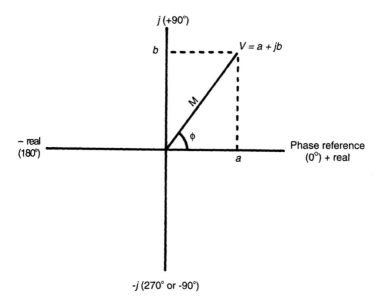

Figure A4 The j-plane

Further reading

Should the reader wish to gain greater understanding of any of the topics which this book touches on, I have found the following to be excellent:

Schuler, C. A. and Fowler, R. J. (1993) *Electric Circuit Analysis*, McGraw-Hill, London.
A clear and thorough explanation of circuit analysis techniques which also contains good chapters on electromagnetism and electrostatics.

Marston, R. (1993) *Electronic Circuits Pocket Book*, Vols One and Two, Butterworth-Heinemann, Oxford.
A comprehensive and informed circuit sourcebook that also analyses the ideas used in sufficient depth to make them really useful. Contains many useful circuits.

Horowitz, P. and Hill, W. (1989) *The Art of Electronics*, Cambridge University Press, Cambridge.
Deservedly considered to be the classic all-round electronics text. Clear, concise and entertaining too.

Phillips, G. (1993) *Newnes Electronics Toolkit*, Butterworth-Heinemann, Oxford.
A great deal of useful information in a small space; an ideal source of reference for the workshop or lab.

Index

LaVergne, TN USA
27 May 2010
184099LV00002B/21/A